OPPORTUNITIES IN

Chemistry

TODAY AND TOMORROW

GEORGE C. PIMENTEL
University of California
Berkeley

JANICE A. COONROD
Lawrence Hall of Science
Berkeley

NATIONAL ACADEMY PRESS
Washington, D.C.   1987

NATIONAL ACADEMY PRESS, 2101 Constitution Ave., NW, Washington, DC 20418

NOTICE: The project that is the subject of this report was approved by the Governing Board of the National Research Council, whose members are drawn from the councils of the National Academy of Sciences, the National Academy of Engineering, and the Institute of Medicine. The members of the committee responsible for the report were chosen for their special competences and with regard for appropriate balance.

This report has been reviewed by a group other than the authors according to procedures approved by a Report Review Committee consisting of members of the National Academy of Sciences, the National Academy of Engineering, and the Institute of Medicine.

The National Academy of Sciences is a private, nonprofit, self-perpetuating society of distinguished scholars engaged in scientific and engineering research, dedicated to the furtherance of science and technology and to their use for the general welfare. Upon the authority of the charter granted to it by the Congress in 1863, the Academy has a mandate that requires it to advise the federal government on scientific and technical matters. Dr. Frank Press is president of the National Academy of Sciences.

The National Academy of Engineering was established in 1964, under the charter of the National Academy of Sciences, as a parallel organization of outstanding engineers. It is autonomous in its administration and in the selection of its members, sharing with the National Academy of Engineering also sponsors engineering programs aimed at meeting national needs, encourages education and research, and recognizes the superior achievements of engineers. Dr. Robert M. White is president of the National Academy of Engineering.

The Institute of Medicine was established in 1970 by the National Academy of Sciences to secure the services of eminent members of appropriate professions in the examination of policy matters pertaining to the health of the public. The Institute acts under the responsibility given to the National Academy of Sciences by its congressional charter to be an adviser to the federal government and, upon its own initiative, to identify issues of medical care, research, and education. Dr. Samuel O. Thier is president of the Institute of Medicine.

The National Research Council was organized by the National Academy of Sciences in 1916 to associate the broad community of science and technology with the Academy's purposes of furthering knowledge and advising the federal government. Functioning in accordance with general policies determined by the Academy, the Council has become the principal operating agency of both the National Academy of Sciences and the National Academy of Engineering in providing services to the government, the public, and the scientific and engineering communities. The Council is administered jointly by both Academies and the Institute of Medicine. Dr. Frank Press and Dr. Robert M. White are chairman and vice chairman, respectively, of the National Research Council.

Copyright ©1987 by the National Academy of Sciences

No part of this book may be reproduced by any mechanical, photographic, or electronic process, or in the form of a phonographic recording, nor may it be stored in a retrieval system, transmitted, or otherwise copied for public or private use, without written permission from the publisher, except for the purposes of official use by the United States Government.

Library of Congress Cataloging-in-Publication Data

Pimentel, George C.
Opportunities in chemistry.

Bibliography: p.
Includes index.
1. Chemistry—Research—United States. I. Coonrod, Janice A. II. Title. III. Title: Chemistry, today and tomorrow.
QD47.P56 1987 540′.72073 87-24000
ISBN 0-309-03742-5

Printed in the United States of America

Acknowledgments

Support for the original *Opportunities in Chemistry* was provided by the American Chemical Society, the U.S. Air Force Office of Scientific Research under Grant No. AFOSR-83-0323, the Council for Chemical Research, Inc., the Camille and Henry Dreyfus Foundation, Inc., the U.S. Department of Energy under Grant No. DE-FGO2-81ER10984, the National Institutes of Health under Grant No. CHE-8301035, the National Bureau of Standards under Contract No. NB835BCA2075, and the National Science Foundation under Grant No. CHE-8301035. Support was also provided by the following industrial companies: Aluminum Company of America, AT&T Bell Laboratories, Calgon Corporation, Celanese Research Company, Dow Chemical Company, Eastman Kodak Company, E.I. du Pont de Nemours and Company, Inc., Exxon Corporation, General Electric Company, GTE Laboratories, Inc., Johnson and Johnson Company, Mobay Chemical Company, Mobil Research and Development Corporation, Monsanto Company, Pfizer, Inc., Phillips Petroleum Company, PPG Industries, Inc., Proctor and Gamble Company, Shell Development Company, Standard Oil Company (Ohio), Stauffer Chemical Company, TRW, Inc., and U.S. Steel Corporation.

Additional financial support has made this volume possible. The National Academy of Sciences contributed funds to subsidize the writing; and the American Chemical Society, Council for Chemical Research, National Science Foundation, Robert A. Welch Foundation, and a number of industrial companies provided support for distribution of copies of *Opportunities in Chemistry: Today and Tomorrow* to high school teachers, libraries, and selected students across the country.

All of this generous support is gratefully acknowledged. A special thanks is given to Julian Systems, Inc., for their assistance. Finally, William Spindel's efforts and encouragement in every step of this project were essential to its successful completion.

Preface

This book is based upon *Opportunities in Chemistry,* which described the contemporary research frontiers of chemistry and the opportunities for the chemical sciences to address society's needs. To accomplish that ambitious task, a committee of 26 eminent scientists was selected under the auspices of the National Research Council. The committee was broadly representative of the major subdisciplines of chemistry, of geographic areas, and of the full range of academic, industrial, and government research. These scientific leaders then called upon more than 350 chemical researchers to suggest topics and prepare commissioned papers on research at chemistry's frontiers. After 3 years of thoughtful deliberation, *Opportunities in Chemistry* was completed in October 1985 and published by the National Academy Press.

Now, we have revised *Opportunities in Chemistry* in an effort to make such a comprehensive survey of modern chemistry more widely available. Our primary goal has been to make the volume valuable to a different audience by reorganizing the content, adjusting the technical vocabulary, and adding explanatory material and supplementary reading suggestions. We believe that this new book, *Opportunities in Chemistry: Today and Tomorrow,* will provide interesting resource material and supplementary reading for high school advanced placement chemistry courses and for college sciences courses developed for nonscience majors. We are confident that it will provide, as well, important background reading for chemistry teachers at all precollege levels.

Finally, we hope that all those who want to look ahead to the promising future of chemistry, who are intrigued by the multitude of doors to be opened by advances in chemistry, and who are concerned about finding the difficult balance between maximizing benefits and minimizing problems, will find this volume illuminating and relevant.

George C. Pimentel
Janice A. Coonrod
Berkeley, California

Contents

OPPORTUNITIES IN
Chemistry
TODAY AND TOMORROW

CHAPTER I
Introduction

This is a book about chemistry. It tells how chemistry fits into our lives. It tells of new chemical frontiers that are being opened, and what benefits may flow from them. It tells how much chemistry contributes to our existence, our culture, and our quality of life. This book shows how central chemistry is among the sciences as they are applied to human needs. It shows how important chemicals are to our survival.

And just what is a chemical? Perhaps you have your answer ready—DDT, Agent Orange, and dioxin are chemicals. Yes, indeed they are, just as much as sugar and salt, air and aspirin, milk and magnesium, protein and penicillin are chemicals. We ourselves are made up entirely of chemicals.

But what of the changes that we see around us: iron nails rust, grass grows, wood burns? Here we see one set of chemicals turning into another set of chemicals. *Chemistry* is the science concerned with such changes. Without these changes, that is, chemical reactions, Earth would be a lifeless planet. A bean plant takes carbon dioxide from the air and water from the soil to produce carbohydrates through a wondrous series of chemical reactions called photosynthesis. All living processes are chemical reactions. And all of the things we use, wear, live in, ride in, and play with are produced through controlled chemical reactions. That is the business of chemists: to design reactions that will convert chemical substances we find around us into chemical substances that serve our needs. We want silicon transistors for our computers, but we find no silicon as such in nature. Instead, we find silica, in the form of sand, on every beach. Through chemistry, silica is converted to elemental silicon. We want a chemical substance that will be effective against Parkinson's disease. Chemists respond by synthesizing the chemical carbidopa, a substance not found in nature, but extremely effective in medical therapy. Drivers want to burn millions of gallons of fuel every day with minimal exhaust contamination of the atmosphere. A part of the answer is found in your automobile's catalytic exhaust converter, and the rest in marvelous chemical manipulations of the raw materials we have at hand, crude oils, which are converted on a gigantic scale into refined chemicals that burn efficiently in your auto engine.

Chemists manage to answer so many needs of our society through a deep understanding of the factors that govern and furnish control of chemical reactions. These understandings are rooted in a powerful concept called the *atomic theory* in which all substances are considered to be made up of submicroscopic particles called atoms. There are a hundred or so chemical types of atoms, each type with its own behavioral characteristics. These are the "elements." Atoms have characteristic abilities to combine with each other to form identifiable groups of atoms, called molecules. Each of these molecules has its own set of properties. These are "compounds."

How well this atomic theory works is demonstrated by the accomplishments of chemists guided by it. Over seven million different molecular compounds have been synthesized, and the rate of discovery is increasing every year. Only a small fraction of these are found in nature—most of them were deliberately designed and synthesized to meet a human need or to test an idea. This book tells about the impact of this growing capability on our society. It shows that chemistry plays a critical role in man's attempt to feed the world population, to tap new sources of energy, to clothe and house humankind, to provide renewable substitutes for dwindling or scarce materials, to improve health and conquer disease, and to monitor and protect our environment. It shows that there are rich opportunities in chemistry for advances through basic research that will help future generations deal with their evolving needs.

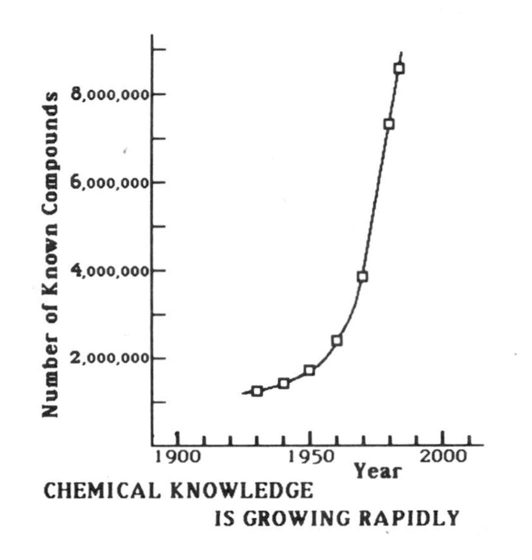

CHEMICAL KNOWLEDGE IS GROWING RAPIDLY

Because of this responsiveness to human needs, chemistry has become a crucial factor in the nation's economic well-being. Equally important, our culture believes that learning about our place in the universe is enough reason for encouraging scientific inquiry. For example, nothing concerns humans more than questions about the nature of life and how to preserve it. Because all life processes are brought about by chemical changes, understanding chemical reactivity is a necessary foundation for our ultimate understanding of life. Thus, chemistry, along with biology, contributes to human knowledge in areas of universal philosophical significance.

Fortunately, we find ourselves in a time of special opportunity for advances on the many fronts of chemistry. The opportunity comes from our developing ability to probe the basic steps of chemical change and, at the same time, to deal with the extreme molecular complexity of biological molecules. A recurrent theme will be the use, by chemists, of the most sophisticated and advanced instrumental techniques that contribute significantly to chemistry's accelerating progress. We hope that your reading will be enjoyable, that it will give you a new view of the beneficial role of chemistry in your life, and that some of you will look to chemistry for a fulfilling and rewarding career.

No Deposit, No Return, No Problem

Each year in this country, you and I dump millions of tons of plastics into the environment. A high percentage of that is spewed directly into the oceans. In fact, 9 million tons of the solid waste produced in the United States each year goes directly into the sea. Merchant ships alone dump 6.6 million tons of trash overboard each year—that's enough garbage to fill 440,000 classrooms!

Contrary to the beliefs of many, plastic debris does eventually degrade—but it may take up to 50 years. A lot of litter can accumulate in that stretch of time. Marine environments are particularly sensitive to this problem—plastic trash floats and is mistaken for jellyfish or eggs or other dinnertime treats by marine animals. In addition, sea animals become entangled in plastic waste, including the 150,000 tons of plastic fishing gear that is discarded in the ocean each year. Another unhappy twist to this problem can be seen in the arctic regions, where litter accumulates, but is inhibited from biodegradation because of the very low temperatures.

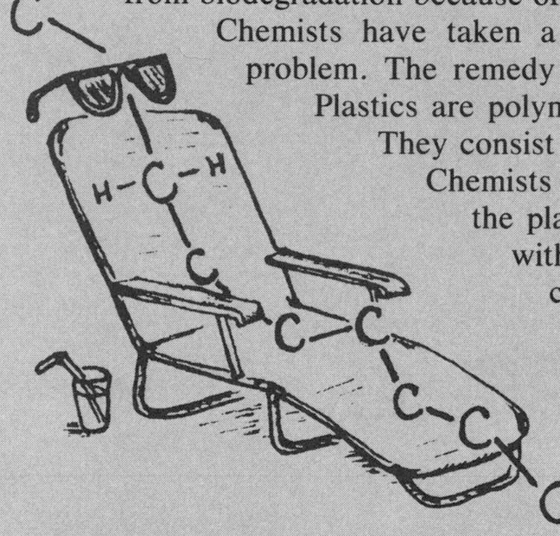

Chemists have taken a big step toward alleviating this distressing problem. The remedy lies in the construction of the plastic itself. Plastics are polymers made from petroleum-based compounds. They consist of long chains of repeating molecular groups. Chemists have found several ways to make changes in the plastic molecules so that they are more in tune with our environmental needs. One way is to chemically attach light-sensitive molecular groups at regular intervals in the macromolecular chain. When plastic made from this polymer is exposed to sunlight, these light-sensitive groups absorb radiation and cause the polymer to break apart at these points. Then Nature does the rest. The small segments which result are easily biodegradable. Presto! Photodegradable plastics! Insertion of ketone groups into common polymers (such as polystyrene and polyethylene) has been of particular interest as photodegradable materials. Such ketone-substituted polymers are stable in artificial light and only undergo photochemical reactions when irradiated with shorter wavelengths of light—like that from the sun.

Another way to tailor long plastic molecules to suit Nature's needs has been to introduce molecular groups that are considered delectable by certain microorganisms in the environment. These microscopic munchers then do the work of breaking the long molecules into smaller bits. Hopefully, with innovations like these, our plastic waste problem will someday be going, going, gone. . . .

CHAPTER II

Environmental Quality Through Chemistry

Every society tries to provide itself with adequate food and shelter and a healthful environment. When these elemental needs are assured, attention turns to comfort and convenience. The extent to which all of these wishes can be satisfied determines the quality of life. Generally, however, choices are required because one or another of these qualities is most easily attained at the expense of others. Today we find our desires for more abundant consumer goods, energy, and mobility in conflict with maintenance of a healthful environment. A major concern of our times is the protection of our environment in the face of increasing world population, continued concentration of population (urbanization), and rising standards of living.

Environmental degradation—with accompanying threats to health and disruption of ecosystems—is not a new phenomenon. Human disturbance of the environment has been noted from the earliest recorded history. The problem of sewage disposal began with the birth of cities. Long before the twentieth century, London was plagued with air pollution from fires used for heating and cooking. An early example of an industrial hygiene problem was the shortened lifetime of chimney sweeps due to cancer, which we can now attribute to prolonged exposure to soot with its trace carcinogen content (polynuclear aromatic hydrocarbons).

There is small consolation, though, in the fact that environmental pollution is not a new invention. The global population becomes ever larger, while cities grow even faster. Per capita consumption and energy use continue to increase. Pollution problems are becoming increasingly obvious, and we are recognizing subtle interactions in the world around us and discovering secondary effects that went unnoticed before. A number of environmental disturbances have begun to appear on a global scale. Occasional industrial accidents, like those at Bhopal, India and Seveso, Italy, remind us that large-scale production of needed consumer products may require handling of large amounts of potentially dangerous feedstock substances. The tragic Bhopal accident highlights the dilemma. This occurred in a country plagued by starvation—the toxic substances were being used to manufacture products that annually saved many thousands of lives by increasing the food supply.

On the positive side, there is high public awareness of the importance of maintaining environmental quality. In the United States, a large majority of citizens from across the political spectrum have indicated they are prepared to pay more for products (e.g., lead-free gasoline), and pay more taxes, to improve their environment. These attitudes are spreading abroad, an essential aspect of environmental protection for the more global problems.

Effective strategies for safeguarding our surroundings require adequate knowledge and understanding. We must be able to answer the following questions:

- What potentially undesirable substances are present in our air, water, soil, and food?
- Where did these substances come from?
- What options are there—alternate products and processes—to reduce or remove known problems?
- How does the degree of hazard depend on the extent of exposure to a given substance? How shall we choose among the available options that offer corrective action?

Plainly, chemists play a central role in the first three crucial questions. To find out what substances are present in the environment, we need analytical chemists to develop more and more sensitive and selective analytical techniques. To track pollutants back to their origin, again we look to analytical chemists acting as detectives, usually in collaboration with meteorologists, oceanographers, volcanologists, climatologists, biologists, and hydrologists. Finding origins can require detailed chemical understandings of reactions that take place between the source of the pollution and the final noxious or toxic product. Thus, development of options calls on the full range of chemistry. If the world's mortality rate due to malaria is not to be reduced with DDT because of its environmental persistence, what substances can be synthesized that are equally effective in saving lives but are spontaneously decomposed? If we must use lower-grade energy sources to satisfy our energy needs, what catalysts and new processes can be developed to avoid making worse the existing problems of acid rain and carcinogen release from coal-fired power plants?

Thus, our society must assure the health of its chemistry enterprise if it wants early warning of emerging environmental damage, understanding of the origins of that degradation, and economically feasible options from which to choose solutions. Other disciplines make their own particular contributions, but none plays a more central and essential role than chemistry.

The fourth question, concerning how much exposure to a substance must be considered hazardous, is the province of the medical profession, toxicologists, and epidemiologists. These scientific disciplines face serious challenges now that society has recognized the inverse relationship between how small a risk can be made and the cost to society to attain it. The medical profession must refine its knowledge of risks associated with substances such as lead in the atmosphere, chloroform in drinking water, radiostrontium in milk, benzene in the workplace, and formaldehyde in the home. A qualitative statement that a certain class of substances might be carcinogenic is no longer sufficient. We must be able to weigh risks and costs against the benefits that would be lost if use of that class of substances were restricted. We must be able to compare those risks with those already present because of natural background levels. More importantly, society cannot afford to pay the excessive cost to eliminate *all* risk, since, as the desired degree of risk approaches zero, the cost escalates toward infinity.

Finally, the choice among options must move into the public arena. Chemists and scientists in the other relevant disciplines carry a special and important informational responsibility here. Every political decision deserves the best and most objective scientific input available. There is nothing more frustrating to our citizens and our government than to be faced with making decisions without having all the facts and a useful understanding of the science involved. Scientists, including chemists, must carry responsibility for providing the public, the media, and the government with a factual picture expressed in language free of technical jargon. That picture must establish the scientific setting for a given decision and indicate the options that lie before us.

TURNING DETECTION INTO PROTECTION

All of our environmental protection strategies should be based on realistic hazard thresholds and on our ability to detect a particular offending substance well before its presence reaches that threshold. Chemists must continue to sharpen their analytical skills so that, even at tiny concentrations well below the hazard threshold, a given substance can be monitored long before frantic corrective action is necessary. When this is possible, we see that *detection can be equated to protection*.

Unfortunately, the media, the public, and our governmental agencies have too often equated detection with hazard. This is based on the common assumption that a substance that is demonstrably toxic at some particular concentration will be toxic at any concentration. There are innumerable examples to prove that this is not generally true. Consider carbon monoxide. This ever-present atmospheric compound becomes dangerously toxic at concentrations exceeding 1,000 parts per million and is considered to have negative health effects for prolonged exposure to concentrations exceeding 10 parts per million. We do not, however, leap to the conclusion that CO must be completely removed from the atmosphere! This would be foolish (and impossible) because we live and thrive in a natural atmosphere that always contains easily detectable CO, about one part per million. Plainly, our task is to decide where we should begin cleanup action between the known toxicity threshold and the known safe range (as the Environmental Protection Agency has attempted to do).

Selenium presents another interesting example. Certain plants growing in soils with relatively high selenium content tend to concentrate the element to levels such that grazing animals are poisoned. *Astragalus* is an example—it has the common name "locoweed." Wheat can do the same, and while humans are not noticeably affected, chickens fed high-selenium wheat produce deformed embryos. On the other hand, it is now well established that selenium is nutritionally *essential* in the diets of rats, chickens, and pigs. Furthermore, it has been found that selenium at proper levels is a natural anticarcinogen; it is a component of glutathione peroxidase, an enzyme that breaks down injurious hydroperoxides. In China, children in populations with low blood levels of selenium suffer from multiple myocarditis (Keshan disease), and the adults display high cancer death rates and a high incidence of liver cancer. Plainly, selenium is an element that is essential to human and animal health at appropriate levels and it becomes toxic at excessive levels. The daily dietary intake of selenium for adults recommended by the National Research Council is 50 to 100 micrograms per day. Presently, the permissible level of selenium in drinking water, as fixed by the EPA, is 10 parts per billion. This level, set to avoid possible toxicity, may be 10-fold below the level needed for optimum health. The example shows vividly that trace-level detection in the environment of a substance that might be toxic at high concentrations does not imply that a hazard exists. Quite the opposite, such early detection allows time for deliberate decisions concerning sources, trends, and levels at which corrective action will be timely. *Detection is protection.*

Some people have forcefully called for a "zero-risk" approach to environmental pollution. Zero risk means achieving absolute and complete freedom from any conceivable hazard. In the carbon monoxide example above it would mean removal of every single molecule of CO from the entire atmosphere. In recent trends, this unattainable zero-risk approach is gradually being replaced by a more sophisticated risk assessment/risk management philosophy. In both assessment and management, a major theme is the crucial importance of being able to analyze complex air, water, soil, and biological systems that may contain hundreds of natural chemical compounds. Conclusions regarding the sources, movements, and fates of pollutants depend upon adequate environmental measurements, whether the issue is acid rain, global climate change, ozone layer destruction, or toxic waste disposal. Enormously costly decisions about how to protect the quality of our air, water, and land resources are sometimes based on environmental information that is dangerously inadequate and inaccurate. Crash projects (like the "Superfund") to remedy crises caused by ineffective strategies of the past have been expensive. The best future investment would be in long-term fundamental environmental science and monitoring techniques to avoid the need for costly patch-up programs.

Increased effectiveness of environmental measurements requires improved tools. The challenge is to measure trace levels of a particular compound present in a complex mixture containing many harmless compounds. The principal objectives of research in environmental analysis and monitoring are improved sensitivity, selectivity, separation, sampling, accuracy, speed, and data interpretation. For example, an active research area is connected with separation techniques to allow rapid and reliable analysis of complex mixtures of pollutants and pesticides found in toxic wastes, polluted streams and lakes, and biological samples. A success story in analytical selectivity can be seen in the development of analytical methods to allow separation and quantitative measurement of each of the individual 22 isomers of tetrachlorodioxin at the parts per trillion level (one part in 10^{12})!

Highly reactive species in the atmosphere cannot be carried to the laboratory for analysis. These substances pose special challenges; they require research aimed at remote sensing techniques that are capable of measuring them where they are originally formed. Past successes include the measurement of formaldehyde and nitric acid in Los Angeles smog by infrared spectroscopy in which absorption due to these pollutants was measured over a one-kilometer distance. With these experiments it was possible to determine the simultaneous concentrations of formaldehyde, formic acid, nitric acid, peroxyacetyl nitrate, and ozone in the air at the parts per billion level. Notice that one part per billion (one part of a pollutant in 10^9 parts of air) is a tiny concentration, too small to be an irritant. However, it can be sufficient to be significant in atmospheric reactions. Scanning laser devices based on radar-like technology (called "lidar") have been used successfully to measure sulfur dioxide at the part per million level in the smoke plumes found downwind of coal-fired power plants. Tunable diode lasers are also capable of providing immediate detection of pollutants from internal combustion engines right at the exhaust pipe.

THERE ARE 22 DIFFERENT TETRACHLORODIOXINS

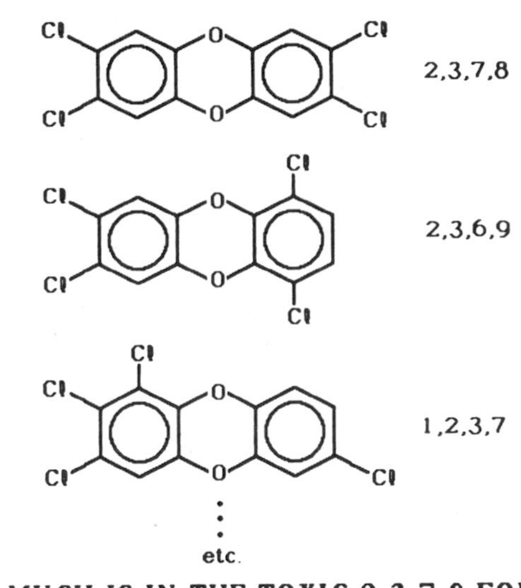

HOW MUCH IS IN THE TOXIC 2,3,7,8 FORM?
AN ANALYTICAL CHALLENGE

Several laser techniques, including absorption, fluorescence, coherent Raman, and two-laser methods, need to be examined more extensively for possible use in atmospheric analysis. One goal of such research should be better measurements in the troposphere (the layer of the Earth's atmosphere closest to the surface) and in the stratosphere above. Rapid, reliable, accurate, and less-expensive methods are needed for measuring concentrations of trace species, such as OH radicals, that play key roles in atmospheric chemistry.

Research directed at fixing the chemical state of environmental constituents is gaining importance because we now recognize that both toxicity and ease of movement from place to place vary markedly with the particular chemical form. Chromium in the hexavalent oxidation state is toxic, while in the trivalent form it is much less so, and for some living systems, it is probably a trace element essential to life. Arsenic in some forms can move rapidly through natural underground water supplies, while other forms are held tightly, adsorbed on rock or soil surfaces. Of the 22 distinct structural arrangements of tetrachlorodioxin, one of them is a thousand times more toxic to test animals than the second most toxic form. These examples illustrate the importance of analytical methods that allow identification of chemical form as well as quantity of potential pollutants. Electrochemistry, chromatography, and mass spectrometry are among the powerful tools used for such studies.

The complexity of environmental problems requires analysis of massive amounts of data. Research is needed to assist in the interpretation and wise utilization of the accumulated information. Developments in the field of artificial intelligence that use pattern recognition should provide a valuable interpretive aid. Recent advances in microprocessors and small computers are coming into use as "intelligent" measuring devices. Attention should be given to organizing, storing, and collecting environmental data.

OZONE IN THE STRATOSPHERE

The possibility of polluting the stratosphere to the point of partially removing the protective ozone layer was first raised only about a dozen years ago. This seemingly improbable notion found much scientific support and has become one of the best examples of a potentially serious environmental problem of global extent. It is a problem, furthermore, that points up chemistry's central role in its understanding, analysis, and solution.

Why do we need to worry about stratospheric chemistry? Ozone in the stratosphere is the natural filter that absorbs and blocks the Sun's short wavelength ultraviolet radiation that is harmful to life. The air in the stratosphere, a cloudless, dry, cold region between about 10- and 50-km altitude, mixes very slowly in the vertical direction, but rapidly in the horizontal. Consequently, harmful pollutants, once introduced into the stratosphere, not only remain there for many years but are also rapidly distributed around the Earth across borders and oceans, making the problem truly global. A large reduction of our ozone shield would result in an increase of dangerous ultraviolet radiation at the Earth's surface.

To understand how easily the ozone layer might be disturbed, it is useful to recognize that ozone is actually only a trace constituent of the stratosphere; at its maximum concentration ozone makes up only a few parts per million of the air molecules. If the ozone layer were concentrated into a thin shell of pure ozone gas surrounding the Earth at atmospheric pressure, it would measure only about 3 millimeters (1/8 inch) in thickness. Furthermore, ozone destruction mechanisms are based on chain reactions in which one pollutant molecule can destroy many thousands of ozone molecules before being transported to the lower atmosphere and removed by rain.

Chemistry's crucial role in understanding this problem has emerged through the identification of several ozone-destroying chain processes. Fifty years ago, the formation of an ozone layer in the midstratosphere was crudely described in terms of four chemical and photochemical reactions involving pure oxygen species (O, O_2, and O_3). Today, we know that the rates of at least 150 chemical reactions must be considered in order to approach an accurate model that will describe the present stratosphere and predict accurately the changes that would result from the introduction of various pollutants. The chemistry begins with absorption of ultraviolet radiation from the Sun by O_2 molecules in the stratosphere. Chemical bond rupture occurs and ozone, O_3, and oxygen atoms, O, are produced. Then, if nitric oxide, NO, is somehow introduced into the stratosphere, an important chemical chain reaction takes place. The NO reacts with ozone to produce NO_2, and this NO_2 reacts with an oxygen atom to regenerate NO. These two reactions together furnish a true catalytic cycle in which NO and NO_2 are the catalysts.

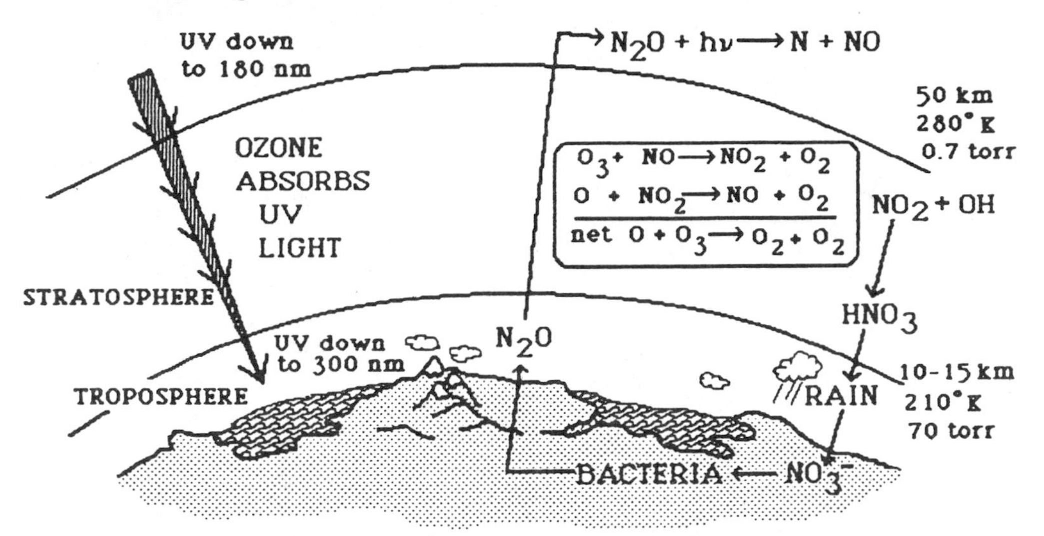

NITROGEN OXIDES REDUCE STRATOSPHERIC OZONE

Neither species is consumed since each is regenerated in a complete cycle. Each cycle has the net effect of destroying one oxygen atom and one ozone molecule (collectively called "odd oxygen"). This catalytic cycle is now believed to be the major mechanism of ozone destruction in the stratosphere. In the natural atmosphere the oxides of nitrogen are provided by natural biological release of nitrous oxide, N_2O, at the Earth's surface by soil and sea bacteria. This relatively inert molecule slowly mixes into the stratosphere where it can absorb ultraviolet light and then react to form NO and NO_2.

Of course, oxides of nitrogen directly introduced into the stratosphere are expected to destroy ozone as well, and this was the basis of the first perceived threat to the ozone layer—large fleets of supersonic aircraft flying in the stratosphere and depositing oxides of nitrogen through their engine exhausts. Nuclear explosions also produce very large quantities of oxides of nitrogen, which are carried into the stratosphere by the buoyancy of the hot fireballs; a significant depletion of the ozone layer in the event of a major nuclear war was forecast in a 1975 study by the National Academy of Sciences, although this environmental effect of nuclear war may be insignificant in comparison with the recently suggested potential of a "nuclear winter." Both effects underscore the delicacy of the atmosphere and its sensitivity to chemical transformations.

Then, in 1974, just as the possible impact of stratospheric planes was reaching the analysis stage, concern was raised about other man-made atmospheric pollutants. Halocarbons such as $CFCl_3$ and CF_2Cl_2 (chlorofluoromethanes, or CFMs) had become popular as spray-can propellants and refrigerant fluids, mainly because of their chemical inertness. This absence of reactivity meant absence of toxicity or other harmful effects on living things. Ironically, this meant that there was no place for the CFMs to go but up—up into the stratosphere where ultraviolet photolysis could occur. Chemists then recognized that if this occurred, the resultant chlorine species, Cl and ClO, could enter into their own catalytic cycle, destroying ozone in a manner just like the destruction caused by the oxides of nitrogen. Once this

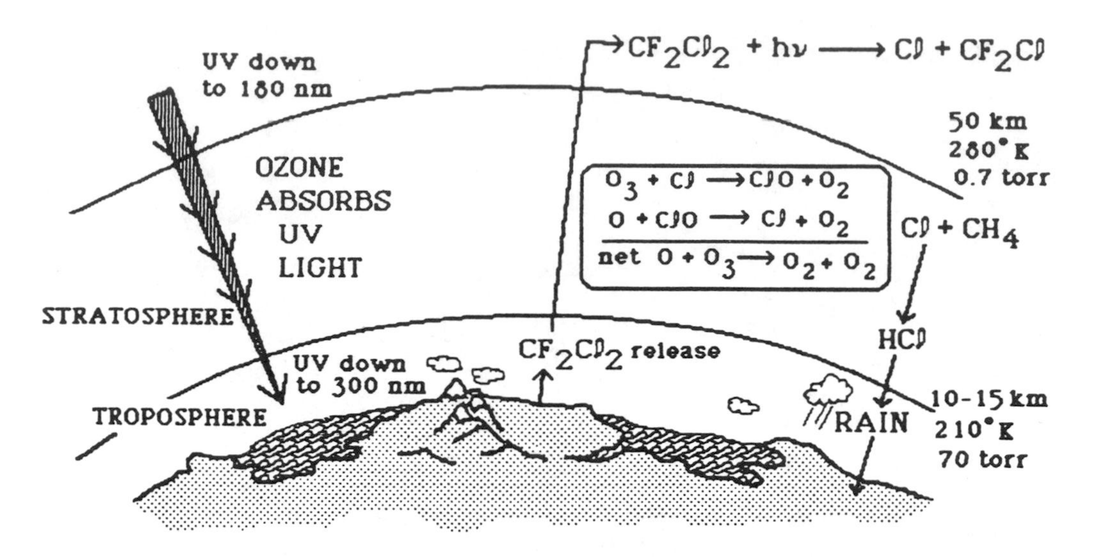

CHLORINE ALSO CAN REDUCE STRATOSPHERIC OZONE

possibility was recognized, analysis of the whole stratospheric ozone chemistry began in earnest. An international committee of scientific experts assembled by the National Academy of Sciences examined in detail the state of our knowledge in every aspect of the problem. It became clear that the additional chemistry introduced into the stratosphere added not just these two catalytic chemical reactions to the roster but a total of about 40 new reactions involving such species as Cl, ClO, HCl, HOCl, $ClONO_2$, the halocarbons, and several others. Most of these reactions had never before been studied in the laboratory.

Chemists have responded to this challenge by measuring in the laboratory reliable reaction rate constants and by clarifying the photochemistry of the suspect compounds, using the growing array of modern experimental methods. Recent progress here has been remarkable. It has become possible to generate nearly any desired reactive molecular species in the laboratory and to measure their rates of reaction with other atmospheric constituents. Such direct measurements of these extremely rapid reactions, only a distant goal a decade ago, are now becoming a reality.

Finally, field measurements of minor atmospheric species have been revolutionized by some of the recent advances in analytical chemistry. Methods originally developed for ultrasensitive detection of extremely reactive species in laboratory studies have been modified and adapted to measure such constituents as O, OH, Cl, and ClO at parts per trillion concentrations in the natural stratosphere. This has been accomplished recently in experiments in which a helium-filled balloon carries an elaborate instrument package to the top of the stratosphere, where the package is dropped while suspended by a parachute. As the instrument travels through the stratosphere, it measures concentrations of several important trace chemical species and relays the information to a ground station. Very recently, the first successful reel-down experiment was performed in which the instrument package was lowered 10 to 15 km from a stationary balloon platform and reeled back up again, as if it were on a giant yo-yo. This method results in a huge increase in the amount of data that can be obtained in a single balloon flight. Also, it allows study of the variability of the stratosphere, both in space and in time.

Much has been accomplished in the past 10 years. Many of the needed 100 to 150 photochemical and rate processes have been measured in the laboratory and many of the trace species have been measured in the atmosphere. Yet, two of the important chemical species containing chlorine, HOCl and $ClONO_2$, have yet to be measured anywhere in the stratosphere. Refinements in the reaction rates for many of the important processes are still required and exact product distributions for many of the reactions are still lacking. Nevertheless, the original NAS study, the research programs that it gave birth to, and the subsequent follow-up studies provided firm and timely bases for legislative decisions about regulation of CFM use. Industrial chemists produced alternative, more readily degradable substances to replace the CFMs in some applications such as aerosol use, in air-conditioning, and in refrigeration systems. Monitoring programs are in place so that trends or changes in the stratospheric composition can be watched. The stratospheric ozone issue provides a showcase example of how science can examine, clarify, and point to solutions for a potential environmental disturbance. Premature initiation of regulation was avoided because the problem was recognized early enough to permit deliberate, objective analysis and focused research to narrow the uncertainty ranges. From first recognition on, chemists played a lead role.

REDUCING ACID RAIN

Acid rain is one of the more obvious air quality problems facing us today. Acidic substances and the compounds that lead to them are formed when fossil fuels are burned to generate power and provide transportation. These are principally acids derived from oxides of sulfur and nitrogen. There are some natural sources of these compounds such as lightning, volcanos, burning biomass, and microbial activity, but, except for rare volcanic eruptions, these are relatively small compared with emissions from automobiles, power plants, and smelters.

The effects of acidic rainfall are most evident and highly publicized in Europe and the northeastern United States, but areas at risk include Canada and perhaps the California Sierras, the Rocky Mountains, and China. In some places precipitation as acidic as vinegar has occasionally been observed. The extent of the effects of acid rain is the subject of continuing controversy. Damage to aquatic life in lakes and streams was the original focus of attention, but damage to buildings, bridges, and equipment have been recognized as other costly consequences of acid rain. The effect of polluted air on human health is the most difficult to determine quantitatively.

The greatest damage is done to lakes that are poorly buffered. When natural alkaline buffers are present, the acidic compounds in acid rain, largely sulfuric acid, nitric acid, and smaller amounts of organic acids, are neutralized. However, lakes lying on granitic (acidic) strata are susceptible to immediate damage because acids in rain can dissolve metal ions such as aluminum and manganese. This can cause a reduction in plant and algae growth, and in some lakes, the decline or elimination

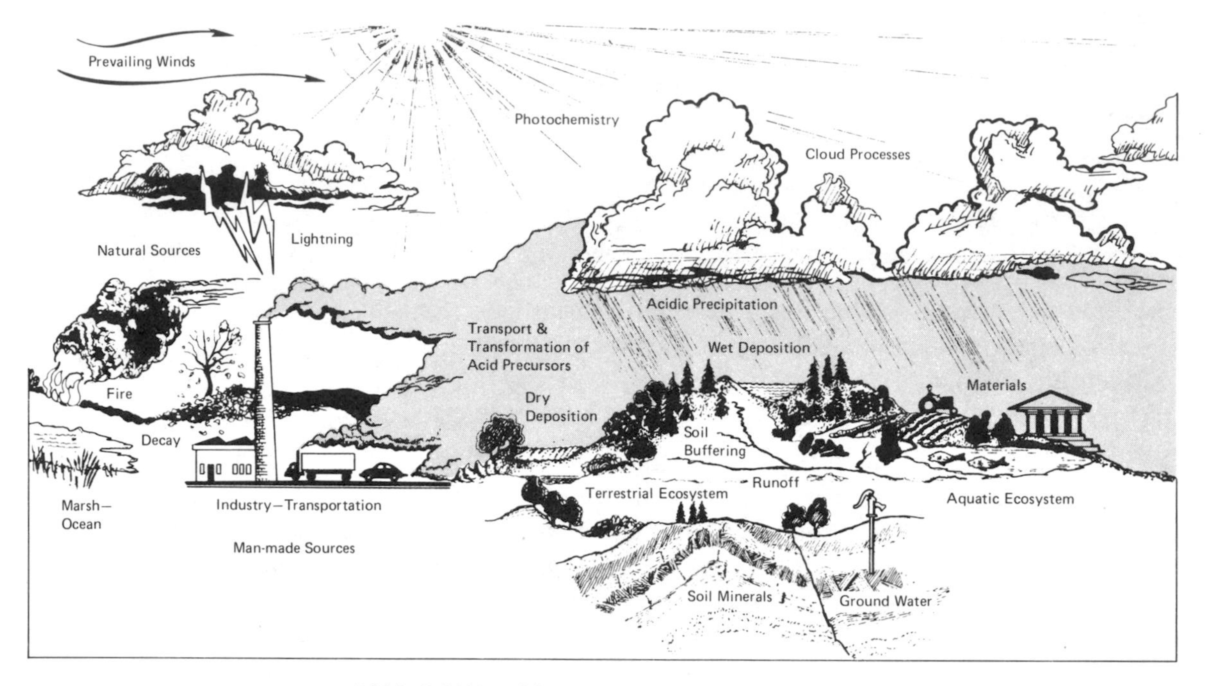

ACID RAIN — SOURCES HERE, IMPACT THERE

of fish populations. Damage to plants from this form of pollution ranges from harmful effects on foliage to destruction of fine root systems.

In a region such as the northeastern United States, the principal candidates for pollutant reduction are the power plants burning coal with high sulfur content. Chemical scrubbers that prevent the emission of the pollutants offer one of the possible remedies. A chemical scrubber is a device that processes gaseous effluents to dissolve, precipitate, or consume the undesired pollutants. Catalysts that reduce oxides of nitrogen emissions from both stationary and mobile sources offer yet another example of the role that chemistry can play in improving air quality.

The various strategies for reducing acid rain involve possible investments of billions of dollars annually. With the stakes so high, it is essential that the atmospheric processes involving transport, chemical transformation, and the fate of pollutants be very well understood.

Acid deposition consists of both "wet" precipitation (as in rain and snow) and dry deposition (in which aerosols or gaseous acidic compounds are deposited on surfaces such as soil particles, plant leaves, etc.). What ends up being deposited has usually entered the atmosphere in a quite different chemical form. For example, sulfur in coal is oxidized to sulfur dioxide, the gaseous form in which it is emitted from smokestacks. As it moves through the atmosphere it is slowly oxidized and reacts with water to form sulfuric acid—the form in which it may be deposited hundreds of miles downwind.

The pathways by which oxides of nitrogen are formed, react, and are eventually removed from the atmosphere are also very complex. Nitrogen and oxygen, when heated at high temperatures in power plants, home furnaces, and auto engines, form nitric oxide, NO, which reacts with oxidants to form nitrogen dioxide, NO_2, and eventually nitric acid, HNO_3. Quantitative estimates of the global budget for the oxides of nitrogen—where they come from and where they go—are still quite uncertain.

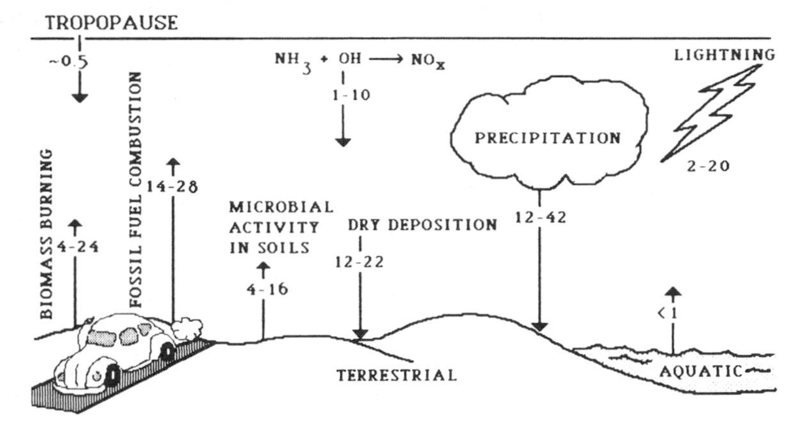

LARGE UNCERTAINTIES REMAIN IN THE GLOBAL NO_x BUDGET

It can readily be seen that until we have a thorough knowledge of the biogeochemical cycles for the various chemical forms of nitrogen, sulfur, and carbon, and of the global origins and fates of these species, it will be difficult to select air pollution control strategies with confidence. Atmospheric and environmental chemistry are central to a clearer and more healthful environment. Development of reliable methods of measurement of trace species in air, kinetics of important atmospheric reactions, and the discovery of new, more effective chemical processes for reducing pollutant emissions are goals that must receive a national commitment for the coming decade.

GUARDING AGAINST CLIMATE CHANGE: THE GREENHOUSE EFFECT

In the quest for food, consumer goods, heat for homes, and energy for our industrial society, we have increased the concentrations of many trace gases in the atmosphere. Some of these absorb solar energy and convert it into heat that might eventually cause climate changes with catastrophic consequences. If the release of these gases to the atmosphere from man's activities causes significant global warming, then the results could be flooding due to melting polar ice, loss of productive farmland to desert, and eventually famine. The most publicized of the gases that trap solar energy is carbon dioxide, but the combined effect of increases in nitrous oxide, methane, and other gases could equal that of carbon dioxide.

Approaches used to reduce emission of other pollutants are not sufficient in the case of carbon dioxide because of the enormous quantities generated in the burning of fossil fuels and biomass. Here the biogeochemical cycling of carbon assumes great importance. What impact will the "slash and burn" clearing of forests and jungles in the Third World countries have? What role does methane play as it is produced by termites and other species? Are atmospheric solid particles and liquid droplets coming from human activities likely to block sunlight and offset the effects of increases in carbon dioxide, methane, and nitrous oxide? Large concentrations of soot and other aerosols have been observed in Arctic regions. The origin, composition, radiative properties, fate, and effects of these aerosols constituting "Arctic Haze" all need to be understood.

The behavior of soot in the atmosphere takes on even greater significance in light of the uncertainties about the possible atmospheric effects of nuclear warfare. It was not until 1982 that the hypothesis of global cooling from soot generated by nuclear war was advanced. This has since been termed "nuclear winter," because even limited nuclear wars have been predicted to cause the injection into the atmosphere of enough soot to black out the Sun so that crops would freeze in summertime. Great uncertainties exist concerning the length of time aerosols remain in the atmosphere and the effects of soot on radiation balance.

Unlike local pollutants, the global pollutant problems are perplexing because they require action on a global scale and the citizens of different countries view their priorities differently. Whether individual countries have emphasized fossil fuel versus nuclear fuel in the past has been based primarily on economic factors such as whether that nation had abundant coal reserves. As global threats such as carbon dioxide buildup (which is accelerated by coal burning) become more clearly defined, we may be forced to reevaluate the costs and benefits of nuclear power. It takes years to develop the knowledge to allow a wise choice. We must accumulate that knowledge base so that we can weigh wisely the real threat posed by carbon dioxide buildup in the light of the options before us, including the environmental safety and waste disposal problems of nuclear energy.

CLEANER WATER AND SAFE DISPOSAL OF WASTES

Our surface and subsurface waters are precious resources. Most of us take it for granted that when we want a drink of water or to go swimming or fishing, our streams, lakes, and aquifers will be safe to use. Yet our progress in protecting bodies of water from contamination has not generally been as successful as efforts in air pollution cleanup. Nonetheless, some important progress has been made. Lake Erie, once thought doomed to die biologically from reduced oxygen content (eutrophication) induced by phosphates and other nutrients, is making a comeback. Improved water treatment, coupled with more rigorous attention to hazardous waste treatment and disposal, holds the key to future advances. To recognize and control the sources of pollution, we must understand the intricacies of pollutant movement and conversion.

Nearly half of the citizens of the United States depend upon wells for their drinking water. A recent NAS assessment of groundwater contamination estimated that about 1 percent of the aquifers in the continental United States may be contaminated to some extent. Evidence of subsurface migration of pollutants makes it increasingly important to protect, with the best science and technology available, the aquifers feeding those wells.

A number of ground-burial disposal practices and waste deposit sites (repositories) have been used for many years with only minimal groundwater contamination. Procedures have been predicated on the assumptions that the waste material was unlikely to migrate and that, over time, the compounds would be oxidized, hydrolyzed, or microbially decomposed to harmless products. Now, however, some instances of serious groundwater contamination have appeared.

Some compounds have proven to be more stable and mobile than expected, while some of them are bacterially converted into more toxic and mobile forms.

Proposals currently under consideration for recovering seriously contaminated aquifers are soberingly expensive. For example, estimated costs for "containment" efforts at the Rocky Mountain Arsenal near Denver, Colorado, are about \$100 million and for "total decontamination" up to \$1 billion. Such enormous prospective cleanup costs require thoughtful weighing of the cost/benefit trade-offs to society in deciding what to do. More relevant here is the inescapable conclusion that it would be wise to invest the much smaller amount of public funds into research that will better define cleanup options and lessen the chances of other such incidents.

If the subsurface of the planet is to be used as a place for depositing our wastes, we must have much more thorough understandings of the physical/chemical/biological system it presents. We must be able to predict the movement and fate of

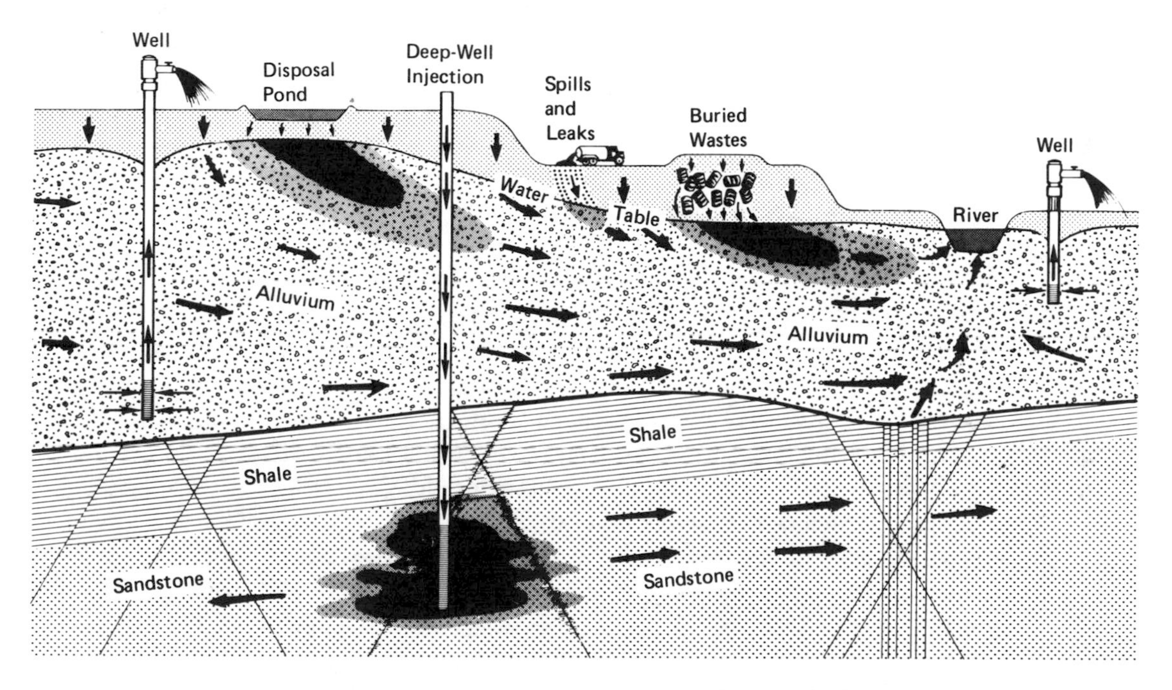

IT GOES IN HERE – BUT IT COMES OUT THERE

waste compounds with much greater confidence than is now possible. Laboratory and field studies must examine migration of compounds and ions through subsurface strata, and we must develop new analytical techniques for detecting and following the movement of polluted subsurface plumes (e.g., by measuring subsurface soil gases).

Groundwater quality can also be improved by developing improved methods for treating wastewaters, including industrial wastewaters that contain especially stable contaminants. Conventional wastewater treatment depends upon combinations of chemical and biological processes. While this is effective for some types of wastes, research is needed on advanced techniques such as exposure to ozone (ozonization), "wet air oxidation" (aqueous oxidation under high temperature and pressure), high- temperature incineration, and the use of adsorbants and resins for pollutant removal.

Innovative methods for recapture and recycling of valuable substances such as metals that would otherwise contribute to water pollution are also needed. Solvent extraction, ion exchange, reverse osmosis, and other chemical separation processes deserve study. Mines pose special problems. Acid mine drainage and movement of radioactive mine tailings are subjects of continuing studies that should reduce unwanted environmental effects.

Agriculture has depended increasingly on pesticides to control disease and insects and to increase food production. An undesired result has been gradual contamination of water supplies in some areas. Assessment of the chemical fate of pesticides and the development of acceptable, degradable alternatives are important research objectives.

It is clear that chemists, geologists, and environmental engineers will need to address these problems in water and waste treatment to safeguard our water resources.

RADIOACTIVE WASTE MANAGEMENT

At present, it is thought that the best place to store radioactive waste is underground rather than, for example, in the oceans, in space, or in accessible surface sites. This choice means that an understanding of the fundamental geochemistry of the underground storage areas is required. We must be able to make reliable predictions concerning possible radionuclide movement through the earth surrounding the storage site. However, mathematical modeling of such movements to indicate the suitability of a given site requires knowledge in several key areas. First, we must understand how much the radiation and heat release associated with the stored radioactivity will affect the local geochemistry (the groundwater chemistry and mineralogy, for example). Next, we have to understand the manner in which radioactive elements are carried through the soil. Do they form water-soluble complexes? Are they adsorbed onto the surfaces of colloidal particles that are then carried along in suspension? We must also look for chemical behaviors that would cause the radioactivity to stay forever wherever we put it. Chemical conversion of the radioactive elements to compounds of very low solubility in water is an example. Adsorption onto stationary solid surfaces is another.

Perhaps most difficult is the need to make predictions that will be reliable far into the future. Here we may find guidance in observations from the geologic record, including those observations connected with the natural reactor discovered at Oklo, in West Africa (see Section IV-C). This need for long-range predictability implies that we should be looking at other means of dealing with radioactive waste besides underground storage, means that will permit easier access to and monitoring of waste deposits. Perhaps this way the risks can be more clearly defined and controlled. Most important, we must have the knowledge base needed to be sure we aren't missing any options and that we understand the relative advantages and risks of each.

SUPPLEMENTARY READING

Chemical & Engineering News

"Tending the Global Commons: Nations Struggle for Ways to Check Global Warming and Depletion of Stratospheric Ozone" by L. Ember (C.& E.N. staff), vol. 64, pp. 14-64, Nov. 24, 1986.

"Chemistry in the Thermosphere and Ionosphere" by R.G. Roble, vol. 64, pp. 23-38, June 16, 1986.

"Incineration of Hazardous Wastes at Sea" by P. Zurer (C.& E.N. staff), vol. 63, pp. 24-35, Dec. 9, 1985.

"Dioxin, A Special C.& E.N. Issue," vol. 61, pp. 7-63, June 6, 1983.

"Federal Food Analysis Program Lowers Detection Limits" by W. Worthy (C.& E.N. staff), vol. 61, pp. 23-24, Mar. 7, 1983.

"Chemistry in the Troposphere" by W.L. Chamedies and D.D. Davis, vol. 60, pp. 39-52, Oct. 4, 1982.

Science

"Treatment of Hazardous Wastes" by P.H. Abelson, vol. 233, p. 509, Aug. 1, 1986.

"Inorganic and Organic Sulfur Cycling in Salt-Marsh Pore Waters" by G.W. Luther III, T.M. Church, J.R. Scudlark, and M. Cosman, vol. 232, pp. 746-749, May 9, 1986.

Scientific American

"Dioxin" by F.H. Tschirley, vol. 254, pp. 29-35, February 1986.

ACS Information Pamphlets

"Acid Rain," 8 pages, December 1985.

"Chemical Risk: A Primer," 12 pages, December 1984.

"Hazardous Waste Management," 12 pages, December 1984.

"Ground Water," 13 pages, December 1983.

Pamphlets available from:
American Chemical Society
Department of Government Relations & Science Policy
1155 16th Street, NW
Washington, DC 20036

CHAPTER III

Human Needs Through Chemistry

In this chapter, we will see how chemistry is essential in our eternal struggle for survival, freedom from toil, and, finally, for comfort. All of the tangible human needs will be considered: food, energy, materials, health, products that raise the quality of life, and economic vitality. We will begin with the most fundamental of these needs, adequate food supply for an ever-increasing world population.

Whipping a Wicked Weed

The plant *Striga asiatica* is one of the most devastating destroyers of grain crops in the world. This wicked weed restricts the food supply of more than 400 million people in Asia and Africa. It is a parasite that nourishes itself by latching onto and draining the vitality of a nearby grain plant. The results are stunted grain, a meager harvest, and hungry people.

Basic research on *Striga asiatica* by chemists and biologists has revealed one of the plant world's incredible host-parasite adaptations. The parasite seed lies in wait until it detects the proximity of the host plant by using an uncanny chemical radar. The give-away is provided by specific chemical compounds exuded by the host. *Striga asiatica* can recognize the exuded compounds and use them to trigger its own growth cycle. Then the parasite has an independent growth period of 4 days, during which it must locate the nearby host.

Researchers trying to solve the mystery of this recognition system faced formidable obstacles; they were seeking unknown, complex molecules produced only in tiny amounts. But, by extending the sensitivity of the most modern instruments, chemists have been able to deduce the chemical structures of these host-recognition substances, even though the agricultural scientist could accumulate the active chemicals in amounts no larger than a few bits of dust (a few micrograms). One method used, nuclear magnetic resonance (NMR), depends upon the fact that the nuclei of many atoms have magnetic fields that respond measurably to the presence of other such nuclei nearby. Thus precise NMR measurements reveal molecular geometries, even of ornate molecules. A second, equally sophisticated approach is high-resolution mass spectometry. In a high vacuum, molecules are given an electric charge, then accelerated with a known energy. By measuring the velocities at which these molecules and fragments from them are traveling (or their curved paths in magnetic fields), chemists can measure the masses and decide the atomic groupings present. These are critical clues to the molecular identities.

Now, the complex host-recognition (xenogistic) substances have been identified and their detailed structures are known. With this information in hand, we may be able to beat this wicked weed at its own game. Chemists can now synthesize the substances and give agricultural scientists enough material for field tests designed to trick the parasite into begining its 4-day growth cycle. It will die out never having found its host. A few days later, grain can be planted safely.

With this success for guidance, similar host-parasite relationships are being sought—and found—here in the United States. In addition to grains, bean crops have similar parasite enemies. Thus in collaboration with agricultural and biological scientists, chemists play a crucial role in our efforts to increase the world's food supply and eliminate hunger.

III-A. More Food

Agriculture, discovered 12,000 years ago, was the beginning of man's attempt to enhance survival by increasing the food supply. The human population at that time was about 15 million, but agriculture helped it rise to 250 million 2,000 years ago. By 1650, it had doubled to 500 million. But then it took only 200 years, until 1850, for the world population to double again, to one billion. Eighty years later, in 1930, the 2 billion level was passed. The acceleration has not abated: by 1985, the number of humans to be fed had reached 5 billion. If the growth were to continue at the 1985 level of 2 percent per year, the world population in 2020 would be about 10 billion. The rate of natural increase in population seems to be slowing worldwide (Table III-A-1), with the industrial countries adding only 80 million up to the year 2000. This is not the case for Africa, however, where population growth has been accelerating at an alarming pace.

Table III-A-1 Population Growth-Rate, 1960-1980

Area	Annual Percent Increase 1960-1965	1975-1980	Percent Change
World	1.99	1.81	−9.0
Industrialized	1.19	0.67	−43.7
Asia	2.06	1.37	−33.5
Latin America	2.77	2.66	−4.0
Africa	2.49	2.91	+16.9

SOURCE: W.P. Mauldi. 1980. *Science* 209:148-157.

In 1983, about 20 million human beings starved to death—about 0.5 percent of the world's population. Moreover, an additional 500 million were severely malnourished. Estimates indicate that by the end of the century, the number of severely undernourished people will reach 650 million.

Plainly, one of the major and increasing problems facing the human race will be providing itself with adequate food and nourishment and, ultimately, to limit its own population growth. And whose problem is this? In the most elemental way, it is the problem of those who are hungry, those who are undernourished, those who are least able to change the course of events on more than a personal and momentary scale. But human hunger is also the problem, indeed, the *responsibility,* of those who *can* affect the course of events. Any attempt to fulfill that responsibility will surely need the options that can come from science, and among the sciences that can generate options, chemistry is seen to be one of the foremost. It can do so, first, by increasing food supply and, second, by providing safe methods by which individuals can limit population growth (see Section III-E).

Food production cannot be significantly increased simply by cultivating new land. In most countries, the farmable land is already in use. In the heavily populated developing countries, expansion of cultivated areas requires huge capital investments and endangers the local ecology and wildlife. To increase the world food supply, we need improvements in food production; food preservation; conservation of soil nutrients, water, and fuel; and better use of solar energy through photosynthesis. Science is providing such improvements and chemistry is playing a central role by clarifying the actual chemistry involved in biological life cycles. We are developing an understanding at the molecular level of the factors that can be controlled to aid in the fight for more food. These factors include the hormones, pheromones, self-defense structures, and nutrients at work in our animal and plant food crops and also those of their natural enemies.

We can best address these problems by using our present understanding of living systems. Pest control, for example, is an essential element of efficient food production. The emphasis in this area has been on the use of chemicals that attempt to eliminate insects or other pests by killing them (biocidal agents). This method risks upsetting nature's balance and introduces foreign substances into the environment. But we want to *control* insect pests, not exterminate them. Then we can avoid the potentially devastating effects that may accompany profound ecological disturbances. By understanding the biochemistry of the organisms themselves, we can limit the impact of pests on food production in ways that can be used indefinitely without harmful effects on nature. Increasingly, such fundamental questions about biological systems have become questions about molecular structures and chemical reactions.

The following examples vividly display the key role chemistry plays in our current attempts to expand the world food supply.

PLANT HORMONES AND GROWTH REGULATORS

Growth regulators are chemical compounds that work in small concentrations to regulate the size, appearance, and shape of plants and animals. They include natural compounds produced within the organism and also some natural products that come from the environment. However, many similar compounds (analogs) that have been synthesized in the laboratory have been found to function as growth regulators. They are usually patterned after compounds found in nature, and some of them work just as effectively but without undesired side effects. The chemicals already present in plants or animals that exert regulatory actions are called hormones (e.g., growth hormones and sex hormones). A hormone can be said to be a chemical message sent between cells. The so-called plant hormones include growth substances (e.g., auxins, gibberellins, and cytokinins) and growth inhibitors (e.g., abscisic acid and ethylene) that seem to be structurally unrelated.

These growth regulators are surely of immense social (and economic) importance for the world's future because they influence every phase of plant development. Unfortunately, even though we know the structures of many plant growth regulators, we have little insight concerning the molecular basis for their activity. Since chemical interactions and reactions are involved, chemistry must play a central and indispensable role in the development of this insight.

Below are some typical growth regulators. Notice the variety of molecular structures nature has developed for this function. By establishing the exact makeup of these structures, scientists have taken an essential step toward understanding and controlling the growth processes being regulated.

Indole Acetic Acid (IAA), an Auxin (1)

This compound was the first plant hormone to be characterized. It promotes plant growth, the rooting of cuttings, and the formation of fruit without fertilization.

The synthesis of numerous IAA analogs led to the first commercial herbicide, 2,4-dichlorophenoxyacetic acid, or 2,4-D.

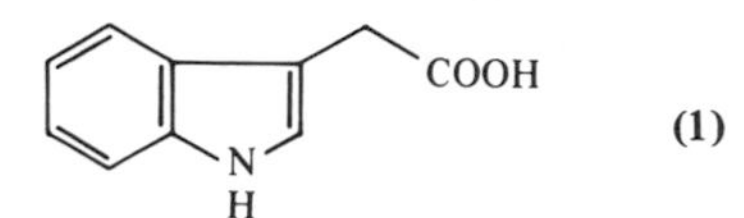

(1)

INDOLE ACETIC ACID (IAA)
Promotes plant growth

Gibberellic Acid (GA) (2)

More than 65 compounds related to gibberellic acid have been characterized from plants and lower organisms since their landmark discovery in the fungus *Gibberella fujikuroi*. Commercially produced by large-scale cultures of this fungus, GA has extensive use in agriculture. Its applications range from triggering formation of flower buds to growing seedless grapes and manufacturing malt in the beer industry.

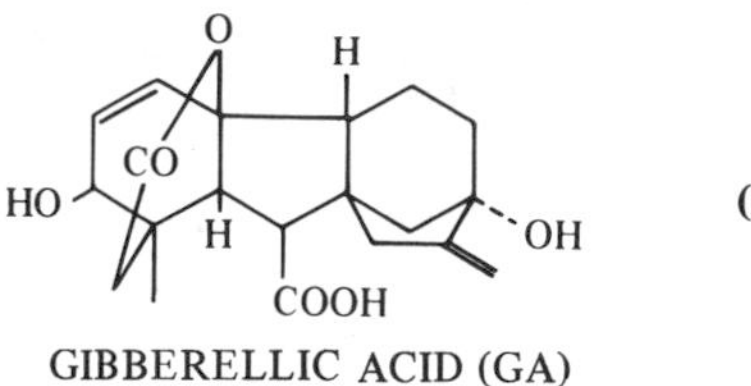

(2)

GIBBERELLIC ACID (GA)
Triggers flowering

Cytokinins (3)

The first cytokinin that was isolated is a compound that enhances cell division. Many analogs, including *trans*-zeatin, have since been isolated from DNA, transfer RNA, and other sources, and quite a number have been synthesized. They promote cell division, enhance flowering and seed germination, and inhibit aging.

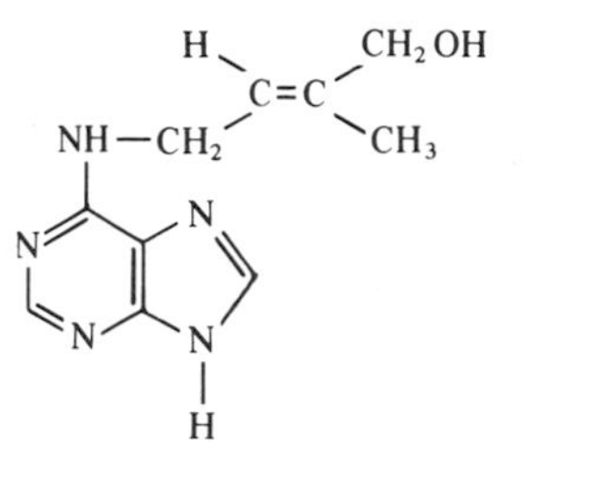

(3)

TRANS-ZEATIN
Promotes seed germination

Ethylene (4)

This simple gas behaves like a hormone since it encourages fruit ripening, leaf-drop, and germination, as well as growth of roots and seedlings. Currently, a substance that generates ethylene above pH 4 is used widely as a fruit ripener. It is suggested that ethylene regulates the action of the growth hormones auxin, GA, and cytokinin.

H H
C=C
H H

(4)

ETHYLENE
Ripens fruit

Strigol (5)

The seeds of witchweed (*Striga*) lie in the soil for years and will only sprout when a particular chemical substance is released by the root of another plant. The weed then attaches itself to the root of the plant and lives off of it (as a parasite). The active substance strigol now has been isolated from the root area of the cotton plant and its structure identified. Now it has been synthesized. Strigol and its synthetic analogs are proving effective in removing these parasitic weeds by causing them to sprout and die off before a crop is planted.

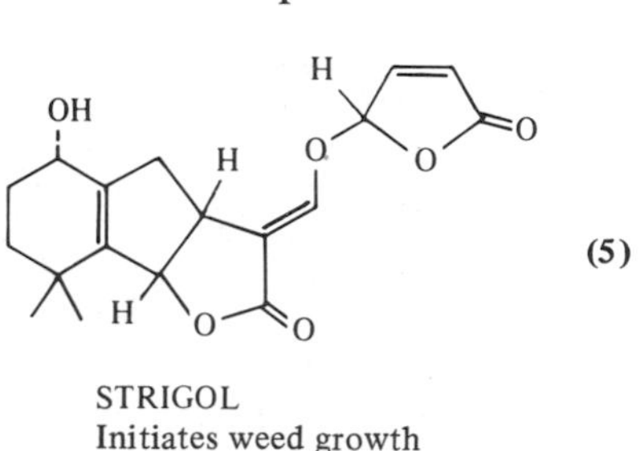

(5)

STRIGOL
Initiates weed growth

G2 Factor or Trigonelline (6)

This compound was discovered to be involved in one stage of a four-step cycle in plant cell reproduction. This stage, called G2 or "gapZ," is characterized by a pause in growth activity. In isolating this compound, the "first leaves" or cotyledons of 15,000 garden pea seedlings provided only one quarter of a milligram of the G2 factor. This compound may be of particular importance because of a link between the G2 stage and the formation of root bumps (called nitrogen nodules) that have the power to convert elemental nitrogen from the soil to nitrate, which enriches the soil.

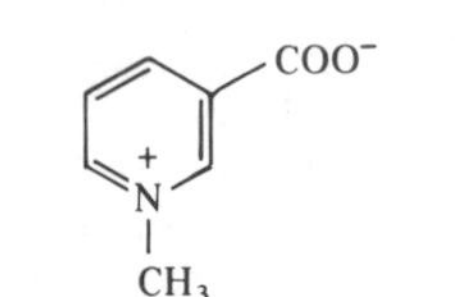

TRIGONELLINE G2 FACTOR
Influences nitrogen fixation

Glycinoeclepin A (7)

Nematodes are tiny worms that inflict huge damage on such crops as soybean and potato. The nematode eggs can rest unchanged in the soil for many years until the root of a nearby host plant releases a substance that will promote hatching. The first such hatching initiator was isolated and understood recently. During a span of 17 years, a total area corresponding to 500 football fields was planted with soybeans to give 1.5 mg of the active substance, glycinoeclepin A, which was then shown to have the unusual structure **(7)**. Synthetic analogs may someday be applied agriculturally to force nematode eggs to hatch before a crop is planted.

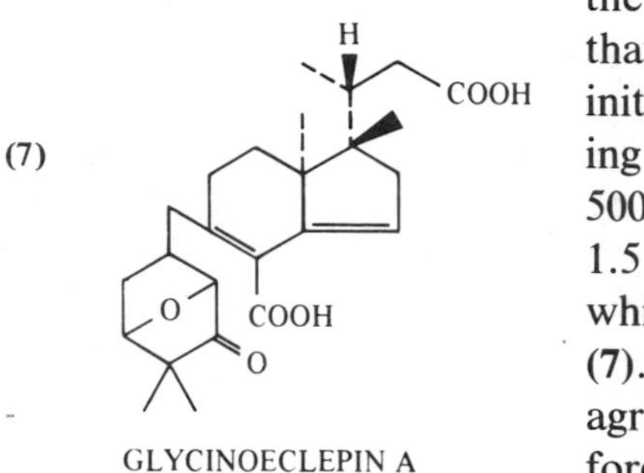

GLYCINOECLEPIN A
Promotes worm egg hatching

Hundreds of natural plant products are now known to exert growth regulatory activity of one sort or another. These compounds represent a surprising range of structural types. Recognition of these structures is the first step toward their systematic use to increase the world's food supply. We are only at the beginning of this important process.

INSECT HORMONES AND GROWTH REGULATORS

Insects that attack food-bearing plants reduce crop yields and, thereby, limit food supplies. The ability to understand and control these natural enemies provides another dimension by which the world's food supply can be increased. The desire to reduce malnourishment and starvation around the globe is not incompatible with the strong element of environmental concern in our society. Pests can be controlled without being exterminated. Furthermore, with the sensitivity of detection methods constantly improving, we can be assured that pest control can eventually be monitored to give ample early warning of unexpected side effects. Certainly, knowledge of the basic chemistry involved in the growth and increase of insect populations should be extended to provide options that may be needed to save human lives.

Molting Hormones (MH) (8)

Two types of hormones are directly involved in the development of insects (known as metamorphosis)—molting hormones and juvenile hormones. The molting hormones cause insects to shed their skins; an example is 20-hydroxyecdysone

(**8**). Nine milligrams of this complex substance (a steroid) were painstakingly extracted from one ton of silkworm pupae (a cocoon-like stage of insect development). It was also shown to be the active molting hormone of crustaceans, using two milligrams isolated from one ton of crayfish waste. It was discovered that molting hormones are widely distributed in plants and are probably produced as a defense against insects. Approximately 50 such steroids with insect MH activity have been identified.

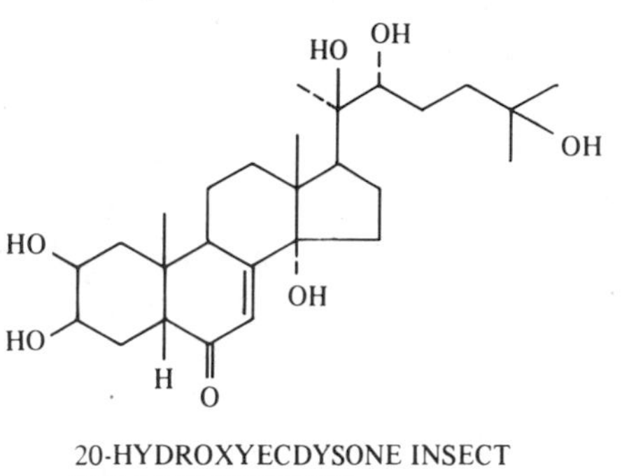

20-HYDROXYECDYSONE INSECT MOLTING HORMONE
Causes insects to shed their skins (8)

Juvenile Hormone (JH) (9)

These hormones tend to keep insects in the juvenile state. The first JH (**9**) was identified using 0.3 milligrams of sample isolated from the butterfly *Lepidoptera*. Several JH analogs are now known, the most universal being JH-III with three methyl groups on carbons 3, 7, and 11. Their importance has stimulated syntheses of thousands of related compounds, one of which is methoprene (**10**). This biodegradable compound imitates (mimics) the natural hormone, and therefore, insects will not readily become resistant to it; it is widely used to kill the larval stage of fleas, flies, and mosquitoes. Because it produced oversized larvae and pupae by prolonging the juvenile stage in silkworm, it has been widely used in China to increase silk production.

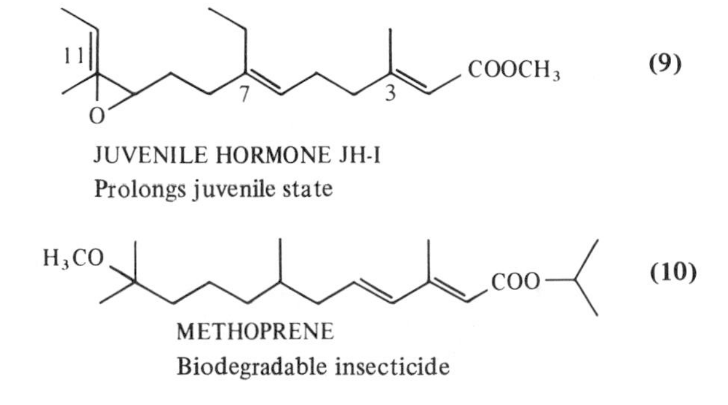

JUVENILE HORMONE JH-I
Prolongs juvenile state (9)

METHOPRENE
Biodegradable insecticide (10)

Anti-Juvenile Hormones

These are substances, both natural and man-made, that somehow interfere with normal juvenile development. Systematic screening of plants has led to the identification of a number of compounds with anti-JH activities. They are called precocenes (**11**). Some insects develop prematurely into tiny sterile adults when they are treated with precocenes.

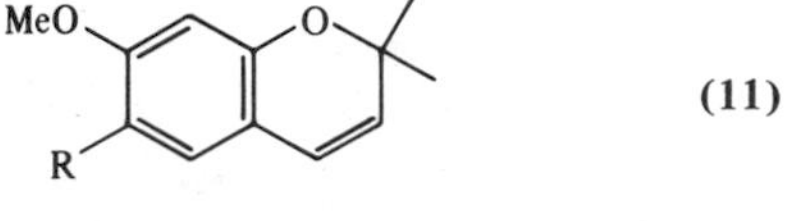

PRECOCENES R=H OR OCH_3
Causes early maturation (11)

Natural Defense Compounds: Antifeedants

Plants produce and store a number of chemical substances used in defense against insects, bacteria, fungi, and viruses. One category of such defense substances is made up of chemical compounds that interfere with feeding. Many antifeedants have been characterized and show a wide variety of structure. Among them, azadirachtin **(12)** is probably the most potent antifeedant isolated to date. It is found in the seeds of the neem tree *Azadirachta indica,* which is known for its use in folk medicine. An amount of only 2 ng/cm^2 (2×10^{-9} g/cm^2) is enough to stop the desert locust from eating. Although **(12)** is far too complex for commercial synthesis, it might be possible to isolate it in useful amounts from cultivated trees. It is known that **(12)** is not poisonous because twigs from the neem tree have been commonly used for brushing teeth, its leaves are used as an antimalarial agent, and the fruit is a favorite food of birds.

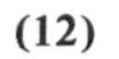
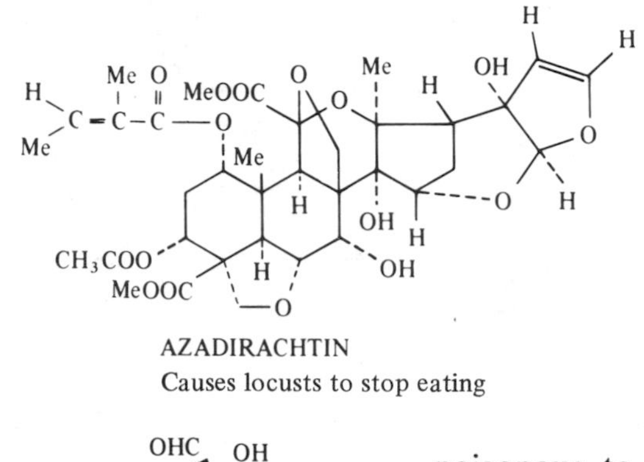

AZADIRACHTIN
Causes locusts to stop eating

Warburganal **(13)** seems to be specifically active against the African army worm. An insect kept for 30 minutes on corn leaves sprayed with warburganal will permanently lose its ability to feed. The plant from which warburganal has been isolated is also commonly used as a spice in East Africa and therefore cannot be highly poisonous to humans. Practically all antifeedants are isolated from plants that are resistant to insect attack. While no antifeedant has yet been developed commercially, they offer an intriguing new avenue for control of insect pests.

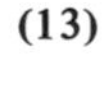

OHC OH CHO

WARBURGANAL
Causes worms
to stop feeding

INSECT PHEROMONES

Pheromones are chemical compounds released by organisms in order to trigger specific behaviors from other individuals of the same species. Pheromones function as communication signals in mating, alarm, territorial display, raiding, nest mate recognition, and marking. They have attracted great interest as a means to monitor and perhaps control pest insects.

The first insect pheromone to be identified was from the female silkworm, and it was shown to be an unbranched C_{16} alcohol containing two double bonds, structure **(14)**. Since then, hundreds of pheromones have been identified, including those for most major agricultural and forest pests. The isolation and full identification always involve handling extremely small quantities. Characterization of the four pheromones for cotton boll weevil pheromones **(15A-D)** required over 4 million weevils and 215 pounds of waste material (feces). It took over 30 years to clarify the structure that

(14)

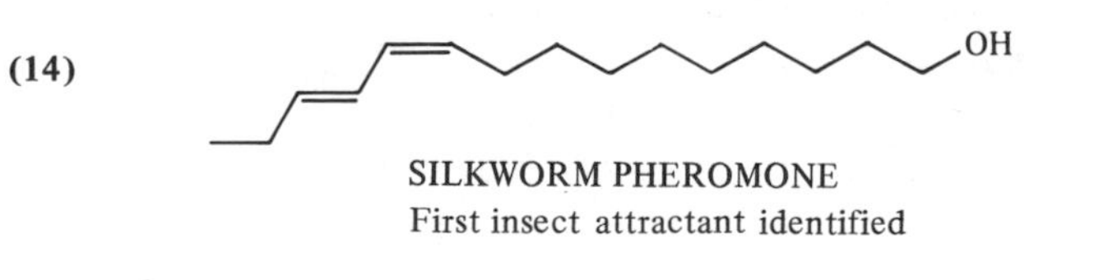

SILKWORM PHEROMONE
First insect attractant identified

stimulates mating in the American cockroach (**16**). It required processing of 75,000 female cockroaches, which finally resulted in 0.2 mg of one compound and 0.02 mg of another.

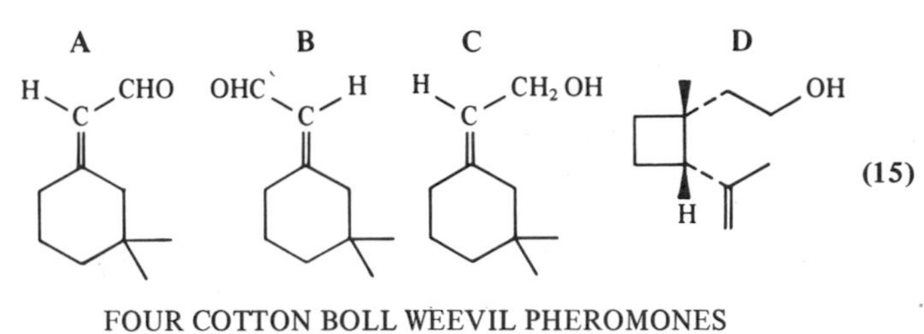

(15)

FOUR COTTON BOLL WEEVIL PHEROMONES
A tiny bit is enough

Special methods for collecting and analyzing these compounds had to be developed to cope with the tiny quantities being investigated. It is now possible to extract a single female moth gland, remove the intestines of a single beetle, collect airborne pheromones on glass wool, and analyze the pheromone from a single insect. One of the most important developments in this area is the electroantennogram technique, in which a single sensory unit from an olfactory antenna hair (used for smell by the insect) is used by researchers to detect the presence of these compounds.

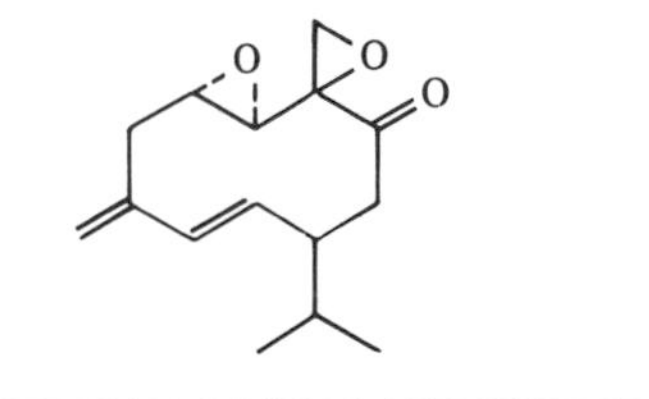

(16)

AMERICAN COCKROACH PHEROMONE
Stimulates mating

In addition to natural pheromones, chemists continue to synthesize artificial pheromones. Pheromone-baited traps have been used worldwide to monitor and survey pest populations. They assist in precise timing of insecticide application, thus reducing the amount of spray, and in insect trapping operations. For example, more than one million traps have been recently deployed for a period of 4 years in the Norwegian and Swedish forests, resulting in spruce bark beetle captures of 4 billion a year. Another commercial use is pheromone distribution throughout an area to confuse the insects. In 1982, pheromones were used on 130,000 acres of cotton to control pink bollworms, on 2,000 acres of artichokes to control plume moths, and on 6,000 acres of tomato to fight pinworms.

Many questions about the basic chemistry and biology of pheromones remain to be answered. In the long run, it is clear that research on pheromones will yield useful benefits to agriculture and to health.

PESTICIDES

Pesticides—insecticides, herbicides, and fungicides—are essential to our attempts to improve food and fiber production and to control insect-transmitted diseases in humans and livestock. Although major changes have recently occurred in pesticide use, environmental concerns make it increasingly difficult to introduce better pesticides into practical use in this country. The time and cost of developing a new compound currently run about 10 years and $30 million. Over 10,000 new compounds normally have to be synthesized and tested before a single acceptably safe—and therefore marketable—pesticide is found.

Insecticides

Most potent insecticides discovered recently are modeled on natural products and act on the nervous system of insects. They include deltamethrin **(17)** and cartap **(18)**, which are based on compounds found in chrysanthemum flowers and marine worms. Another compound still on the drawing board is pipercide **(19)**, which includes an unusual cyclic diether unit. Chemical synthesis and testing programs have led to other novel structures that act as nerve poisons, inhibitors of chitin synthesis, and growth disruptors (e.g., **(20)**). This new range and variety of insecticide classes has helped immensely in the pest control battle.

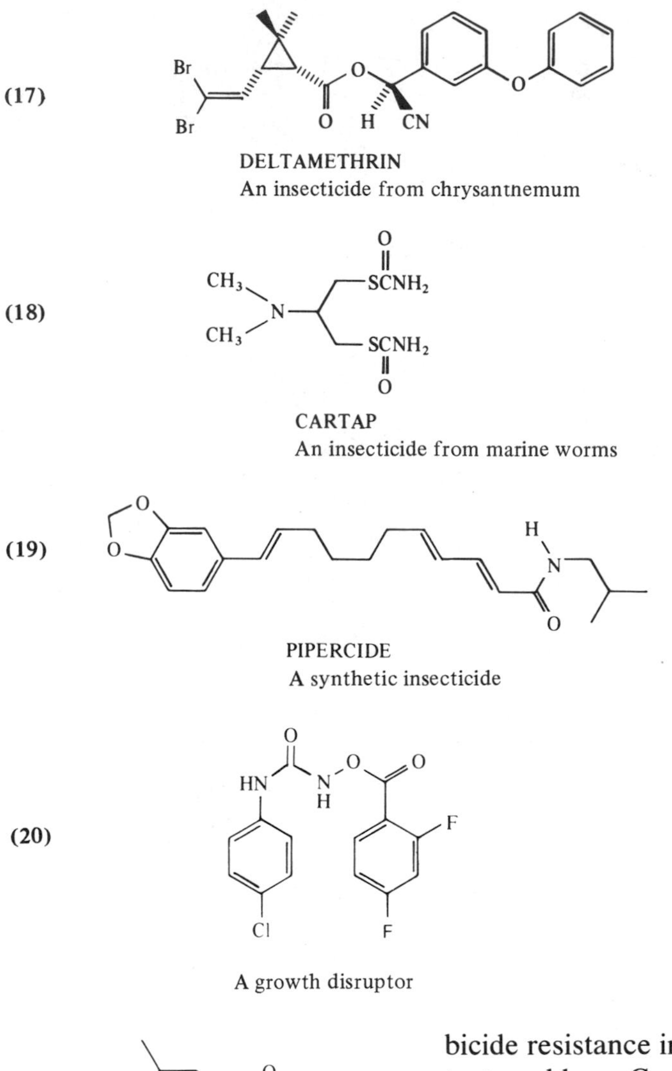

DELTAMETHRIN
An insecticide from chrysantnemum

CARTAP
An insecticide from marine worms

PIPERCIDE
A synthetic insecticide

A growth disruptor

BUTYLATE
A weed control agent

Herbicides

These are substances that work to control weed pests. Highly novel structures derived through chemical synthesis have provided a variety of new herbicides in recent years. The butylates **(21)** work on the weed before it emerges from the soil, while atrazine **(22)** blocks photosynthesis by the weed. Still others interfere with seed germination or block formation of chlorophyll. Herbicide resistance in weeds is an increasingly important problem. Genetic research currently directed toward improved crop tolerance suggests that we should transfer to the crop the gene that a weed has developed to make itself herbicide resistant.

Fungicides

Major advances have been made in fungicides and antibiotics to control plant diseases caused by fungal and bacterial microorganisms. Some fungicides, such as triadimefon **(23)**, work by slowing RNA synthesis. Other compounds block cell division or formation of cell walls, as in benomyl **(24)**. New fungicides are needed that are not only highly selective of their targets but that may also disrupt more than one biological function so that resistance is less likely to develop.

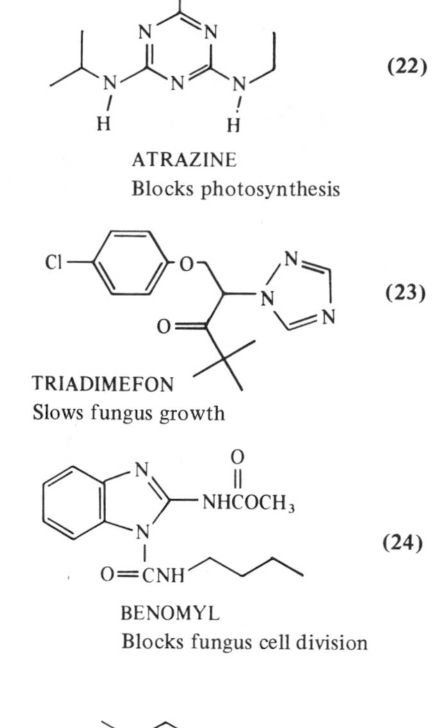

ATRAZINE
Blocks photosynthesis

TRIADIMEFON
Slows fungus growth

BENOMYL
Blocks fungus cell division

DIPROPYLNITROSAMINE
A herbicide impurity

Special Techniques

Specialized techniques, instrumentation, and facilities are required to solve the multidisciplinary problems encountered in pesticide chemistry. The quantities of pesticide that can be used on crops are restricted so that crops will be free of hazardous leftover chemicals. The chemical by-products of pesticide use are also being evaluated for environmental impact and safety levels. Some hazardous impurities have been placed under strict control, such as tetrachlorodibenzodioxin (''dioxin,'' an impurity in the herbicide 2,4,5-T) and nitrosamines **(25)** that occur in some other useful herbicides. The fact that pesticide research involves many scientific disciplines requires increased cooperation on a local, national, and international basis between industrial, government, and university scientists.

Research into pesticide chemistry can provide farmers and public health officials with safe and effective means to control pests. The research permits replacement of compounds that may be highly toxic or that have unfavorable long-term effects with better and environmentally safe pesticides. Because pest control problems are complex, but of extreme importance to society's well-being, long-term commitments to pesticide research are necessary and will be rewarding.

FIXATION OF NITROGEN AND PHOTOSYNTHESIS

All of our food supply ultimately depends upon the growth of plants. Hence, a fundamental aspect of increasing the world's food supply is to deepen our knowledge of plant chemistry. Because of special promise, two frontiers deserve special mention—nitrogen fixation and photosynthesis.

Nitrogen Fixation

Nitrogen is a crucial element in the chemistry of all living systems and one that can limit food production. Since nitrogen is drawn from the soil as the plant grows, restoring nitrogen to the soil is a primary concern in agriculture. This concern accounts for the centuries-old practice of crop rotation, and it figures importantly in the choice and amounts of fertilizers used by farmers. Ironically, nitrogen is abundant—air is 80 percent nitrogen—but it is present in the elemental form that is difficult to convert into useful compounds. Some plants know how to convert this elemental nitrogen into compounds they can use; we'd like to know just how they do it.

Certain bacteria and algae are able to reduce nitrogen in the air to ammonia (nitrogen fixation), which is then converted into amino acids, proteins, and other nitrogenous compounds by plants. A rather diverse group of organisms has the capability of reducing nitrogen. A group of plants called legumes, which includes soybeans, clover, and alfalfa, has the capability of fixing nitrogen, with the assistance of bacteria that live on their roots. About 170 species of nonleguminous plants also fix nitrogen in this manner. Additional nitrogen fixers in nature are certain free-living bacteria and the blue-green algae.

Nitrogen fixation involves an enzyme called nitrogenase which consists of two proteins. One protein (dinitrogenase) has a molecular weight of approximately 220,000. It contains 2 molybdenum atoms and about 32 atoms each of iron and reactive sulfur atoms. The other protein (dinitrogenase reductase) is made up of two identical subunits of 29,000 molecular weight, each containing 4 iron and 4 sulfur atoms.

The sequence of events involving this enzyme complex in the reduction of elemental nitrogen to ammonia has been partially resolved by spectroscopic and purification techniques. Many critical aspects are not yet understood. Research with other compounds that can also be reduced with these enzymes (e.g., acetylene, cyanide, hydrogen ion, and cyclopropane) may provide clues. In another direction, a number of novel metal organic compounds are showing promise as soluble catalysts for nitrogen fixation.

On another active frontier, genetic studies are being applied to nitrogen fixation in plants. Recombinant DNA techniques might permit control of a plant's aging to extend its period of nitrogen fixation or development of more efficient nitrogen-fixing strains of bacteria. A still more adventurous goal would be to transfer genetically the nitrogen fixation ability to food-bearing plants so that they become self-fertilizing.

Photosynthesis

Photosynthesis will be discussed in Section III-C in its relevance to the world's energy supply. Since all of our food supply ultimately depends upon the growth of plants, we see that photosynthesis is also the key to the world's food supply. Photosynthesis is the process in nature by which green plants, algae, and photosynthetic bacteria use the energy from sunlight to stimulate chemical reactions in plants. These reactions convert carbon dioxide and water into organic building block molecules used by the plant cells which act as chemical factories that satisfy the plant needs. Since 10^{11} tons of carbon are annually converted into organic compounds by photosynthesis, determination of the mechanisms of photosynthesis remains an important goal. Despite the rapid progress described in Section III-C, we are still far from duplicating natural photosynthesis in the laboratory. Nevertheless, chemists hope and expect to add to the world's food supply (as well as its energy supply) by developing an artificial photosynthetic system that will use solar energy to produce safe and abundant animal feedstocks.

FOOD FROM THE SEA

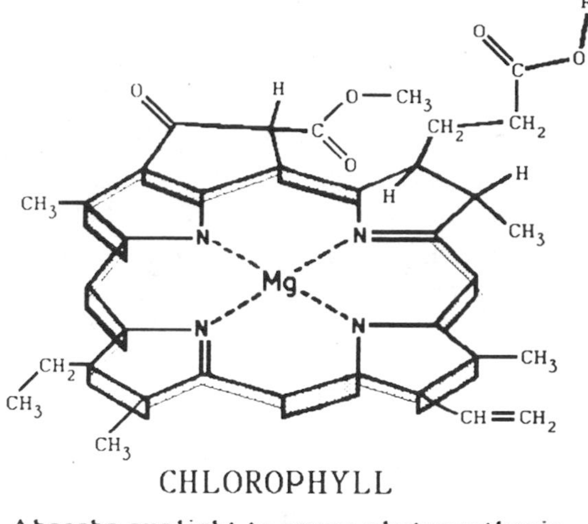

CHLOROPHYLL
Absorbs sunlight to power photosynthesis

Seventy-one percent of the Earth's surface is covered by water, so more than two-thirds of the solar energy potentially available for photosynthesis is absorbed in our oceans and seas. Yet, on a global scale, food from the waters has not been as important as that from terrestrial sources. Of the total of 3.3 billion tons of food harvested in 1975, only 2 percent came from the ocean and inland waters. Moreover, the harvest of fish, mollusks, and crustaceans has leveled off in recent years. Significant advances can be made, for example, in aquaculture technology and in the cultivation of algae, fish, and crustaceans. Knowledge of the chemistry of biological life cycles in marine species is an important requirement for such advances.

ISOLATION AND CHARACTERIZATION TECHNIQUES OF BIOACTIVE MOLECULES

All of the advances discussed above are the more remarkable in view of the tiny amount of quite complex molecular compounds that are available for isolation and identification. While a successful purification can take years of work, it is a necessary first step in explaining a biological behavior on the basis of molecular structure. In addition, new and unique methods are being developed to determine precisely how much and what kind of a chemical has been isolated. In the case of biologically active molecules, these methods themselves are often biological in nature and are allowing chemists to work effectively with extremely small amounts of material, in the range of a thousandth of a millionth of a gram (i.e., a nanogram, 10^{-9} g).

These separation and isolation techniques have been a major factor responsible for the launching of genetic engineering. A key process in genetic engineering is the ability to cut the DNA strand at specific sites. This cutting is called "cleavage." Discovering what the cleavage products are requires that they be separated in reasonable purity. Chromatography and electrophoresis techniques fill this need.

None of the molecular structures displayed in this section could have been determined without the use of the most modern spectroscopic methods. The instrument with the widest impact has undoubtedly been nuclear magnetic resonance (NMR), which permits clarification of the local molecular neighborhood of individual atoms in a large molecule.

Mass spectrometry has also been extremely effective, as measured by our ability to identify larger and larger molecules. Currently, solids with molecular weights up to 23,000 and without noticeable vapor pressure can be measured, and under favorable conditions, as little as 10^{-13} g of the solid is needed. Computer-aided infrared and Raman spectroscopy show the vibrational motions that characterize certain chemical groupings. Diffractive methods (X-ray, neutron, electron microscope) can now clarify structures and shapes of nonrigid biopolymers, including flexible proteins. These powerful instruments have played a central role in the advances already mentioned in this section; they are essential to our continued progress in the science that underlies modern methods of food production.

CONCLUSION

Food supply and efficient use of energy are rapidly emerging as major concerns for the world's future. The theme "more food" requires understanding of the basic principles of nature so that wise choices can be made. The traditional disciplinary classifications of biology, chemistry, biochemistry, physics, physiology, and medicine are becoming less distinct; and cooperative effort among scientists with broad and overlapping interests is becoming common as research moves into topics dealing with the nature of life. In these cross-disciplinary collaborations, chemists are essential because we need to know the structures and shapes of molecules, their reactivities, and how to synthesize molecules of biological importance. Chemistry will play a central role in the search for options that will help us feed and limit the world's population in the decades ahead.

SUPPLEMENTARY READING

Chemical & Engineering News

"First Tunichrome Isolated and Characterized" by R.J. Seltzer (C.&E.N. staff), vol. 63, pp. 67-69, Sept. 16, 1985.

"Plants Natural Defenses May Be Key to Better Pesticides" (C.&E.N. staff), vol. 63, pp. 46-51, May 27, 1985.

"Proteinaceous Pheromones Found in Golden Hamsters" by R.J. Seltzer (C.&E.N. staff), vol. 62, pp. 21-23, Oct. 22, 1984.

"Pesticide Chemists Are Shifting Emphasis from Kill to Control" by W. Worthy (C.& E.N. staff), vol. 62, pp. 22-26, July 23, 1984.

"Cutting Carbonyl Group Stabilizes Weedkiller" (C.&E.N. staff), vol. 62, pp. 26-27, Apr. 23, 1984.

"Lemon Odor Helps Identify Male Moth Pheromone" (C.&E.N. staff), vol. 61, pp. 34-36, Sept. 19, 1983.

"Ultraviolet-Active Compounds Kill Insect Pests" (C.&E.N. staff), vol. 61, p. 334, Apr. 11, 1983.

"Allelopathic Chemicals, Nature's Herbicides in Action" by A.R. Putnam, vol. 61, pp. 34-45, Apr. 4, 1983.

"Herbicides" by H.E. Sanders (C.&E.N. staff), vol. 59, pp. 20-35, Aug. 3, 1981.

"Photosynthesis and Plant Productivity" by I. Zelitch, vol. 57, pp. 28-48, Feb. 5, 1979.

Beauty Is Only Skin Deep

Ever think of going into the gold-brick business? Just take a big hunk of gold and a hacksaw and you've got a good-looking brick with a nice heft. Unfortunately, one such brick and you're talking $140-150,000! There's no room for mark-up. But suppose you get an ordinary brick (wholesale in South Jersey, 17¢!) and just coat the surface with gold—the cost will come down a lot. And you'll have a beautiful brick—well, at least "skin-deep."

So how much would such a surface coating cost? For openers, put a one-atom-thick layer of gold atoms over the entire surface of the brick. Let's see, 2 inches by 4 inches by 8 inches—gold at $320 an ounce—one atom thick—that'll be . . . 0.3¢ worth of gold. Wow! There, we've got an attractive product at a total material cost of 17.3¢ (not including packaging).

That's pretty impressive. It means that the outermost layer (the surface) of a $150,000 piece of gold involves so few atoms that they would cost less than a cent. Yet that miniscule fraction of atoms on the surface of a piece of metal controls the chemistry of that piece. For instance, these surface atoms are the ones that determine whether the metal surface acts as a catalyst or not. And catalysts account, one way or another, for about 20 percent of our gross national product.

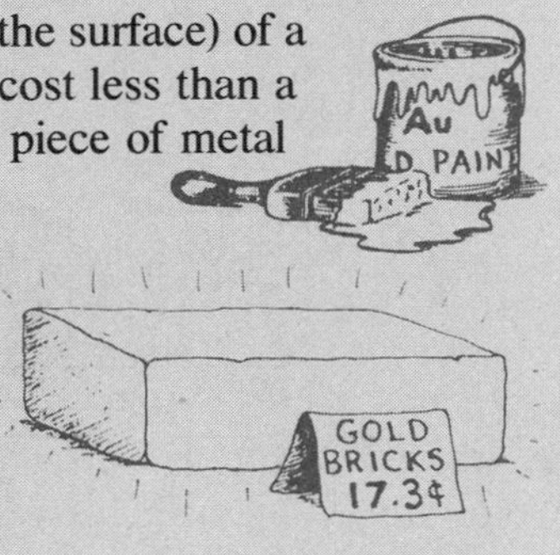

So what is a catalyst? It's a chemical substance that speeds up a chemical reaction without itself getting into the act (i.e., it is not consumed while doing its thing). A solid catalyst merely furnishes its surface as a meeting place for gaseous molecules. For instance, when a molecule of methanol lands on a rhodium catalyst surface, it usually sticks for a while (becomes adsorbed). Now, if a carbon monoxide molecule happens to arrive, zingo, it reacts with the adsorbed methanol molecule and they leave the surface as acetic acid. When methanol and carbon monoxide meet in the gas phase, they won't even give each other the time of day. But because of the special environment provided by that thin layer of surface atoms on the rhodium catalyst, methanol and carbon monoxide react so rapidly that 500,000 tons of commercial acetic acid are made every year this way! This kind of speed-up might be anywere from a thousand-fold to a million-fold when things are working.

Because of such successes, chemists care a lot about how these catalytic gold bricks do their job. What actually happens to that thin layer of adsorbed molecules as they come and go on a catalytic metal surface? Unfortunately, that's where the skin-deep principle works against us. If there isn't much on that surface, there isn't much to see.

But nowadays, we have several powerful instruments with which we can learn about the special properties of the skin of a metal. These instruments also let us watch molecules as they lodge on the surfaces of catalysts like platinum and rhodium and many others. We can see how the molecules are chemically changed by the metallic skin to make them more reactive when a suitable reaction partner comes along. So chemists are begining to understand how to design these catalytic gold bricks to do whatever we want. Right now, every gallon of your gasoline began as a bunch of molecules sure to make your engine knock and then some chemist catalytically converted them into other molecules that make your engine purr. But now we are looking ahead to new energy feedstocks with more sulfur and metallic contaminants that will require much better catalysts so that we can keep your engine purring and the air clean at the same time. We'll do it by learning how those catalytic gold bricks work so we can tailor them to our needs. This is a case where skin-deep beauty really pays off!

III-B. New Processes

A prime reason for wishing to understand and control *chemical reactions* is so that we can convert abundant substances into useful substances. When this can be done on an economically significant scale, the reaction (or sequence of reactions) is called a *chemical process*.

Many employed chemists are engaged in perfecting existing chemical processes and developing new ones. Their success is reflected in the present vitality and strength of the U.S. chemical industry. The industry makes billions of pounds of chemicals at low cost, in high yield, and with minimum waste products. For example, every year we produce 9.8 billion pounds of synthetic fibers (such as polyesters), 28 billion pounds of plastics (such as polyethylene), and 4.4 billion pounds of synthetic rubber. To sense the magnitudes of these annual production figures, imagine 10 Astrodome-size football stadiums—9.8 billion pounds of polyester would fill up all 10!

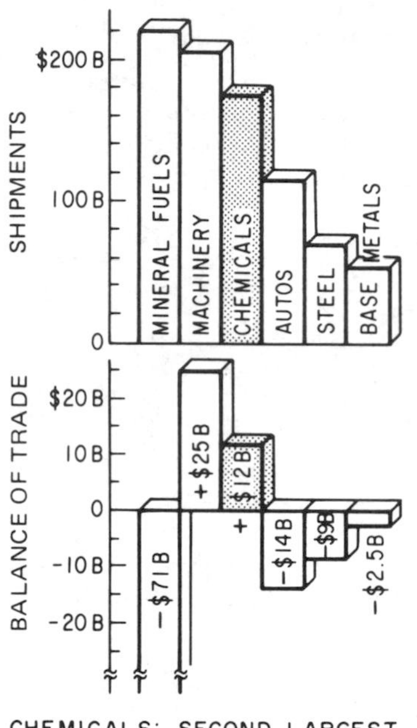

CHEMICALS: SECOND LARGEST POSITIVE TRADE BALANCE

Our current position of world leadership in this wide-ranging industry can be attributed to our strength in the field of chemical catalysis. The major role of catalysis in industry is indicated by estimates that 20 percent of the gross national product is generated through the use of catalytic processes. On the horizon, new catalysts will help us tap new energy sources (the subject of Section III-C).

A catalyst is a substance that speeds up chemical reactions without being consumed. Some reactions can be speeded up by a factor as large as 10 billion (10^{10}). A selective catalyst can have the same dramatic effect but working on only one of many competing reactions. A stereoselective catalyst not only controls the end product, it also favors a particular molecular shape, often with remarkable effects on the physical properties (such as tensile strength, stiffness, or plasticity) and, for biologically active substances, on the potency. Catalysis can be subdivided according to the physical and chemical nature of the catalytic substance.

- In *heterogeneous catalysis,* the catalyzed reaction occurs at the surface interface between a solid and either a gaseous or a liquid mixture of the reactants.
- In *homogeneous catalysis,* reaction occurs either in a gas mixture or in liquid solution in which both catalyst and reactants are dissolved.
- In *electrocatalysis,* reaction occurs at an electrode surface in contact with a solution but assisted by a flow of current. Thus, electrocatalysis is like heterogeneous catalysis, but it adds the opportunity to put in or take out electrical energy.
- In *photocatalysis,* reaction can take place at a solid surface (including electrode surfaces) or in liquid solution, but in these reactions energy encourage-

ment is provided by absorbed light.

- In *enzyme catalysis,* some characteristics of both heterogeneous and homogeneous catalysis appear. Enzymes are large protein structures that provide a surface, or interface, upon which a dissolved reactant molecule can be held to await reaction. In addition, the enzyme provides a suitable chemical environment that will catalyze the desired reaction when an appropriate partner arrives.

We discuss below aspects of each of these catalytic situations that are related to the development of new chemical processes. Then they will be revisited in Section III-C because of their importance in the development of new energy sources.

HETEROGENEOUS CATALYSIS

A heterogeneous catalyst is a solid prepared with an extremely large surface area (1-500 m^2/gram) upon which a chemical reaction can occur. To appreciate the magnitude of this surface area, consider that a one-gram cube of platinum catalyst would be 4 mm high and it would have a surface area of 1.0 cm^2. If that one-gram cube were sliced into eight equal cubes, the surface area would be doubled. To reach an area per gram of 100 m^2, this process would have to be continued until the one gram has been divided into 10^{18} tiny cubes, each about 40 Å on a side and containing only 2,750 platinum atoms.

Table III-B-1 shows the fruitful commercial outcome of developments in heterogeneous catalysis in recent years. The potential economic significance is displayed in the last column, the total U.S. production (in metric tons) by all processes.

TABLE III-B-1 New Processes Based on Heterogeneous Catalysis

Feedstocks	Catalyst	Product	Used to Manufacture	1982 U.S. Production (metric tons)[a,b]
Ethylene	Silver, cesium chloride salts	Ethylene oxide	Polyesters, textiles, lubricants	2,300,000
Propylene, NH_3, O_2	Bismuth molybdates	Acrylonitrile	Plastics, fibers, resins	925,000
Ethylene	Chromium titanium	High-density polyethylene	Molded products	2,200,000
Propylene	Titanium, magnesium oxides	Polypropylene	Plastics, fibers, films	1,600,000

[a] Production by all processes, including the innovative process; U.S. Tariff Commission Report.
[b] One metric ton = 1,000 kg.

With the new measurement techniques of *surface science,* we can now begin to understand how such solid catalysts work. Because surface atoms have unused bonding capacity, they change the chemistry of molecules stuck on that surface. Hence, when two reactants A and B meet in this two-dimensional reaction zone, their chemistry can differ greatly from when they meet in solution or in the gas phase. To understand this different chemistry, we must know the molecular structures for A and B as they exist on the reactive catalyst surface. Fortunately, we now have laboratory

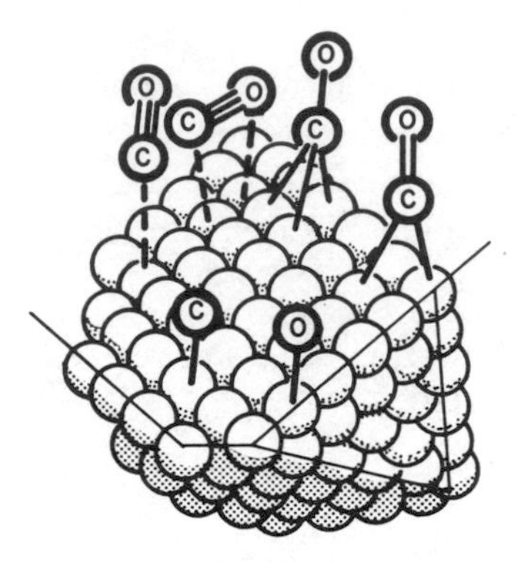

HOW DOES CARBON MONOXIDE BIND TO A METAL SURFACE?

tools with which chemists can "see" what these molecular structures are. Then our knowledge of reactions in familiar settings can be applied and the door begins to open to understanding, control, and design of catalysts.

Here are four examples of heterogeneous catalysis where the fruits of such understanding will have a major impact on new technologies that benefit our society.

Molecular Sieve Synthesis and Catalysis

Molecular sieves are natural or synthetic solids made of aluminum, silicon, and oxygen (aluminosilicates). A special property of these solids is that they contain tiny holes or channels into which gaseous molecules can enter if they are not too large. Once caught inside such a channel, these molecules can undergo reactions that would require a much higher temperature outside. Thus, the sieve acts as a catalyst. Furthermore, the shape and size of the cavity both select which molecules can react and also limit the size of the product. This means that the sieve is a selective catalyst. They have been used with remarkable efficiencies both for the breakdown of crude oil into smaller, more combustible molecules ("cracking") and for converting methanol (from biological sources) into gasoline.

Metal Catalysis

It has long been known that extremely small metallic particles of certain elements can catalyze hydrocarbon conversions for fuels and catalyze ammonia synthesis from nitrogen for fertilizer production. These elements are from the middle of the Periodic Table; they include cobalt and nickel and elements below them: rhodium, palladium, and platinum. We have already noted that the catalyst particles may contain only a few thousand atoms. We need to know why these tiny particles are so effective and why these particular metals work while other, more abundant metals, do not. Unfortunately, many of these catalytically active metals are rare, and their ore deposits are not located in the United States: e.g., cobalt, manganese, nickel, rhodium, platinum, palladium, and ruthenium. When we understand why these metals work so well, we are on our way to finding more available substitutes. Among the candidates are metal oxides (including rust, iron oxide), carbides, sulfides, and nitrides.

Conversion Catalysts

We must find catalysts to convert abundant and cheap substances to more useful compounds. Thus, we would like to convert nitrogen to nitrates (for fertilizer use); coal to hydrocarbons (for fuels); and one-carbon compounds, like carbon monoxide, carbon dioxide, methane, and methanol, to two-carbon compounds, like ethylene, ethanol, acetic acid, and ethylene glycol (for industrial feedstocks).

Catalysts to Improve the Quality of Air and Water

We have many environmental pollution problems that must and can be solved, the way the catalytic converter helped clean up automobile exhaust gases. Thus, we would like to have catalysts that remove sulfur oxides from the smokestacks of factories, purify water, and prevent acid rain.

HOMOGENEOUS CATALYSIS

Homogeneous catalysts act in the gas or liquid state in the absence of a surface. Important among these are the soluble catalysts that are active in a liquid solution. Often they are complex, metal-containing molecules whose structures serve to fine-tune reactivity and achieve highly selective end results. The largest industrial-scale process using homogeneous catalysis is the partial oxidation of *para*-xylene to terephthalic acid, with U.S. production of 6.2 billion pounds in 1981. The process uses salts of cobalt and manganese dissolved in acetic acid at 215°C as the catalyst system. Most of the product is then copolymerized with ethylene glycol to give us polyester clothing, tire cord, soda bottles, and a host of other useful articles.

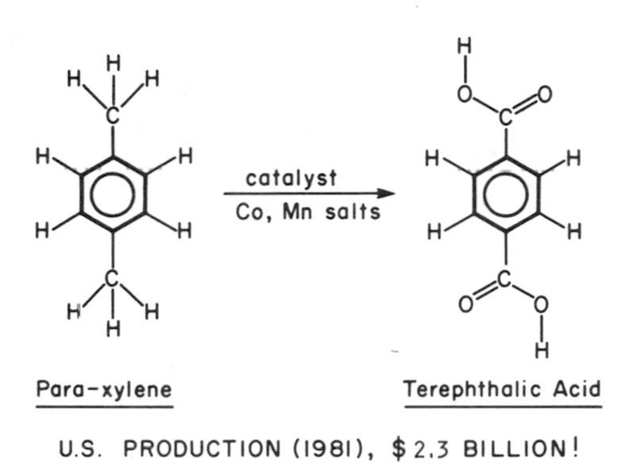

The chemical industry in the United States has been repeatedly strengthened by the introduction of new processes based upon homogeneous catalysts. Table III-B-2 lists six such processes, whose 1982 production figures were valued at over one billion dollars.

TABLE III-B-2 New Processes Based on Homogeneous Catalysis

Feedstocks	Catalyst[a]	Product	Used to Manufacture	Start-up Date	1982 U.S. Production (metric tons)[b]
Propylene, oxidizer	Mo^{VI} complexes	Propylene oxide	Polyurethanes (foams) Polyesters (plastics)	1969	303,000
Methanol, CO	$[Rh(CO)_2I_2]^-$	Acetic acid	Vinyl acetate (coatings) Polyvinyl alcohol	1970	495,000
Butadiene, HCN	$Ni(L_1)_4$	Adiponitrile	Nylon (fibers, plastics)	1971	220,000
α-Olefins	$RhH(CO)(L_2)_3$	Aldehydes	Plasticizers Lubricants	1976	300,000-350,000
Ethylene	$Ni(L_3)_2$	α-Olefins	Detergents	1977	150,000-200,000
CO, H_2 (from coal)	$[Rh(CO)_2I_2]^-$	Acetic anhydride	Cellulose acetate (films)	1983	[225,000, Capacity]

[a] L = Ligand, L_1 = Triaryl Phosphite, $L_2 = PPh_3$, $L_3 = OOCCH_2PPh_2$, Ph = C_6H_5.
[b] One metric ton = 1,000 kg.

An important branch of homogeneous catalysis has developed from research in organometallic chemistry. For example, in the second reaction in Table III-B-2, rhodium dicarbonyl diiodide catalyzes the commercial production of acetic acid from methanol and carbon monoxide. With this catalyst present, the reaction economically gives more than 99 percent preference for acetic acid over other products. Almost a billion pounds of acetic acid is so produced, a large part of which is used to manufacture such polymeric materials as vinyl acetate coatings and polyvinyl alcohol polymers.

Activation of Inert Molecules

Several abundant substances are inviting as reaction feedstocks, including nitrogen, carbon monoxide, carbon dioxide, and methane. However, these are relatively inert molecules, so catalysts are needed to speed up their chemistry. Soluble organometallic compounds are showing promise here. For example, soluble compounds of molecular nitrogen, N_2, with tungsten and molybdenum have been prepared that permit ammonia production under mild conditions. Additionally, carbon-hydrogen bonds in normally unreactive hydrocarbons like methane and ethane have been split by organorhodium, organorhenium, and organoiridium complexes. Hope for the buildup of complex molecules from one-carbon molecules, such as carbon monoxide and carbon dioxide, is stimulated by recent demonstrations of carbon-carbon bond formation at metal centers bound in soluble metal-organic molecules. Synthesis of compounds with multiple bonds between carbon and metal atoms has been of special importance. These compounds catalyze the interconversion (metathesis) of various ethylenes to make desired polymer starting materials.

Metal Cluster Chemistry

An adventurous frontier of catalysis has been opened by the increasing capability of chemists to synthesize molecules built around several metal atoms bonded together. In size, these cluster compounds lie between the molecular-sized homogeneous catalysts and the particles of bulk metal that are used in heterogeneous catalysts. It is intriguing that many of the metals that are most active as heterogeneous catalysts also form such cluster compounds (e.g., rhodium, platinum, osmium, ruthenium, and iridium). Now the chemistry of these elements can be studied as a function of cluster size. Are small clusters better catalysts than bulk metal? Can they improve on the action of metal-organic catalysts that contain only one or two metal atoms? With our new preparative methods, we will be able to answer these questions.

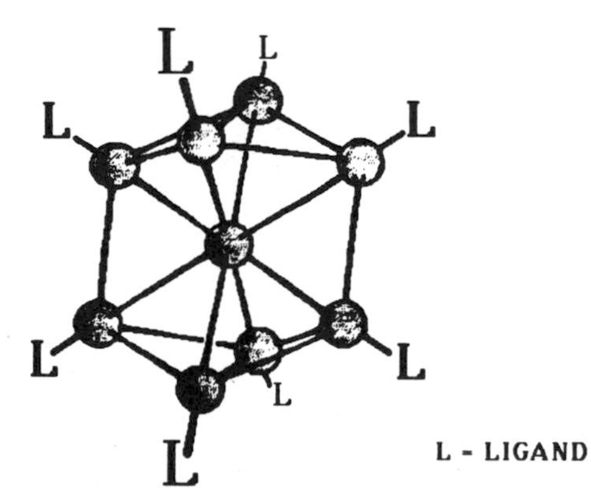

A GOLD CLUSTER COMPOUND

Many metal cluster compounds consist of several metal atoms bound to each other in the "core" of the molecule with carbon monoxide molecules chemically attached on the outside. These metal carbonyls have formulas $M_x(CO)_y$, and x can be made very large. The world's record as of this writing is a platinum compound with $x = 38$, $Pt_{38}(CO)_{44}^{2-}$, which is actually approaching the smallest catalyst particle sizes that have been made from bulk material. This closes the gap between molecular and bulk catalysts. At the same time, very low temperature techniques are revealing the structures and chemistry of small clusters containing *only* metal atoms (''naked clusters''). These are of special interest because heterogeneous catalysts are presumed to consist only of metal atoms.

Still other cluster compounds are called ''cubanes.'' These are molecules built around a unit of four metal atoms and four sulfur atoms at the eight corners of a cube. Such ''cubane'' structures have been made for iron, nickel, tungsten, zinc, cobalt, manganese, and chromium. The iron example, ferrodoxin, has been found to be the functional part of the iron proteins that catalyze electron-transfer reactions in biological systems. Here is a small cluster compound in use by nature as a biological enzyme.

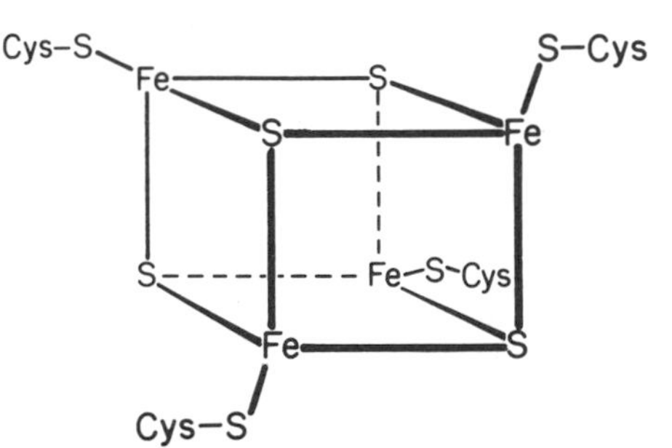

THE BIOLOGICAL ENZYME FERRODOXIN: AN IRON-SULFUR "CUBANE" STRUCTURE

Stereoselective Catalysts

Many biological molecules can have either of two geometric structures, one being the mirror image of the other. These are called ''chiral'' structures. Generally, only one of these structures is functionally useful in the biological system. If a complex molecule has seven such chiral carbon atoms and a synthetic process produces all the mirror-image structures, there would be $2^7 = 128$ structures, 127 of which might have no activity or, worse, some undesired effect. Thus, the ability to synthesize at every chiral center the desired structure with the desired geometry is essential. A catalyst that will do this is called a *stereoselective catalyst*. An example is provided by L-dopa, a particular mirror-image structure of an amino acid that has revolutionized the treatment of Parkinson's disease. This molecule has been made using the stereoselective addition of hydrogen to a carbon-carbon double bond. The catalyst that does this is a soluble rhodium phos-

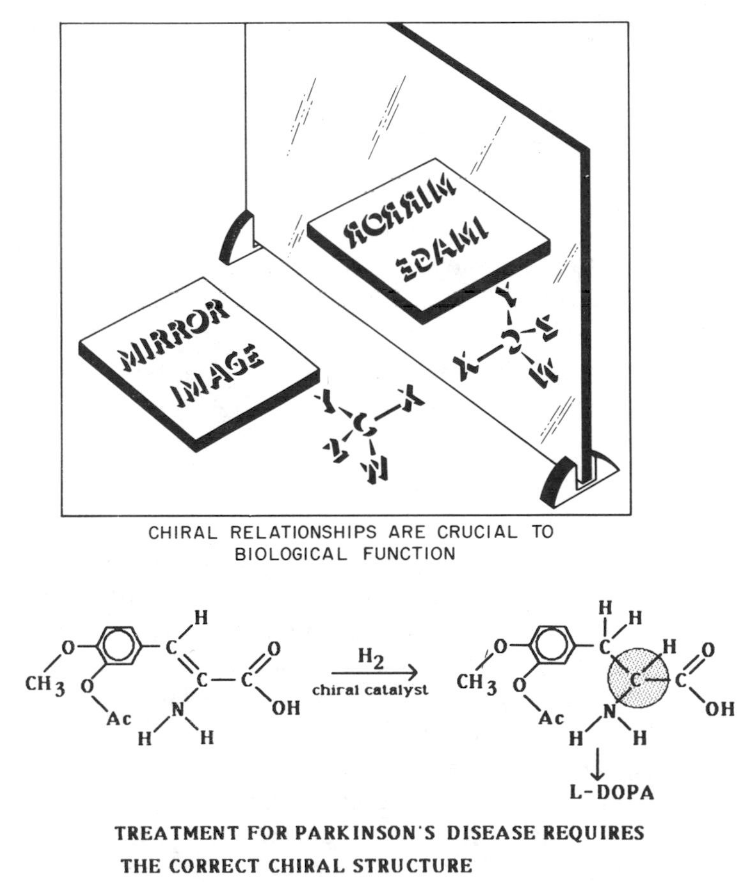

CHIRAL RELATIONSHIPS ARE CRUCIAL TO BIOLOGICAL FUNCTION

TREATMENT FOR PARKINSON'S DISEASE REQUIRES THE CORRECT CHIRAL STRUCTURE

phine compound that gives the correct structure in 96 percent yield. Stereospecific oxidations can also be carried out. The recent discovery of a titanium catalyst to add, in a specific geometry, an oxygen atom across a carbon-carbon double bond has lowered 10-fold the price of the synthetic sex attractant of the gypsy moth. The gypsy moths are ecstatic about this; there are commercial applications as well.

PHOTOCATALYSIS AND ELECTROCATALYSIS

Significant advances have recently been made in controlling the chemistry that takes place at the surface boundaries between liquid solutions and electrochemical electrodes (electrocatalysis). In some applications, absorption of light by a semiconductor electrode initiates the chemistry (photocatalysis). This rapidly moving field depends upon our knowledge of homogeneous catalysis, heterogeneous catalysis, and semiconductor behavior.

Photocatalysis

An electrochemical cell can be built with one or both electrodes made of semiconductor materials that absorb light. In such a cell, the light absorbed by the electrode can be used to promote catalytic oxidation-reduction chemistry at the electrode-solution interface. The same sort of chemistry can be induced in solutions containing small particle suspensions of semiconductor materials but now at the particle-solution interface. Such oxidation-reduction chemistry has significant scientific interest and, without doubt, practical importance as well. For example, photodestruction of toxic waste material, such as cyanide, has been demonstrated at titanium dioxide surfaces. A more popularized concept is that such photocatalytic chemistry, driven by solar energy, could give a process for producing massive amounts of hydrogen and oxygen from water. It is an interesting prospect: to convert from using the dwindling, polluting petroleum fuels to a renewable fuel—hydrogen—that burns to form water and that is made from water using energy from the sun.

Electrocatalysis

Even without light-initiated processes, electrode surfaces with catalytic activity offer new opportunities for chemical syntheses. Recent developments have shown that electrode surfaces can be chemically tailored to promote particular reactions. For example, this research area has benefited from techniques used by the semiconductor industry, such as deposition of chemical vapors on the electrode surface, by combining those techniques with imaginative synthetic chemical techniques for modifying surfaces.

This is demonstrated by the group of electrocatalysts developed for use in making chlorine in chlor-alkali cells. One successful case is based upon a thin layer of ruthenium dioxide—the catalyst—deposited on a common metal electrode. This electrocatalyst has dramatically improved energy efficiency and reduced cell maintenance in the chlor-alkali industry, an industry representing billions of dollars in sales. The savings are enormous because this crucial industry consumes up to 3 percent of all electrical energy produced in the United States.

Chemistry at the Solid/Liquid Interface

Before the technological potentialities of any of the above can be fully realized, we must have a much better understanding of the chemistry going on at the semiconductor/liquid interface. Most of the extraordinary instrumentation so far developed for surface science studies can only be applied at solid/vacuum interfaces. We need similar capability at the solid/liquid boundary, and there is already reason for optimism. For example, when light is scattered by a molecule, it can leave behind energy to excite the vibrational motions of the molecule. Hence, the scattered light contains the "signature" of the molecule and gives us clues to its structure. This behavior, the Raman effect, has been found to be intensified a millionfold when the scattering molecule is adsorbed (held) on a silver metal surface. This intensification permits us to detect the tiny number of molecules at a solid/liquid interface. Other scattering methods that depend upon the very high power of laser light sources (e.g., "surface-enhanced harmonic generation") show that other such discoveries are to be expected.

The possible gains from these areas are considerable. We would like to know how to catalyze multielectron transfer reactions at an electrode surface. That is the chemistry required, for example, to photogenerate a liquid fuel like methanol from carbon dioxide and water. Multielectron transfer catalytic electrodes for oxygen reduction in electrochemical cells would find a welcoming home in the fuel cell industry.

It is also likely that research on semiconductor electrode surfaces will benefit the field of electronics. The integrated circuit technology based on the new material gallium-arsenide depends upon control of its surface chemistry. Already, scientists concerned with design of the tiny chips used in computers are recognizing the importance of the chemistry involved. The crowded circuits on an electronic chip must be chemically etched with great precision and on a microscopic scale.

ARTIFICIAL-ENZYME CATALYSIS

A striking outcome of our expanding chemical knowledge has been the development of our ability to deal with molecular systems of extreme complexity. Using such modern instrumentation as nuclear magnetic resonance, X-ray spectroscopy, and mass spectroscopy, we can now synthesize and control the structure of molecules that come close to the complexity of biological molecules. This control includes the ability to fix the molecular shape, even including the mirror-image properties so critical to biological function.

One intriguing application of these capabilities is to combine them with our growing knowledge of catalysis in the synthesis of artificial enzymes. There are compelling reasons to do this. Without catalysts, many simple reactions are extremely slow under normal conditions. Raising the temperature speeds things up, but at risk of a variety of possible undesired outcomes such as acceleration of unwanted reactions, destruction of delicate products, and waste of energy. Unfortunately, natural enzymes do not exist for most of the chemical reactions in which we have interest. In the manufacture of polymers, synthetic fibers, medicines, and many industrial chemicals, few of the reactions used can be catalyzed by naturally occurring enzymes. Even where there are natural enzymes, their properties are not ideal for chemical manufacture since they are proteins, sensitive substances that are easily broken down and destroyed. In industries that do use enzymes, major effort is devoted to modifying them to increase their stability.

Controlled Molecular Topography and Designed Catalysts

We have a pretty good idea of how enzymes work. Nature fashions a molecular surface shaped to recognize and bind to a specific reactant. This surface attracts the unique molecular type desired from a mixture and holds it in a distinct position that encourages reaction. When the reaction partner arrives, the scene is set for the desired reaction to take place in the desired geometry.

Organic chemists who are trying to make artificial enzymes are making notable progress. Without special control, large molecules usually have convex outer surfaces (ball-like shapes). So a first step toward making shaped surfaces has been to learn to synthesize large molecules with concave surfaces and hollows. Cyclodextrins provide examples: they are shaped like a doughnut. The crown ethers, developed over the last 15 years, have a quite different surface topography. For instance, 18-crown-6 consists of 12 carbon atoms and 6 oxygen atoms evenly spaced in a cyclic arrangement. In the presence of potassium ions, this ether assumes a crown-like structure in which the 6 oxygen atoms point toward and bind a potassium ion. Lithium and sodium ions are too small, and rubidium ions too large, to fit in the crown-shaped cavity; so this ether extracts the intermediate-sized potassium ions from a mixture. More complicated examples now exist. Chiral binaphthyl units can be coupled into cylindrical or egg-shaped cavities. With benzene rings, enforced cavities have been made with the shapes of bowls, pots, saucers, and vases. One of the descriptive names for such compounds is *cavitands*.

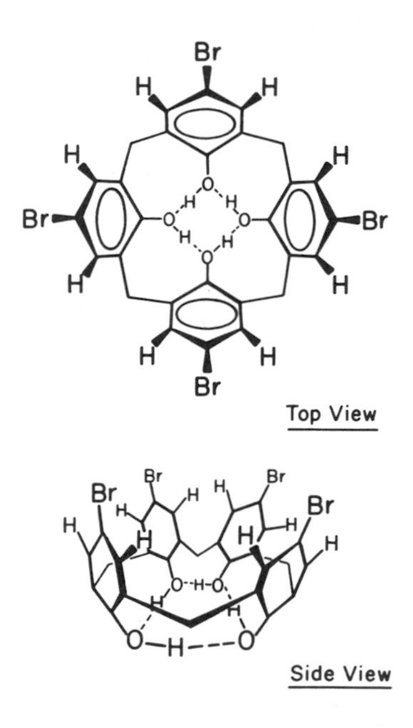

CAVITANDS
WHAT SHAPE DO YOU NEED?

We are clearly moving toward the next step, which is to build into these shaped cavities a catalytic binding site. This will probably be a metal-organic compound already known to have catalytic activity in solution. The earliest successes are likely to be patterned after natural enzymes, but there is no doubt that, in time, artificial, enzyme-like catalysts will go beyond what we find in nature.

Biomimetic Enzymes

A shortcut approach to improved catalysis is to pattern artificial enzymes closely after natural enzymes—sometimes called "biomimetic chemistry." For example, biomimetics, or mimics, have been prepared for the enzymes that biologically synthesize amino acids. Artificial enzymes that structurally resemble such natural enzymes as vitamin B_6 have shown good selectivity and even the correct mirror-image preferences in the product. Mimics have been prepared for several of the common enzymes involved in the digestion of proteins, and substances that catalyze the cleavage of RNA have been synthesized based upon the catalytic groups found in the enzyme ribonuclease. Mimics have also been synthesized that imitate the class of enzymes called cytochromes P-450, which catalyze many biological oxidations, and another that imitates the oxygen carrier hemoglobin.

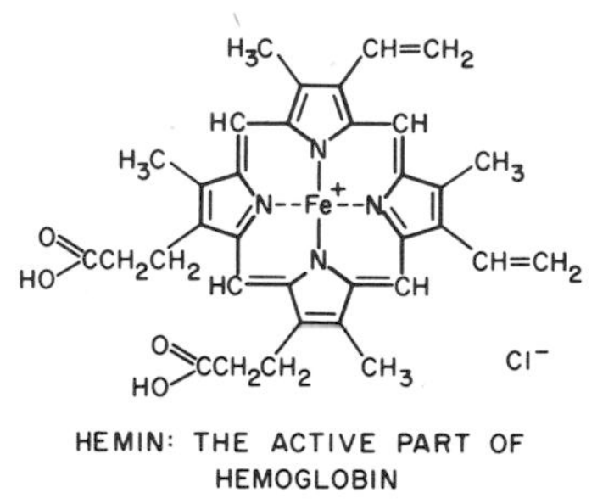

HEMIN: THE ACTIVE PART OF HEMOGLOBIN

The United States is a forerunner in this field, and the Japanese have specifically targeted biomimetic chemistry as an area of special opportunity. Such study is pointed toward a logical approach to catalyst design, an area ripe for development.

CONCLUSION

A significant share of our economy is built upon the chemical industry. The long-range health of this critical industry will depend upon our ability to develop new processes that increase energy and cost efficiency, and create new products for new markets, all the while strengthening our protection of the environment. Today's basic research in all aspects of catalysis will provide the source of such creative invention. It will also produce young scientists working at the frontiers of knowledge with the state-of-the-art instrumental skills needed to recognize and take advantage of the rich potentialities.

SUPPLEMENTARY READING

Chemical & Engineering News

"Stereospecific Routes to Silyl Enol Ethers" by S. Stinson (C.& E.N. staff), vol. 63, p. 22, July 15, 1985.

"New Dow Acrylate Ester Processes Derive From C_1 Efforts" by J. Haggin (C.& E.N. staff), vol. 63, pp. 25-26, Feb. 4, 1985.

"Rice University Chemists Study Reactivity on Metal Clusters" (C.& E.N. staff), vol. 63, pp. 51-52, Jan. 21, 1985.

"Catalysts Selectively Activate C-H, C-C Bonds" (C.& E.N. staff), vol. 63, pp. 53-54, Jan. 14, 1985.

"Organic Electrosynthesis" by R. Jannso, vol. 62, pp. 43-58, Nov. 19, 1984.

"Flame Synthesis of Fine Particles" by G.D. Ulrich, vol. 62, pp. 22-30, Aug. 6, 1984.

"Low-Severity Route to Acrylic Acid Developed" (C.& E.N. staff), vol. 62, p. 32, Apr. 30, 1984.

"Dow Continues Fischer-Tropsch Development" by J. Haggin (C.& E.N. staff), vol. 62, pp. 24-25, Mar. 5, 1984.

"Chemists Detail Catalysis Work with C_1 Systems" by J. Haggin (C.& E.N. staff), vol. 62, pp. 21-22, Feb. 27, 1984.

"Surface Modification Gives Selectivity to

Poisoned Catalysts'' (C.& E.N. staff), vol. 61, pp. 24-25, Sept. 5, 1983.

''Aluminophosphates Broaden Shape Selective Catalyst Types'' by J. Haggin (C.& E.N. staff), vol. 61, pp. 36-37, June 20, 1983.

''Shape Selectivity Key to Designed Catalysts'' by J. Haggin (C.& E.N. staff), vol. 60, pp. 9-15, Dec. 13, 1982.

''Metal Clusters: Bridges Between Molecular and Solid State Chemistry'' by E.L. Muetterties, vol. 60, pp. 28-39, Aug. 30, 1982.

Science

''Enhanced Ethylene and Ethane Production With Free-Radical Cracking Catalysts'' by J.H. Kolts and G.A. Delger, vol. 232, pp. 744-746, May, 9, 1986.

''The Zeolite Cage Structure'' by J.M. Newsom, vol. 231, pp. 1093-1099, Mar. 7, 1986.

A Lithium-Powered Heart

The cardiac pacemaker is a modern miracle of science that many of us take for granted—but not a person who owns one! These pacemakers operate on battery power, and the demands put on the tiny batteries that generate it are awesome! They must start the human machine every morning without fail, and the human lights and radio are running all the time. Yet many, many people are adding healthy years to life by betting on the chemical reactions that occur in these batteries to generate—day in, day out—the electric current that drives their pacemakers.

These batteries have special requirements because they must be implanted in a human body. They must be rugged and leakproof, have long life and minimal weight, and, of course, they must be nontoxic. The first batteries used in pacemakers had a lifespan of only 2 years, and the periodic operations required for replacement meant additional risk and stress for the patient.

Chemists began to tackle this problem, and research efforts in electrochemistry uncovered lithium metal, a new ingredient with the potential to give long life to batteries. Unfortunately, lithium is highly reactive—it burns in air and reacts with water to produce flammable hydrogen gas. If lithium were to be used, it would be necessary to discover new, nonaqueous electrolyte systems.

Electrolytes are substances that dissolve in water to form conducting solutions. They dissolve to produce ions, particles carrying electrical charge. The movement of these charges carries the current as the battery's chemistry releases its stored energy. Conventional batteries that draw on the chemical energy of zinc and mercuric oxide depend upon aqueous electrolytes. So the problem for the chemists to solve was defined—to design a battery that would operate without water.

Extensive investigations into new solvents and new materials for use in high-energy, long-life batteries eventually led to the discovery of a solid electrolyte for use with lithium metal. The solid electrolyte is iodine, and the lithium-iodine battery was born for biomedical applications. These revolutionary batteries are currently in use, and they have an impressive lifespan of 10 years! The benefits to those who must depend upon cardiact pacemakers are incalculable.

The lithium-iodine battery is not the end of the story. It is a vast improvement over its predecessors and extremely useful in pacemakers, but it has a lower power than would be optimum for other uses. On the horizon is the need for new, higher-power batteries for use in other implantable organs like artificial kidneys and hearts. But further electrochemical research will undoubtedly provide the answer. It has in the past, and it will again.

III-C. More Energy

This country's economic development is tied to the growth of its use of energy. For six decades, the Industrial Revolution was fueled primarily by coal. Then, petroleum energy use caught up with coal in 1948. Meanwhile, throughout the twentieth century the 3-fold growth in population has been accompanied by a 10-fold increase in energy use in all its forms. As we look ahead, there can be no doubt that the nation's wealth and quality of life will be strongly linked to continued access to energy in large amounts.

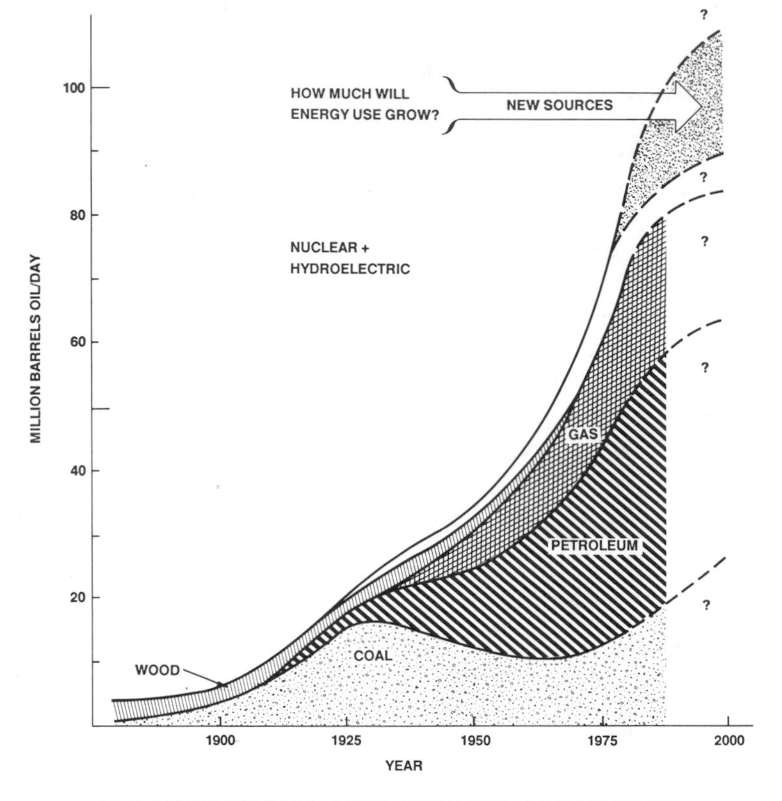

U.S. ENERGY USE: NEW SOURCES ARE NEEDED

Today, about 92 percent of the U.S. energy consumption is based upon chemical fuels. Because of societal concern about nuclear fission as an energy source, this dependence upon chemical technologies will continue well into the twentieth century. Meanwhile, every estimate of future energy use emphasizes the need to conserve and develop every energy source at our disposal. The need is urgent because, first, the planet's petroleum supply is limited and will eventually be exhausted. Second, our desire to protect the environment will result in stiffer restraints on new energy technologies. Chemistry and chemical engineering will play central roles as we develop each of the following energy sources and emerging alternatives:

- Petroleum
- Natural Gas
- Coal, Lignite, Peat
- Shale Oil, Tar Sands
- Biomass
- Solar
- Nuclear Fission
- Nuclear Fusion
- Conservation

PETROLEUM

Petroleum use has increased steeply worldwide; as much petroleum was taken from the ground between 1968 and 1978 as was removed in the 110 years before that. Complex chemical processing is required to convert the raw natural product into chemical forms that meet the demands of modern, high-compression engines.

Challenging research opportunities for chemists and chemical engineers lie in such key areas as *recovery* (getting more oil from the known deposits), *refining* (converting the crude oil into the most useful chemical form), and *combustion* (getting the most energy from the finished fuel).

Recovery

Recovery refers to the amount of oil that is actually removed from a known oil deposit. About 4,000 billion barrels of oil have been discovered worldwide, with about 12 percent of it in the United States. *Most of that oil, however, is not recoverable by presently known methods of removal. Primary recovery,* based upon natural pressure, can usually recover no more than 10 to 30 percent of the oil from its natural reservoir, a complex structure of porous rock. *Secondary recovery,* in which water, gas, or steam is injected to force more oil from the deposit, can raise the recovery efficiency, but even then, only about 35 percent of the known U.S. oil deposits are classified as recoverable. Of that recoverable part, *more than 80 percent has already been extracted and consumed.*

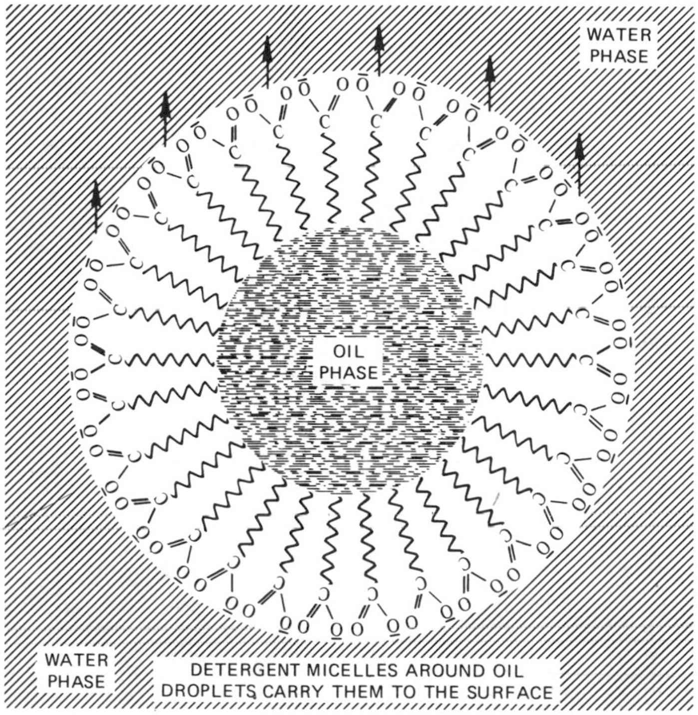

WATER PHASE

OIL PHASE

WATER PHASE

DETERGENT MICELLES AROUND OIL DROPLETS CARRY THEM TO THE SURFACE

Tertiary recovery goes after the rest of this valuable resource; it requires new chemistry and new methods. Two of these methods are the use of detergents (called surfactants) and solution polymers to separate oil droplets from the surrounding water. If tertiary recovery could be made possible, it would have enormous economic significance because it would permit us to tap the remaining 350 billion barrels of U.S. oil already discovered but currently beyond economic reach.

Refining

Crude oil, as it is pumped out of a well, is a liquid solution containing mostly hydrocarbons. The largest component is made up of compounds containing only single bonds; these are called *alkanes*. Most of the alkanes have long carbon chains, but some are branched and some are cyclic. A smaller fraction of the hydrocarbons have one double bond; these are called *alkenes*. Then there are also some molecules containing benzene rings; these are called *aromatics*. The molecular weights range from those of natural gas (methane, CH_4, 14; ethane, C_2H_6, 30;

propane, C_3H_8, 44; butane, C_4H_{10}, 58) all the way up to those of the waxes (a typical wax would have the formula $C_{30}H_{62}$ and a molecular weight of 422).

The first purpose of refining is to extract from this complex solution those hydrocarbons that have the right volatility for use in an auto engine and that burn well. Octane, C_8H_{18}, is about optimum, so we "calibrate" gasolines on the basis of "octane number" (octane equivalent). As far as combustibility is concerned, branched and cyclic alkanes burn smoothly, alkenes and aromatics are fine, but extended alkanes (normal alkanes) tend to explode in the auto cylinder rather than burn (causing your car to "knock"). So the second purpose of refining is to convert molecules that are unsatisfactory into the best molecular weight range and combustibility. This is where the chemistry becomes sophisticated.

Refinement of the crude oil begins with *disillation,* in which different petroleum ingredients are separated from each other according to their boiling points. Then, sulfur may be removed to improve the quality of the product. After that, the large molecules must be broken down into smaller, lower-boiling molecules by *catalytic cracking*. Then, *catalytic reforming* can be used to change the molecular structures to forms that burn better (high octane number). Catalysis is the key.

Chemistry on a Catalytic Surface

The best petroleum catalysts are expensive and rare elements like platinum, palladium, rhodium, and iridium, and they work as catalysts in the metallic state. Metallic crystals can exhibit a variety of surfaces, depending upon the angle of the surface relative to the natural crystal axes. The most stable surfaces tend to be flat and close-packed, with each surface atom surrounded by a large number of nearest neighbors. These are the surfaces we see in a neatly packed tray of oranges in a grocery store. Looking carefully at these trays one can see a variety of packing arrangements. There may be steps in the surface, forming terraces several oranges wide, with the terrace width depending on the tilt of the tray. It is the same for the surface of a metal. When there are terraces on a metal surface, the atoms at the ledges are even more exposed, thus more reactive, than surface atoms embedded in the smooth terraces. Further, there may be kinks in the steps, or the surface may be "rough" with atomic-sized openings between surface atoms that, again, will display special reactivity because of

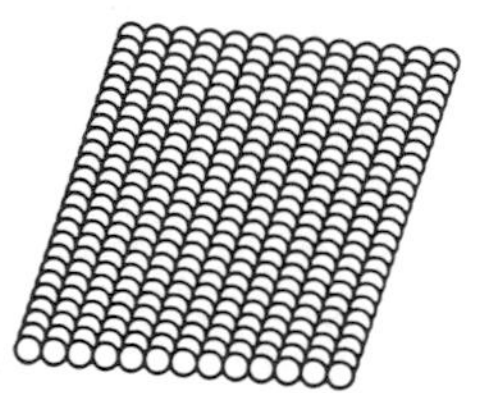

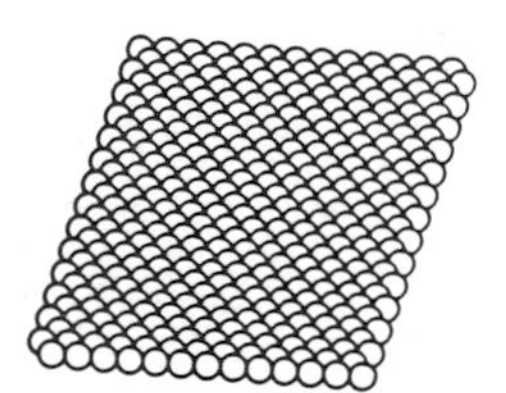

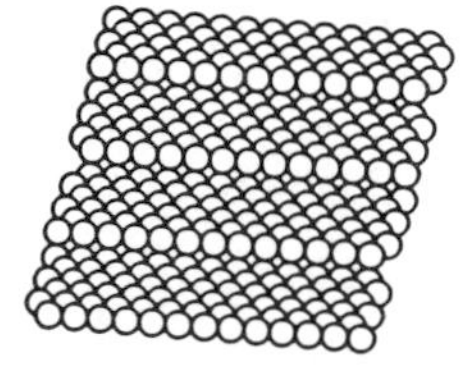

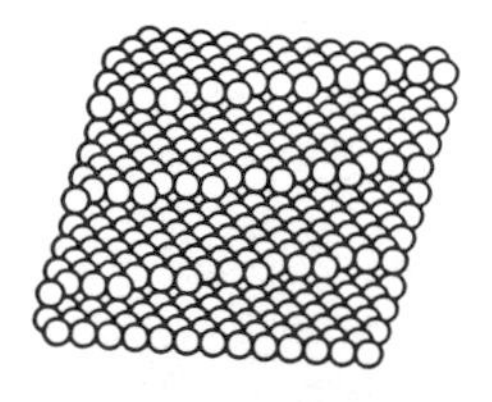

CHEMISTRY ON A PLATINUM SURFACE DEPENDS ON THE SURFACE EXPOSED

unsatisfied bonding capability. These special sites may be crucial in determining the catalytic activity of a metal surface. Fortunately, such chemically important surface irregularities can now be identified by low-energy electron diffraction (LEED) (see Section V-C).

Petroleum refining shows how important these surface structures are in catalysis. Platinum is one of the best catalysts to restructure alkane hydrocarbons to forms with better combustion properties (e.g., octane number, volatility). Now it is possible to determine which catalyst surfaces give the most of the desired products. Thus, *n*-hexane, a stretched-out chain structure alkane with a low octane number, can be converted to forms with higher octane numbers, such as benzene and branched or cyclic alkanes, using a platinum catalyst in the presence of hydrogen. We know now that benzene is favored on the flat (1,1,1) surface or on stepped surfaces with terraces of (1,1,1) orientation, like (7,5,5). In contrast, formation of branched or cyclic alkanes is favored on the flat (1,0,0) surface or stepped surfaces with (1,0,0) terraces. Kinked surfaces, like (10,8,7), tend to produce less desirable products like propane and ethane. Knowing this, we can seek a reagent that will permanently bind to and block ("poison") these active kink sites to eliminate their less desirable products.

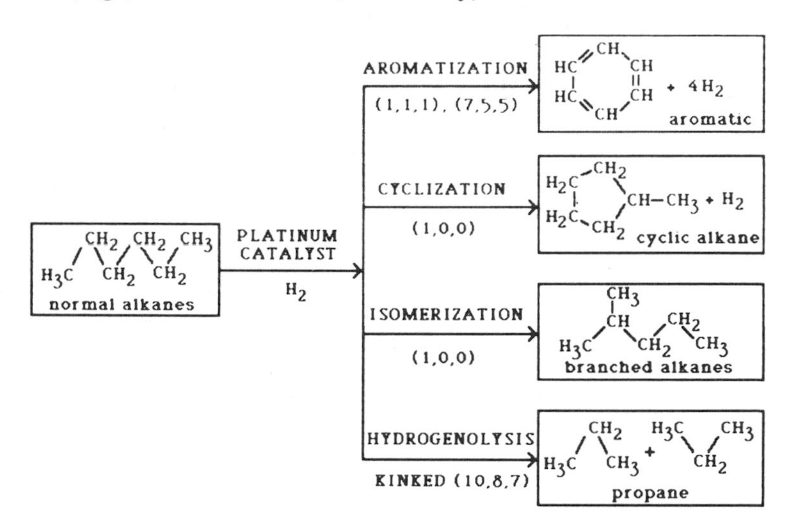

DIFFERENT SURFACES FAVOR DIFFERENT PRODUCTS

You benefit from this understanding of catalysis every time you fill up your tank with the gasoline that is best for your car. Table III-C-1 lists four important catalytic processes recently introduced during a period when our concerns for the environment influenced researchers to develop high-octane lead-free gasoline and reduce hazardous by-products. The need for new discoveries is even greater today

TABLE III-C-1 Heterogeneous Catalysis in the Petroleum Industry

Feedstocks	Catalyst	Product	Used for
C_{16}-C_{24} oils	Zeolite molecular sieves (aluminosilicates)	C_7-C_9 alkanes, alkenes	"Cracking" to high-octane fuels
C_7-C_9 unbranched hydrocarbons	Platinum-rhenium/ platinum-iridium	Aromatics, other hydrocarbons	"Reforming" to high-octane fuels
CO, NO, NO_2	Platinum/palladium/ rhodium	CO_2, N_2	Auto exhaust cleanup
CH_3OH	Zeolite molecular sieves (aluminosilicates)	C_7-C_9 branched hydrocarbons, aromatics	Gasoline production

as we turn to lower quality petroleum sources (called feedstocks) with higher sulfur content, with higher molecular weights (Alaska oils), and containing impurities that interfere with catalysts (e.g., vanadium and nickel in California offshore oils).

It is likely that future refining techniques will differ a great deal from those currently used. Petroleum refining technology is already undergoing an evolution as refineries are being adapted to feedstocks of lower quality. Future developments may be based upon combustion of the low-hydrogen and coke components of these low-quality feedstocks to drive energy-consuming processes. Some of the least desirable portions may be used as fuel for other refinery processes or to produce other useful reactants such as hydrogen.

Combustion

The United States annually spends about $30 billion (10 percent of its GNP) on materials that are burned as fuel. It seems ironic that much remains to be learned about the chemistry of combustion since it is one of the oldest technologies of mankind dating back to the discovery of fire. The need for more knowledge stems from an ever increasing dependence on combustion, from changes in the composition of our fuel, and, most important, from the sudden awareness and concern about the environmental impacts of combustion. In the last 30 years, society has recognized, and begun to grapple with, the undesired side effects of careless combustion of fossil fuels. These side effects include smog from nitrogen oxides, acid rain from sulfur impurities, dioxins from inefficient burning of chlorinated compounds, and a problem almost too unwieldy to deal with, the long-range effect on the global climate of accumulating CO_2.

The combustion process is a tightly coupled system involving fluid flow, diffusion processes, energy transfer, and chemical kinetics. This complexity is shown in an oxyacetylene torch flame and even in a Bunsen burner flame. After 60 years of intensive study, it is only in the last few years that this methane-air flame has been well described in chemical and physical detail.

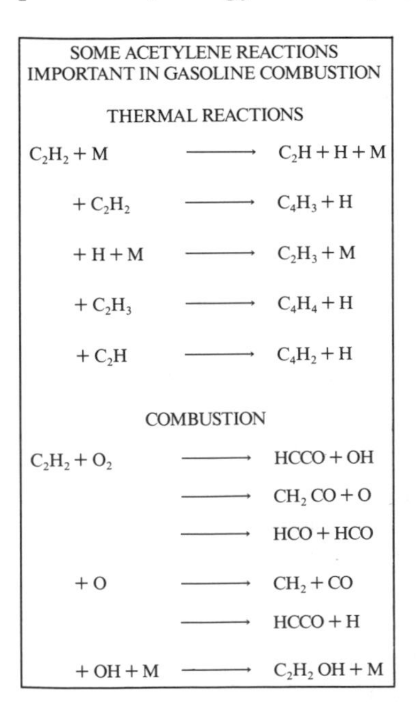

Fortunately, the area of chemical kinetics currently offers great promise. Such optimism comes because of an array of new, sophisticated instrumental techniques that allows us to understand the basic chemical behaviors at work (see Section IV-A). These advances, as they occur, will be quickly taken up by chemical engineers and will mean more efficient combustion and decreased environmental pollution. To indicate their importance, an increase of only 5 percent in the efficiency with which we burn coal, oil, and gas would be worth $15 billion per year to the U.S. economy, and an immeasurable additional value if it also reduces the growing problems of smog and acid rain.

NATURAL GAS

Natural gas is a mixture of low-molecular-weight hydrocarbons, mostly methane, CH_4. In North America, natural gas is typically 60 to 80 percent methane (the rest, ethane, C_2H_6; propane, C_3H_8; and butane, C_4H_{10}, in varying percentages). It contains some sulfur- and nitrogen-containing impurities, but they can be removed to give a clean-burning fuel and a widely useful chemical feedstock. The ethane and propane can be catalytically changed to ethylene, C_2H_4; propylene, C_3H_6; and acetylene, C_2H_2,—all valuable raw materials for products needed by our society.

Natural gas is an important resource as it is easily transported in pipelines and has many applications. Its contribution to U.S. energy use has almost doubled since 1960. The U.S. natural gas reserves are about equivalent to our petroleum reserves, perhaps somewhat larger. However, like petroleum, there is a limited amount of natural gas both worldwide and in this country, and its production will undoubtedly peak one or two decades from now.

COAL

Coal is the most abundant of the fossil fuel energy sources. Estimates of recoverable supplies worldwide indicate 20 to 40 times more coal than crude oil. The contrast is even more dramatic here in the United States where the estimates indicate 50 to 100 times more coal than crude oil. There can be no doubt that dependence on coal must increase during the next two or three decades as petroleum reserves are used up. Fortunately, being aware of this gives us time for the basic research needed to use this valuable resource efficiently and cleanly.

It must be noted, too, that petroleum is not only a fuel; it also provides us with many important fine chemicals and chemical feedstocks. In fact some people hold the opinion that petroleum as a source of other chemicals ought to be classified as "too valuable to burn." If coal can be economically converted on a huge scale into combustible fuels, then we gain the option to save petroleum for more sophisticated uses. Further ahead, we can predict that with creative advances in chemistry, coal itself can provide its own variety of valuable feedstocks, including some that now come from petroleum.

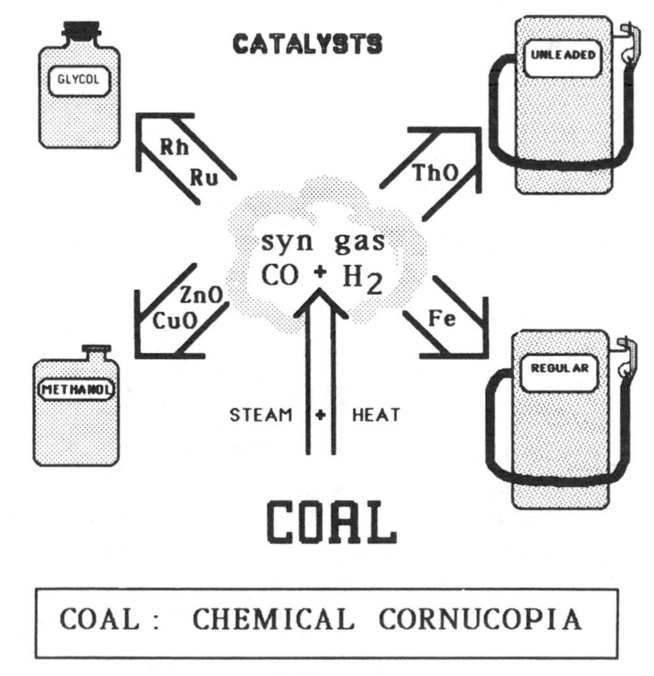

COAL : CHEMICAL CORNUCOPIA

Coal is a carbonaceous rock containing chemically bound oxygen, sulfur, and nitrogen as well as varying amounts of minerals and moisture. As a fuel, it has an undesirably low hydrogen-to-carbon ratio (its H/C ratio is near unity, about half that of gasoline), which makes it burn less efficiently. In order to use coal for anything more sophisticated than simple combustion, its molecular weight must be reduced; sulfur, nitrogen, and minerals must be removed; and its hydrogen content must be increased. These goals can be reached either through processes that convert coal to liquid products, which can then be refined (hydroliquefaction), or by converting coal to a gaseous form called "syn gas" (an abbreviation for "synthesis gas"), which is a mixture of carbon monoxide and hydrogen.

The use of synthesis gas has enormous potential, but so far it is not economically competitive. Table III-C-2 shows some catalysts that are effective with syn gas, the products that result, and the useful applications of those products.

TABLE III-C-2 Catalyst Specificity for Syn Gas Conversion to Useful Products

$$CO + H_2 \xrightarrow{\text{Catalyst}} \text{Product}$$

Catalyst	Product	Useful for
Nickel	Methane, CH_4	Fuel
Copper/zinc oxide/aluminum oxide	Methanol, CH_3OH	Fuels, via zeolite catalysts, chemical feedstock
Iron[a]/cobalt	Straight chain hydrocarbons, $CH_3(CH_2)_NCH_3$, N = 0 to 30	Feedstock for petroleum refineries
Molybdenum/cobalt	Mixed alcohol	Octane booster
Ruthenium complexes (in solution)	C_1 to C_3 oxygenated compounds	Chemical feedstocks
Thorium oxide	Low-molecular-weight branched chain hydrocarbons	High-octane fuel
Rhodium complexes (in solution)	Ethylene glycol	Polyester feedstock

[a] The catalyst developed by Hans Fischer and Franz Tropsch in the early 1920s.

The details of catalytic conversion of CO and H_2 to particular desired products present an active research area. The potentials for the liquefaction processes are equally promising, and more research in this area would clearly be fruitful.

The importance of the things that can be learned from research on both types of coal conversion was strikingly displayed during World War II. Germany, denied easy access to petroleum, was able to produce 585,000 tons of fuel hydrocarbons from coal. While a good fraction came through gasification combined with cobalt catalysts (Fischer-Tropsch chemistry), the larger share was produced through catalytic liquefaction. In a current situation, the Republic of South Africa now produces 40 percent of its gasoline requirements by similarly converting coal into 1,750,000 tons of hydrocarbons annually (using iron catalysts). Its plants and refineries are literally built on top of large coal deposits, and the coal enters the chemical reactors via conveyor belts rising from the mines. However, these examples are economically unique because the countries involved were denied access to petroleum because of political events.

SHALE OIL AND TAR SANDS

Shale is a form of sedimentary rock and is a major potential source of liquid hydrocarbons in Colorado, Utah, and Wyoming. It is estimated that 4,000 billion barrels of hydrocarbons are contained in the shale of these three states alone. If only a third of this enormous reserve could be recovered, it would give us almost 10 times as much fuel as has been removed so far from U.S. oil wells. Complicated new problems of chemistry, geochemistry, and petroleum engineering must be overcome to reach this end.

Shale, which comes from ancient marine deposits of mud and plant life, contains varying amounts of kerogen, a mixture of insoluble organic polymers, and smaller amounts of bitumin, a mixture of organic compounds soluble in benzene. Formidable environmental questions relating to water sources and land reclamation are raised in the development of shale deposits, because a ton of shale may yield only 10 to 40 gallons of crude oil. Shale oil has a favorably high H/C ratio—about 1.5—but it also contains undesired organic nitrogen and sulfur compounds that must be removed. Arsenic compounds can also present a special problem.

In Utah, sands are found containing dense, thick petroleum. Such deposits (called tar sands) exist in amounts equivalent to about 25 billion barrels of petroleum. Problems similar to those discussed for oil shales must be faced, particularly the environmental aspects. Because of its potential impact on the environment, practical use of this potential energy reserve may depend upon whether the rather complicated chemical conversions needed can be handled underground.

BIOMASS

An estimated 500 to 800 million tons of methane (equivalent to about 4 to 7 million barrels of oil and with an H/C ratio equal to 4!) are released each year into the air through the action of bacteria that work without oxygen. This bacterial action, which results in methane gas, is called anaerobic respiration. The obvious possibility of using such anaerobic processes to produce methane from what is called biomass (agricultural by-products, garbage, or other organic wastes) is complicated by the slowness of the process and by its great sensitivity to solution acidity. A detailed understanding of the chemical mechanism of methane production and of the biochemistry of the organisms involved could suggest ways to overcome the problems. Concerning the chemical mechanism, the reduction of carbon dioxide is now believed to occur in a succession of two-electron steps catalyzed by enzymes. Nickel plays a key role in the active enzyme, but its specific action is not known. Research on both synthesis and catalytic activity of metal-organic compounds, artificial enzymes, and natural enzymes should help us weigh the potential of biomass as a source of hydrocarbon fuels or chemical feedstocks. Of course, there is tremendous appeal in the prospect of producing useful energy from garbage, sewage, and plant waste disposal.

A particularly appealing aspect of biomass as a major fuel source is related to the amount of carbon dioxide in our atmosphere. Because carbon dioxide, CO_2, is transparent to visible light but absorbs infrared light, it lets most of the normal solar radiation reach the ground but intercepts infrared light which is radiated from the cooler Earth's surface. Thus, CO_2 "traps" the solar energy, tending to warm up the atmosphere (the "greenhouse" effect). The problem we face is that measurements from the beginning of this century indicate that the amount of CO_2 in the atmosphere is rising, which raises concern that in time the atmospheric temperature might rise enough to melt the polar ice caps and inundate coastal areas all over the world (an average global rise of only 5°C might be sufficient).

It is likely that most of the increase in atmospheric carbon dioxide over the last 60 years has resulted from combustion of fossil fuels. To halt this trend, we should be seeking new energy sources that do not release CO_2. Solar energy is such an alternative. Less widely recognized, however, is that new biomass is an ongoing solar energy use that does not add to the CO_2 problem. While combustion of new biomass does produce CO_2, its carbon content was all taken recently from the atmospheric CO_2 reservoir during growth of the biomass. Hence there is no *net* change in the CO_2 balance.

As mentioned above, this desirable concept can only be put into practice when researchers discover economic chemical methods for converting massive amounts of biomass into combustible substances. Further, there are trade-offs to be considered, such as the need to divert agricultural land from food production to biomass production. With the prospects provided by genetic engineering, even this conflict may be diminished or eliminated. Food *and* energy-producing biomass could possibly be produced by the same plant. Perhaps, as well, we can learn to genetically "engineer" plants that tend to counteract any rise of carbon dioxide in the atmosphere by growing more efficiently when carbon dioxide availability goes up.

SOLAR ENERGY

By far the most important natural process that uses solar energy is photosynthesis—the process by which green plants use the energy of sunlight to manufacture organic (carbon) compounds from carbon dioxide and water, with the simultaneous release of molecular oxygen. To be able to duplicate this process in the laboratory would clearly be a major triumph with dramatic implications. Despite much progress in understanding photosynthesis, we are still far from this goal.

The solar spectrum that drives photosynthesis places about two-thirds of the radiant energy in the red and near-infrared spectral regions. Understanding the way nature manages to carry out photochemistry with these low-energy photons is one of the keys to understanding (and imitating) photosynthesis. In current explanations, the energy of one near-infrared photon sets off a series of electron-transfer reactions (oxidation-reduction steps). While each of these steps uses up some of the absorbed energy, a fraction is stored through the production of adenosine triphosphate (ATP). Further, the chemistry is set up for the absorption of a second infrared photon to produce still more ATP and to begin reduction of atmospheric CO_2. This sequence of events gives the raw materials from which the cellular factory manufactures its high-energy carbohydrate products. This factory is run by the solar energy that was stored in ATP.

Thus, natural photosynthesis is energized by near-infrared light through production of energy-storing intermediate substances with long enough lifetimes to await arrival of a second near-infrared photon. The second photon "stands on the shoulders" of the first, so that their combined energy is enough to make or break the chemical bonds of the plant molecules. Several of the steps in this sequence take place in much less than a millionth of a second, at rates that were, only 15 years ago, too fast to measure. Now we have picosecond laser and nanosecond electron spin resonance techniques with which to probe each successive reaction on its own characteristic time scale. Hence, we are in a period of rapid progress in clarifying the chemistry of the photosynthetic process.

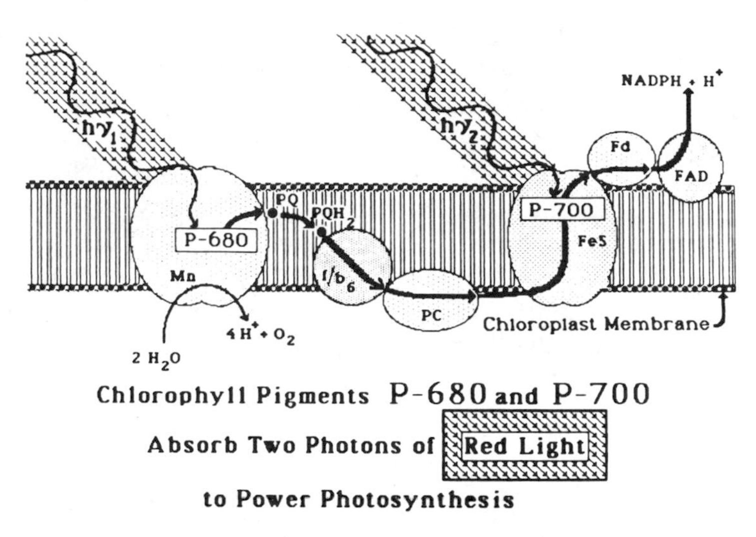

Chlorophyll Pigments P-680 and P-700 Absorb Two Photons of Red Light to Power Photosynthesis

This type of spectroscopic study reveals photosynthesis to be a complex process involving the cooperative interaction of many chlorophyll molecules. The packing arrangement of neighboring chlorophyll molecules has been probed by X-ray spectroscopy and by proton and ^{13}C nuclear magnetic resonance (NMR). Electron spin resonance experiments have shown that an electron is rapidly ejected or transferred from chlorophyll shortly after the light absorption (within nanoseconds). This leaves an unpaired electron shared by two chlorophyll molecules. This observation has led to the idea that the center of photoreaction is a pair of parallel chlorophyll rings held closely together by hydrogen bonding between amino acid groups.

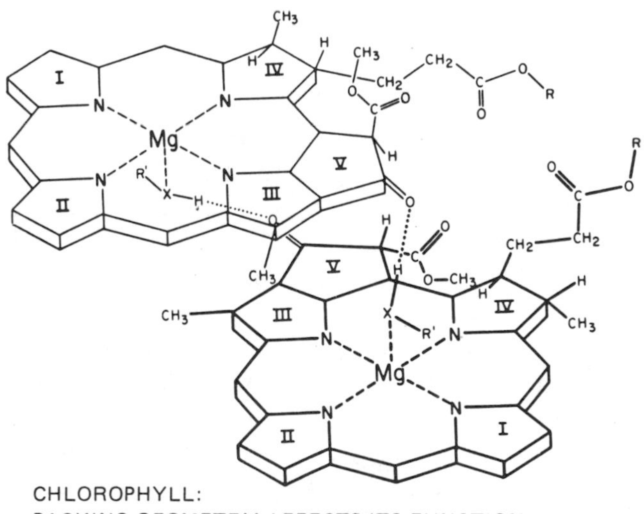

CHLOROPHYLL: PACKING GEOMETRY AFFECTS ITS FUNCTION

Another promising approach to the use of solar energy is the direct conversion of sunlight to electrical or chemical energy with the aid of electrochemical devices. Recent advances in electrochemistry have brought us closer to this goal. In a photoelectrochemical cell, one or both electrodes are made of light-absorbing semiconductors. The light absorption results in oxidation-reduction chemistry at the electrode-electrolyte interface and, hence, current flow in the external circuit. Alternatively, with suitable control, the final products of the oxidation-reduction chemistry can be hydrogen and oxygen. Determination of the thermodynamics and kinetics of light-induced processes at interfaces has, over the past decade, led to a 10-fold increase in efficiency of conversion from light to electrical energy (from 1 percent to better than 10 percent). Development of thin, polycrystalline semiconductor films with high conversion efficiencies to replace the expensive single crystals currently used has been another important achievement. For example, efficiencies of nearly 10 percent have been reported with thin cadmium-selenide-telluride films.

NUCLEAR ENERGY

At the same time that physicists and chemists gave us the atomic bomb, they made available atomic energy, a new source of energy with seemingly unlimited capacity. However, the role of nuclear energy in man's energy future is clouded by long-term risks that are difficult to weigh. Chernobyl has shown plainly the essential need for caution. Whatever course society ultimately chooses, it will rely heavily upon the ingenuity of chemists and chemical engineers to reduce those risks.

Indeed, chemical research is essential to practically all phases of nuclear energy generation and the subsequent management of radioactive waste. To begin with, geochemistry plays a lead role in locating uranium ore deposits. Then, chemical separations are centrally important in the nuclear fuel cycle—from concentration steps at the uranium mill, through reactor fuel manufacture, to the highly automated, remote-control reprocessing of fuel elements from nuclear reactors. This last step has controversial overtones. While recovery of plutonium from fission products puts an appealing "recycling" aspect into the use of nuclear energy, it also makes more widely available plutonium, from which atomic weapons can be made.

Radioactive waste management is also largely based upon chemistry and geochemistry. If these wastes are to be stored underground, we must find appropriately stable ground sites from which dangerous substances will not spread; we must develop more efficient separations of particularly hazardous radioactive elements (e.g., the actinides that pose the major health hazard after a few hundred years); and we must understand fully the geochemistry of potential waste storage sites. If temporary, recoverable containers are used, the problem shifts to the possibility that those containers might corrode and weaken under intense irradiation. Next, our analytical techniques must be made more sensitive for a variety of uses that extend from exploring for new uranium deposits to environmental monitoring that is sufficiently sensitive to reveal potential problems before real hazard has developed. Finally, we must extend our understanding of the unfamiliar chemistry that would accompany a catastrophic reactor accident. We must have useful estimates of release rates for fission products from a decomposing ceramic in the presence of high pressure (up to 150 atmospheres), high-temperature steam (up to 3,000K), and an intense radiation field.

The use of nuclear reactors to generate energy is plainly a controversial and emotionally charged issue. Nevertheless, the scientific setting must be fully understood so that political choices can be made among well-defined and well-informed options. It must be remembered, too, that the use of nuclear energy is a global issue. U.S. decisions about our nuclear future may influence, but they will not determine, the policies of other governments. Plainly, it would be unwise to stop the research efforts that will define these options more clearly.

FUSION ENERGY

Nuclear fusion is a process in which two nuclei join or "melt together" to form a larger nucleus. An example would be the joining of the nuclei of a deuterium atom and a tritium atom to form a helium nucleus (with ejection of a neutron). The product helium nucleus is much more stable than the reactant hydrogen isotopes, though it is not clearly understood why. An enormous energy release would result from this fusion—more than 100 million kilocalories of heat would be released for every gram of helium formed. Thus, nuclear fusion competes with nuclear fission as a possible future source of energy, but it does not produce the myriad of radioactive fission products that present such a troublesome radioactive waste problem. We know that it can work since this same principle is used in the hydrogen bomb.

Very large research efforts have been devoted for the last quarter century to the development of nuclear fusion. (The federal investment for 1985 exceeded $400 million.) The difficulty lies in the need to find an appropriate "match" to light this nuclear fire. The match will have to raise the fuel temperature to about a thousand million degrees before it will ignite. To "light" a hydrogen bomb, an ordinary nuclear fission bomb is used as a match—hardly a practical device at a neighborhood power plant. But even putting aside the question of what match to use (a chemical laser has been considered!), think of the container problem! What could an oven be made of to cook something at 10^9 degrees? The oven walls will be exposed to these solar furnace temperatures, intense ultraviolet radiation, and bombardment by neutrons and chemical ions.

Studies of materials that might be suitable for nuclear reactor components have begun with coated, temperature-resistant solids (refractory and ceramic materials). There is much to be learned, however, about the chemical changes that will occur at the surface of reactor components exposed to the high-temperature gas (which is called a "plasma"). While it is not clear yet that controlled nuclear fusion will ever be practical, it is obvious that everyday application will require significant chemical breakthroughs in the development of high-temperature materials.

CONCLUSION

Nothing is more critical to the long-term health of our technological society than continued access to abundant and clean sources of energy. As we try to look ahead to these needs, we must face these challenging expectations for the next three decades:

- by the year 2000 the U.S. annual energy consumption will probably exceed today's use by 20 to 50 percent;
- during the next three decades, growth in the use of nuclear power will be severely restricted by social concerns already in evidence;
- further increase in hydroelectric power has natural limits and is in conflict with widespread desire to minimize environmental change;
- even those most optimistic about nuclear fusion do not see it providing a large fraction of our energy use before well into the twenty-first century;
- dependence upon high-grade petroleum crudes and high-grade coal deposits must decline as worldwide reserves are used up and as access to foreign crude oil is restricted by political developments beyond our control.

These depressing expectations surely point to the need for expanding the knowledge base upon which new energy technologies can be built. Chemical and electrochemical systems provide some of the most compact and efficient means of energy storage. And we can predict with confidence that foremost among the new energy sources will be low-grade chemical fuels, such as high-sulfur coal, shale oil, tar sands, peat, lignite, and biomass. For none of these alternatives does the technology yet exist that can economically meet the strict demand that environmental pollution be avoided. Enormous chemical challenges must be met—for new catalysts, new processes, new fuels, new extraction techniques, more efficient combustion conditions, better emissions controls, more sensitive environmental monitoring, and many others. Biomass must be further developed to reduce the amount of fossil fuel burned and thus to help check the rate of increase in atmospheric carbon dioxide. Solar energy must be fully investigated and put to use. We must develop artificial photosynthetic and electrocatalytic techniques that completely avoid combustion by converting the light energy directly to electrical or chemical energy. Fortunately, chemistry is ready to respond to these challenges.

SUPPLEMENTARY READING

Chemical & Engineering News

"Photovoltaic Cells" by K. Zweibel, vol. 64, pp. 34-48, July 7, 1986.

"First Methanol-to-Gasoline Plant Nears Startup in New Zealand" by J. Haggin (C.& E.N. staff), vol. 63, pp. 39-41, Mar. 25, 1985.

"New Dow Acrylate Ester Processes Derive from C_1 Efforts" by J. Haggin (C.& E.N. staff), vol. 63, pp. 25-26, Feb. 4, 1985.

"Dow Develops Catalytic Method to Produce Higher Mixed Alcohols" by J. Haggin (C.& E.N. staff), vol. 62, pp. 29-30, Nov. 12, 1984.

"Surface Sites Defined on Synthesis Gas Catalysts" (C.& E.N. staff), vol. 62, pp. 38-39, Sept. 17, 1984.

"Chemical Microstructures of Electrodes" by L.R. Faulkner, vol. 62, pp. 28-42, Feb. 27, 1984.

"New Processes Upgrade Heavy Hydrocarbons" (C.& E.N. staff), vol. 61, pp. 43-44, Apr. 11, 1983.

"Two New Routes to Ethylene Glycol from Synthesis Gas" (C.& E.N. staff), vol. 61, pp. 41-42, Apr. 11, 1983.

Science

"Surface Functionalization of Electrodes

with Molecular Reagents'' by M.S. Wrighton, vol. 231, pp. 32-37, Jan. 3, 1986.

Scientific American

''Molecular Mechanisms of Photosynthesis'' by D.C. Youvan and B.L. Marrs, vol. 256, pp. 42-48, June 1987.

''Materials for Energy Utilization'' by R.S. Claasen and L.A. Girifalco, vol. 255, pp. 102-107, October 1986.

''Photonic Materials'' by J.M. Rowell, vol. 255, pp. 146-157, October 1986.

Chem Matters

''Hydrogen and Helium'' pp. 4-7, October 1985.

''Detergents'' pp. 4-7, April 1985.

''Soap'' pp. 4-7, February 1985.

''The Sun Worshippers'' pp. 4-7, April 1984.

Stone Age, Iron Age, Polymer Age

There was a time when everything from arrowheads to armchairs was made from stones. Other features of those good old days were air-conditioned caves and charbroiled saber-tooth tiger steaks (if you caught him instead of the other way around). Fortunately, this age ended when someone discovered how to reduce iron oxide to metallic iron using coke (carbon) as the reducing agent. That all happened several thousand years ago, so the caveperson chemist who got the patent rights to the Iron Age wasn't educated at MIT or the University of Chicago. But this chemical discovery profoundly changed the way people lived. It led to all sorts of new products like swords and plowshares and the inner-spring mattress. Can you imagine how those stone-agers would have reacted the first time they put on a suit of armor, or went up the Eiffel Tower, or took the train to Chattanooga? Well, brace yourself, because chemists are at it again! This time, we're about to enter the Polymer Age.

You may think we're already there, with your polyester shirt, polyethylene milk bottle, and polyvinylchloride suitcase. We walk on polypropylene carpets, sit on polystyrene furniture, ride on polyisoprene tires, and feed our personal computers a steady diet of polyvinylacetate floppy disks. In just the last 40 years, the volume of polymers produced in the United States has grown 100-fold and, since 1980, actually exceeds the volume of iron we produce. But the best is yet to come.

The structural materials with which we have been building bridges since even before the one to Brooklyn, and automobiles since the Model T, would seem to be the last stronghold of the iron age (pun intended). Would anyone dare to suggest that polymers could compete on this sacred ground? Well, no one perhaps except chemists. Right now, there's talk of an all-plastic automobile, and you're already flying in commercial airliners with substantial structural elements made of composite polymers. One of these, poly(para-phenylene terephthalamide), has a tensile strength slightly higher than that of steel. But where this polymer really scores is in applications where the strength-to-weight ratio matters a lot, as it does in airplanes. Even with its cumbersome name, this polymer has a strength-to-weight ratio six-fold higher than steel! To appreciate this advantage, you should know that a 1-pound reduction in the structural weight of an airplane reduces its take-off weight by 10 pounds (counting the fuel to lift the pound and the fuel to lift the extra fuel). No wonder this polymer, under the trade name Kevlar®, is used to build tail sections for the biggest airliners. Oh, and bullet-proof vests, too.

So what about this all-plastic automobile? Of course, weight reduction is the name of the game in trying to build fuel-efficient cars. Already there are automobile driveshafts made of polymers strengthened with stiff fibers, and similar composites are used for leaf springs (oops, there goes the inner-spring mattress!). Right now, U.S. cars contain about 500 pounds of plastics if you count, as well, the rubber and paint and sealants and lubricants and upholstery.

But what about the engine and the electrical system? What will we do about these in this allegedly all-polymer car? Gee, I'm glad you asked. . . .

III-D. New Products and Materials

Webster: *Material.* noun *The substance or substances out of which a thing is constructed.*

Chemistry. noun *The science that deals with the composition, properties, and changes of properties of substances.*

Expectations are universally high for advances in the materials sciences. What is a *material*? Webster's definition includes all of the substances from which one might construct autos and airplanes, bridges and buildings, dishes and doors, parachutes and pantyhose, raincoats and radios, spacecraft and sewer pipe, tires and transistors, windows and walls, shirts, sheets, and shoes. That incredible range of application is reason enough for the high hopes scientists have for finding new substances and new ways to tailor their properties to fit our changing and diverse needs.

Chemists clearly have a role here, because chemistry is the central science for understanding and controlling what substances are composed of, their structure, and the manner in which those substances behave. When designing a substance to meet a particular need, the chemist's special talent for synthesis and control of composition can play a key role. By no means does that exclude other disciplines. To make this point, we need only mention the remarkable advances made in solid-state physics over the last three decades characterizing and developing semiconductor materials. Because of this we can now fashion calculators as thin as credit cards, and pocket radios to carry along when you go jogging. The fields of ceramics and metallurgy, too, have provided substances to meet special needs, from heat shields on the space shuttle to cylinder heads in automobiles. Equally important are the contributions of engineers in the processing and fabrication of the products we wish to use. There is probably no scientific frontier that is more interdisciplinary.

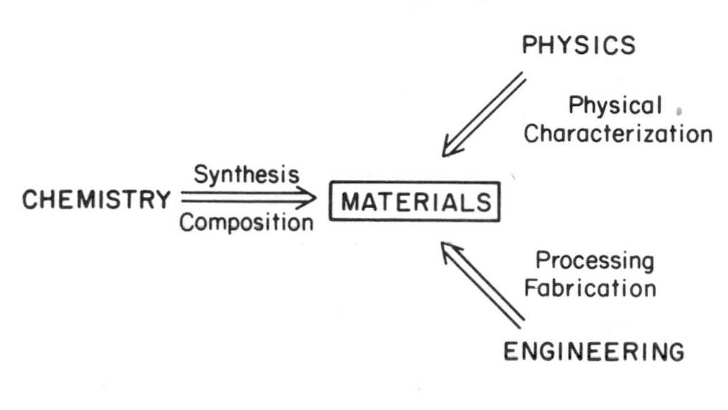

MATERIALS SCIENCE IS INTERDISCIPLINARY

The following analysis will focus on the rich opportunities for chemists to advance materials science to benefit us all. However, realization of these opportunities will often depend upon cooperative interactions with other scientists in the materials science community.

PLASTICS AND POLYMERS

We find natural polymeric materials all around us—in proteins and cellulose, for example. Polymers are long molecules made up of the same chemical unit repeated over and over, linked by covalent bonds into a chain. Chemists probably learned most about how to make polymers through their attempts to imitate nature in synthesizing natural rubber. Today, chemists have designed so many polymers for so many purposes it is difficult to picture a modern society without their benefits. This importance is dramatically displayed in the 100-fold growth of U.S. production of plastics over the last 40 years. Its production, expressed on a volume basis, now exceeds that of steel, whose growth has barely doubled over this same time period. The economic implications of these comparisons are obvious. Furthermore, production of plastics continues upward.

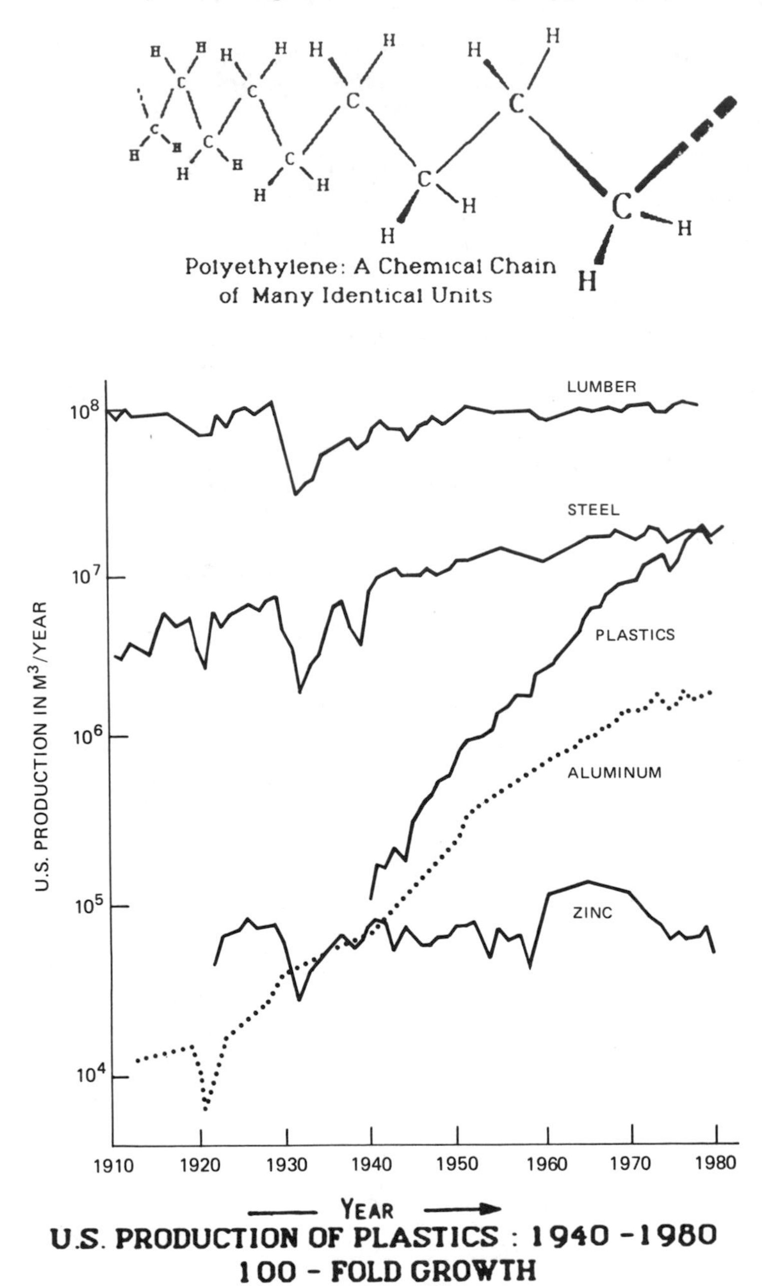

U.S. PRODUCTION OF PLASTICS : 1940 - 1980 100 - FOLD GROWTH

There are many dimensions to polymer chemistry—dimensions that chemists are increasingly able to control. Extremely careful choice of reaction conditions (temperature, pressure, polymerization initiator, concentration, solvent, emulsifiers, etc.) and reactant (monomer) structures can determine a variety of different qualities of a polymer. We can fix the average chain length (molecular weight), extent of chain branching, cross-linking between polymer strands, and, through the addition of carefully chosen functional units, the physical and chemical properties of the final polymer.

By clever manipulation of these factors, chemists can design polymers with tailored properties such as plasticity or hardness, tensile strength, flexibility or elasticity, thermal softening or thermal stability, chemical inertness or solubility, attraction or repulsion of solvents (wetting properties), permeability to water, responsiveness to light (photodegradability), responsiveness to organisms (biodegradability), and viscosity variability under flow (thixotropy). All of these possibilities account for the continuing growth of plastics production and their increasing presence in the things we use, wear, sit upon, ride in, eat from, and otherwise find in our everyday environment.

A FEW SIMPLE POLYMERS SHOW THAT POLYMERS CAN BE TAILORED TO NEED

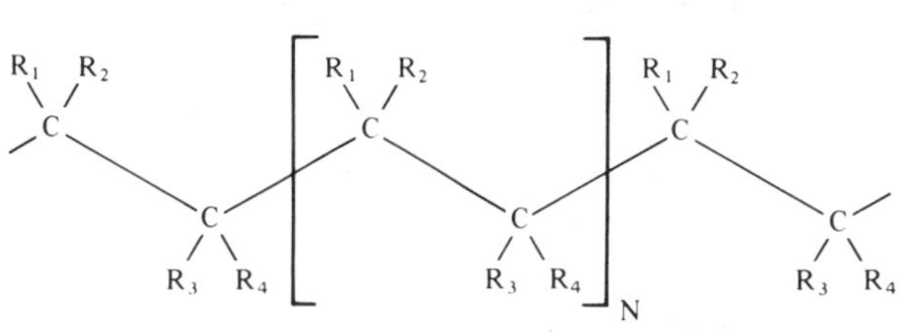

R_1,R_2,R_3,R_4	Name	Product	1986 U.S. Production (tons/year)
H,H,H,H	Polyethylene	Plastic bags, toys, bottles, wire & cable coverings	8,100,000
F,F,F,F	Polytetrafluoro-ethylene	Cooking utensils, insulation (e.g. Teflon®)	
H,H,H,CH_3	Polypropylene	Carpeting (indoor, outdoor), bottles	2,700,000
H,H,H,Cl	Polyvinyl chloride	Plastic wrap, pipe, phonograph records, garden hose, indoor plumbing	3,500,000
H,H,H,C_6H_5	Polystyrene	Insulation, furniture, packaging	2,100,000
H,H,H,CN	Polyacrylonitrile	Yarns, fabrics, wigs (e.g., Orlon®, Acrilon®)	920,000*
$H,H,H,OCOCH_3$	Polyvinyl acetate	Adhesives, paints, textile coatings, floppy disks	500,000*
H,H,Cl,Cl	Polyvinylidine chloride	Food wrap (e.g., Saran®)	
$H,H,CH_3,COOCH_3$	Polymethyl methacrylate	Glass substitute, bowling balls, paint (e.g., Lucite®, Plexiglas®)	

*1982 Production

Polymers as Structural Materials

The potentiality of polymers as structural materials is demonstrated in the hundreds of commercial airplanes flying today that have substantial structural elements made of a substance called Kevlar®. This is a composite material made up in part of the lightweight, ultrastrong organic polymer poly(*para*-phenylene terephthalamide). The widely known Lear jet is largely built of polymer composites. More down to earth, the efforts directed toward an all-plastic and ceramic automobile shows the high expectations for polymers' capacity to reduce weight, eliminate corrosion, and lower costs.

In the past, differences in mechanical properties of polymers were only discussed empirically, that is, in terms of their observed behaviors. Nowadays, much is known about the molecular conformation of these polymeric molecules. Using primary molecular data and the basic principles of chemical bonding, chemists can now predict how each polymer will behave. The elasticity in the polymer chain direction can now be calculated from bond lengths, bond angles, and the vibrational spring constants derived from infrared spectroscopic measurements. The resulting progress is shown in Table III-D-1, which compares the tensile strengths of two organic polymer fibers to those of both aluminum alloy and drawn steel. The two polymers significantly outperform both of the conventional structural metals in a crucial measure, strength per unit weight.

TABLE III-D-1 Polymer Fibers Compete as Structural Materials

	Tensile Strength[a]	Tensile Strength per Unit Weight[a]
Aluminum alloy	(1.0)	(1.0)
Steel (drawn)	5.0	1.7
Poly(*p*-phenylene terephthalamide)[b]	5.4	10.0
Polyethylene[c]	5.8	15.0
Ceramic whiskers	25	50

[a] Relative to aluminum alloy.
[b] Kevlar®.
[c] Highly oriented samples.

Further developments will surely flow from continued research. It is already known, for example, that the elasticity that can be obtained with a zig-zag polymer chain is far higher than with a helical (spiral) structure. Polyethylene has a strength-to-weight ratio that is 10-fold better than that of steel. Calculations show that, in theory, it could be improved by another factor of five. Research is needed to tell us how to take advantage of these possibilities.

Liquid Crystals and Polymer Liquid Crystals

Though known for over a century, liquid crystals flared into prominence only a decade ago. Now, liquid crystal display (LCD) devices provide an industry second only to television cathode ray tubes in the world market of display technology. Nothing comes close to matching the LCDs in low-power consumption for small-area displays.

Liquid crystals are organic molecules that have been constructed to possess geometric and/or polar characteristics that will encourage one- or two-dimensional order. Because at least one dimension remains disordered, the substance remains fluid, and thus appears to be a liquid. However, the optical properties of these compounds give proof of their degree of order on the molecular level. Long, slender molecules that are highly rigid line up like logs floating down a river (such one-dimensional order is called a ''nematic phase''). More complex shapes, such as large but flat molecules, can give layered structures, like the successive sheets in a piece of plywood (such two-dimensional order is called a ''smectic phase''). The actual behavior is determined by a balance between the effects of molecular shape and electrical charge distribution as the molecule interacts with its local environment. This balance can be affected by a small electric field, which provides a ready means of switching from one optical behavior to another (e.g., from transparent to opaque).

Plainly, design of liquid crystals is an exciting research area for chemists. Their ability to synthesize new molecules of spherical, rod-like, or disc-like shape, containing functional groups placed as desired, is crucial to progress. In fact, one of the most promising frontiers of liquid crystal chemistry is the application of this knowledge to the preparation of polymers. Combining the molecular ordering of a nematic liquid with polymerization chemistry permits the order to be built into the polymer, with dramatic effects on physical (and optical) properties. It is just this control that lies behind the production of fibers of exceptionally high tensile strength which can replace steel in products ranging from airframe construction to bullet-proof vests.

Block Polymers and Self-Organized Solids

Another area of research that is destined to lead to entirely new types of materials is connected with *block polymers*. These polymers exploit the fact that long molecules of suitable structure will organize themselves into clusters. These organized clusters can take the shape of spheres or alternating layers or rods in a continuous pattern.

	Stiffness	Tensile Strength
GRAPHITE	(1.00)	(1.00)
KEVLAR poly-(p-phenylene terephthalamide)	0.66	2.25
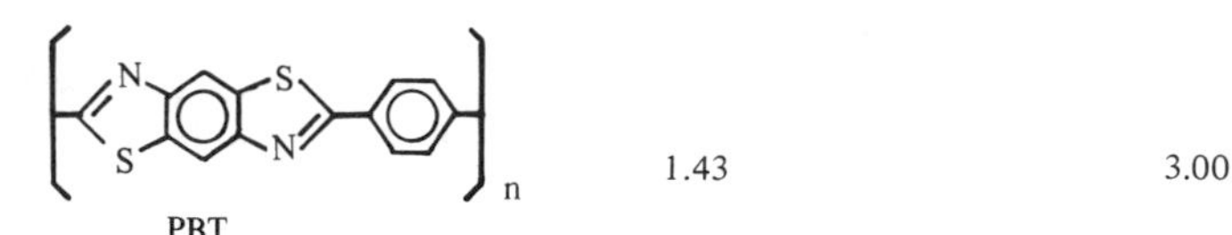 PBT poly-(p-phenylene benzobisthiazole)	1.43	3.00
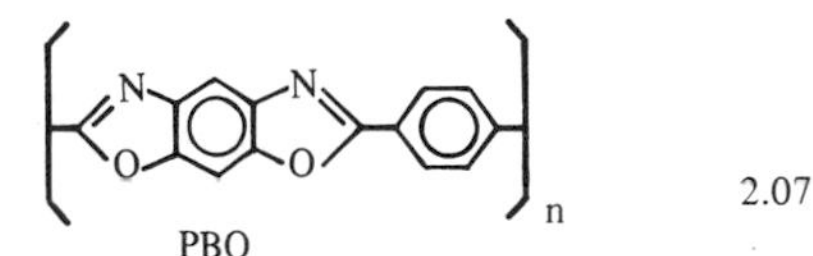 PBO poly-(p-phenylene benzobisoxazole)	2.07	3.00
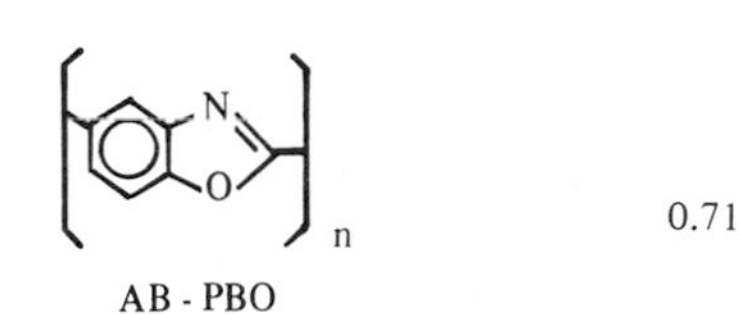 AB - PBO	0.71	3.25

EXPERIMENTAL POLYMERS FROM LIQUID CRYSTALS STRONGER AND STRONGER!

A "triblock" polymer is constructed of two polymers, $\overline{A}$ and $\overline{B}$, so that one polymer $\overline{B}$ is sandwiched between two segments of a different polymer $\overline{A}$. The resulting material, $\overline{A}$-$\overline{B}$-$\overline{A}$, has the properties of A at its ends and the properties of B at its middle. If A and B are chemically designed to be unfriendly to each other, one polymer will try to reject the other. This chemical conflict can result in a molecule in which the A ends curl up into a ball in order to avoid contact with B. The result is a polymer in which spheres of A molecules are found distributed fairly regularly in a continuous matrix of B molecules. The values

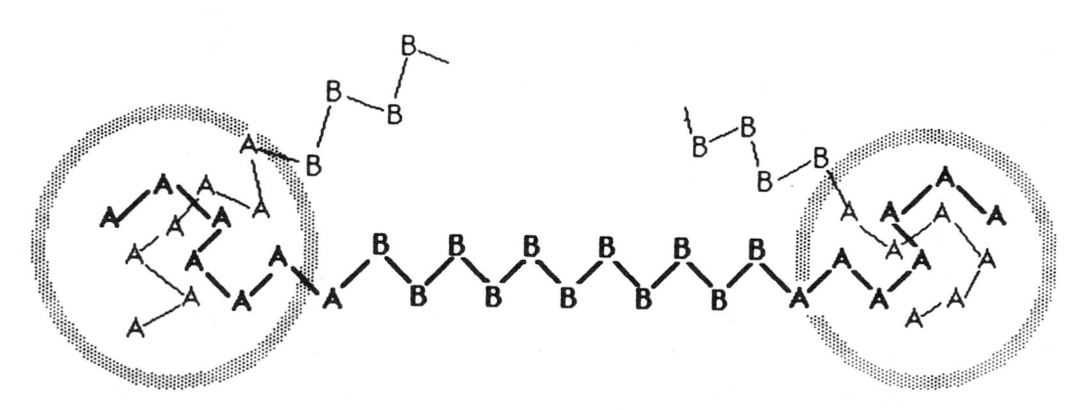

BLOCK POLYMERS CAN SELF-ORGANIZE

of such molecular design are shown dramatically by comparing the tensile strengths of the two types of triblock polymers that can be made from butadiene (B) and styrene (A). With $\overline{B}$ chains containing 1,400 B molecules and $\overline{A}$ chains with 250 A molecules, the triblock polymer $\overline{A}$-$\overline{B}$-$\overline{A}$ has a useful tensile strength. If the polymers are hooked together in the reverse triblock arrangement, $\overline{B}$-$\overline{A}$-$\overline{B}$, the polymer is a syrupy liquid, showing no real tensile strength at all. The first of these two, $\overline{A}$-$\overline{B}$-A, can be shaped to any desired form at high temperature. On cooling to room temperature, it becomes rigid and behaves like a cross-linked rubber. However, unlike conventional rubber, the $\overline{A}$-$\overline{B}$-$\overline{A}$ block polymer can be warmed again and reshaped. Such "thermoplastic" behavior has many useful applications.

This is only the beginning, however. The ability of block polymers to self-organize into microdomains of 10 to 100 Å size and of different shapes (spheres, rods, planes) is sure to provide new materials with novel combinations of properties. The self-organization can give directional properties (anisotropic behavior) to mechanical, optical, electrical, magnetic, and flow characteristics. As research advances give us control of these various dimensions, new applications, new devices, and new industries will be seen.

NOVEL OPTICAL MATERIALS

Optical Fibers

Just as the vacuum tube has been replaced by the transistor in modern electronics, copper wires are being replaced by hair-like silica fibers to transmit telephone conversations and digital data from one place to another. Instead of a pulse of electrons in a copper wire, a pulse of light is sent through the transparent fiber to convey a bit of information. The critical development that made this optical technology possible was the production of highly transparent silica fibers through a new process known as chemical vapor deposition (CVD). Essentially, a silicon compound is burned in an oxygen stream to create a "soot" of pure silica that is deposited inside a glass tube. The tube and its silica deposit are melted and drawn out to give a glass-coated silica fiber about one-tenth the diameter of a human hair. The CVD process made it possible, in less than a decade, to vastly improve the performance of fiber optics, reducing fiber light losses 100-fold. A new class of materials, the fluoride glasses, may result in fibers that are even more transparent. In contrast to traditional glasses, which are mixtures of metal oxides, fluoride glasses are mixtures of metal fluorides. Although many practical problems remain to be resolved, these new glasses would, in principle, permit transmission of an optical signal across the Pacific Ocean without any need for relay stations.

Optical Switches

In addition to chemistry's role in development of new materials and processes for optical fibers, it also has a major part in synthesis of materials for optical devices that switch, amplify, and store light signals. The possibilities in this area are quite remarkable since an optical switch might be able to operate in a millionth of a millionth of a second (a picosecond). Current optical devices are based on lithium niobate and gallium aluminum arsenide, which are spin-offs from the electronics industry. In new directions, mirror-image organic molecules, liquid crystals, and polyacetylenes can display desirable optical effects greater than those of lithium niobate. The potential for discovery and practical applications in this field are great.

NOVEL ELECTRICAL CONDUCTORS

Semiconductors

The modern age of solids was launched during the 1950s by brilliant advances of solid-state physicists as they developed deep understanding of pure semiconductor materials. There were early challenges to chemists, too, as it became clear that elemental silicon and germanium were needed in single-crystal form with impurity levels as low as one part in 100 million. Thereafter, similar behaviors were found in compounds consisting of two elements, one from the third group of the Periodic Table (e.g., gallium) and one from the fifth group (e.g., arsenic). A typical "III-V" compound is the mixed semiconductor indium antimonide, which has for 15 years provided one of the most sensitive detectors known for near-infrared light. Lately much attention has been directed toward single crystals of the III-V compound gallium-arsenide grown on single-crystal substrates of indium phosphide, another III-V compound. There may be as many as half a dozen gallium-arsenide layers with differing impurity compositions and thicknesses. This class of materials forms the basis for lasers and laser display devices for long-wavelength optical communications.

As the range of materials used in semiconductor technologies has broadened, more and more chemists have joined the physicists in such work. This upswing of the chemist's participation came about the same time as the startling discovery that amorphous silicon (silicon that is not crystalline) also can demonstrate semiconductor behavior. Because the prevailing and extremely successful textbook theory of semiconductor behavior is based upon the properties of perfectly ordered solids, such amorphous semiconductors were neither predicted nor comfortably described by theory. The language and concepts of chemistry are now being used to explain this puzzle (e.g., "dangling bonds" in amorphous silicon).

We are on the verge of a new era in the solid-state field, one in which physicists will continue to expand their success in characterizing new solids, but now chemists will play an increasingly important role. The reason is that entirely new families of electrically conducting solids are being discovered—families susceptible to a chemist's ability to control local structures and molecular properties. As will be seen, some of these new families are inorganic solids and some are organics.

Conducting Stacks

The field of organic conductors had its beginning in the late 1960s-early 1970s with the synthesis of organic crystals that conducted electricity. The first examples were formed by the reaction of compounds such as tetrathiafulvalene (TTF) with tetracyanoquinodimethane (TCNQ). Both of these molecules are flat, and in their mixed crystal they are found alternately stacked like poker chips. The interaction between two neighbor molecules is a familiar one to chemists—a charge-transfer complex is formed. Such an interaction always includes an electron donor, a molecule from which electrons are readily removed, and an electron acceptor, a molecule that has a high electron affinity. These two roles are filled, respectively, by TTF and TCNQ. The surprise is that this charge transfer between two neighbors in the crystalline stack provides a mechanism for current flow up and down the stack, the length of the crystal.

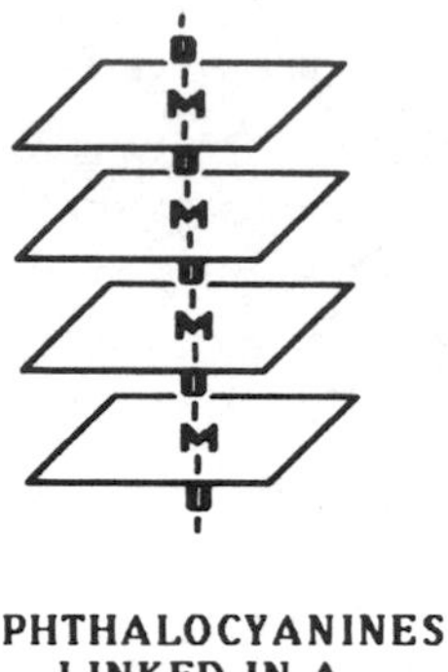

PHTHALOCYANINES LINKED IN A "CONDUCTING STACK"

The bright future for conducting stacks has recently been assured by the imaginative synthesis of polymeric conductors with charge transfer properties. Again large, flat molecules furnish the elements of the conducting stack (metallomacrocycles). The clever innovation lies in lacing them together with a string of covalently bound oxygen atoms. The fact that this chemically designed molecule is, indeed, an electrical conductor is quite a breakthrough. Plainly, the metal atom and the surrounding groups in the flat metallomacrocycle can be substituted and altered in great variety. Then these units can be connected by an intervening atom chosen to give the desired spacing. The result is a polymer in which carefully chosen macrocycles are held in a molecular stacking that is rigidly enforced by the covalent bonding and designed to fit the desired function.

Organic Conductors

Polyacetylene is one of the simplest organic polymers. It has a carbon skeleton of alternating single and double bonds. Chemists call this bonding situation "conjugation," which means that electric charge is especially mobile along the skeletal chain. Nevertheless, it came as a surprise, half a dozen years ago, when the unusual electrical properties of polyacetylenes were discovered. Such polymers, when exposed to suitable chemical agents such as bromine, iodine, and arsenic pentafluoride (which physicists call "dopants"), become shiny, like metals, and they display electrical conductivities higher than those of many metals (though not yet as good as copper).

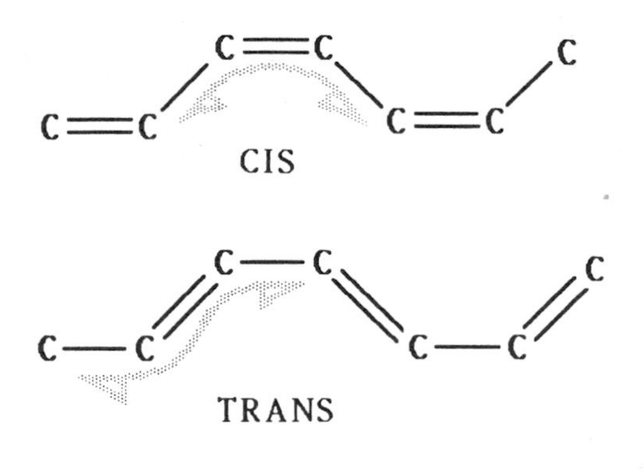

POLYACETYLENE
DIFFERENT STRUCTURES ⇒ DIFFERENT PROPERTIES

Plainly, the gates are now open, and other conducting polymers are rapidly appearing. The polymer poly(*para*-phenylene) has been shown to become a conductor when exposed to the correct chemicals. The same is true of poly(*para*-phenylene sulfide) and poly-pyrrole. Organic chemists can use their creative skills to fashion compounds that combine electrical conductivity with the various beneficial properties of polymers, such as structural strength, thermoplasticity, or flexibility. Electrochemical methods are under study that make polyacetylene photovoltaic cells possible. Because response to light can be designed to match the solar spectrum, these polymers give us hope for cheap organic photovoltaic cells with which to convert solar energy to electricity. Extensive research is in progress to develop lightweight, high-power density, rechargeable batteries with polymeric electrodes.

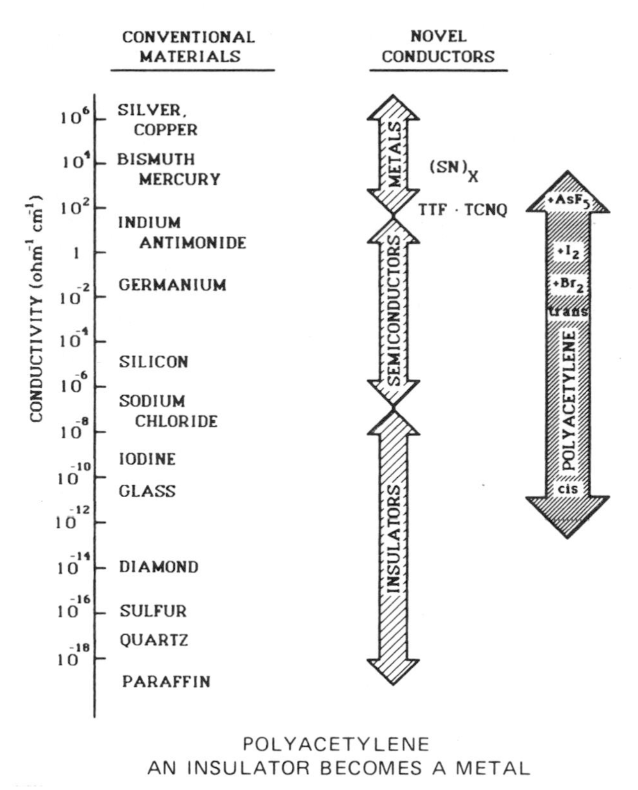

POLYACETYLENE
AN INSULATOR BECOMES A METAL

Superconductors

Another discovery as significant as polyacetylene was the synthesis of pure, single crystals of the *inorganic* polymer poly(sulfur nitride), $(SN)_x$. This material not only showed metallic conductivity, it was found to become *superconducting* (having no detectable electrical resistance) at about 0.3K! It was the first covalent polymer with metallic conductivity (ahead of polyacetylene by 4 or 5 years), and also the first covalent polymer comprised of nonmetals to show superconductivity. To solid-state scientists, this opened a whole new world of possibilities of chemical candidates for electrical behavior. For example, chemists have used the conducting stack design involving tetrathiafulvalene (TTF), mentioned earlier, to develop a superconducting polymer. They synthesized an analogous compound by replacing the sulfur atom on each TTF molecule with a selenium atom. Like TTF, this selenium analog also forms conducting salts, but in addition it displays superconductivity and at much higher temperatures than poly(sulfur nitride), $(SN)_x$.

Inorganic compounds involving three elements are also under systematic study, and materials with relatively high superconducting temperatures have been discovered among this family of ternary compounds known as Cheveral phases. An example is $PbMo_6S_8$ which can remain superconducting in the presence of magnetic fields of several thousand gauss. This is a crucial property because construction of compact, high-field magnets is one of the most important applications of superconductors.

Late in 1986, however, an enormous breakthrough was made when certain copper oxide solids were found to become superconducting at temperatures above 90K. These solids have layers of copper oxide with various metal atoms sandwiched between in the so-called *perovskite* crystal structure. A typical composition is $YBa_2Cu_3O_x$ where x is about 7.5, indicating an oxygen-deficient lattice. However, the yttrium atom can be replaced by almost any other lanthanide atom, and the barium atom can be partially replaced by calcium or strontium. These many substitutions have little effect on T_c, the temperature at which the material becomes superconducting (all have T_c in the range 85 to 98K), so it seems that the electrical behavior is a property of the oxygen-deficient, hence strained, copper-oxygen layers.

The implications of this landmark discovery are mind-boggling. It provides zero-loss electrical conductivity at temperatures easily maintained with inexpensive liquid nitrogen coolant (77K). This makes many applications practical, ranging from loss-free energy transmission over long distances, tinier computer integrated circuits that are not limited by heat generation, and trains levitated with superconducting magnets to make them virtually friction-free. But most remarkable is the fact that after 75 years since the discovery of superconductivity, the highest T_c recorded was only 23K. Then, in a period of a few months, this record reached 100K. We are irresistibly drawn to the expectation that other materials will be discovered that raise superconductivity upward further toward room temperature. The impact of such a development would likely have as dramatic effect on our culture as that caused by the transistor.

Solid-State Ionic Conductors

Solid materials with ionic structures are now known with ionic charge mobilities approaching those in liquids. Investigations of such materials over the last decade have already led to their use in memory devices, display devices, chemical sensors, and as electrolytes and electrodes in batteries. Thus, sodium beta-alumina provides the conducting solid electrolyte in the sodium/sulfur battery.

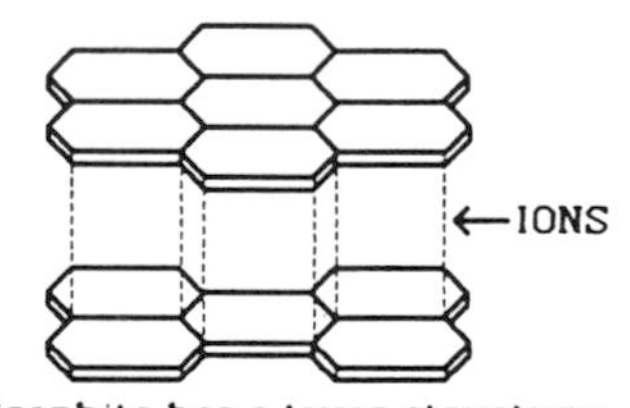

Graphite has a layer structure
Ions can move between the layers

Normally, an ionic solid like sodium chloride has a fixed composition and is an electrical insulator. The new solid electrolytes are produced by carefully manipulating defects in crystals and by deviating from integer chemical formulas. In a process called *intercalation,* charges are inserted between the weakly bound layers of a crystal lattice that encourages charge migration. The mobile charge carriers might be small ions like lithium ion or hydrogen ion. Substances with layered molecular structures—like graphite—provide excellent crystal hosts for such manipulations. This method places charges in a two-dimensional zone where movement can be exceptionally high. Many such layered structures are known, so significant opportunities for new discoveries lie ahead.

In a practical example of this ionic conduction, zirconium dioxide is used as a sensing element in the oxygen analyzer of an automobile emission control system. The electrical conductivity of this solid changes with the oxygen content of the exhaust gases.

Acentric Materials

Materials with directional properties (such as magnets, ferroelectrics, and pyroelectrics) are under active development and include a wide variety of ionic crystals, semiconductors, and organic molecular crystals. Both electrical and optical applications are probable: optical memory devices, display devices (for digital wrist watches), capacitors for use over wide temperature ranges, pyroelectric detectors (for fire alarm systems and infrared imaging), and nonlinear optics (second harmonic generation and optical mixing). To cite an example, the polymer of vinylidene chloride, (CH_2CCl_2), changes shape in an electric field (it is piezoelectric), and has found use in sonar detectors and microphones.

Conducting Glasses

Both metallic and semiconducting glasses can be prepared by rapidly freezing a liquid, by condensing gases on a very cold surface, or by ion-implantation in ordinary solids. Thus, noncrystalline, semiconducting silicon can be prepared by rapidly condensing the products from a glow discharge through gaseous silane, SiH_4. Low-cost solar cells can be made of such material, and their performance depends critically upon hydrogen impurities chemically bound to the silicon atoms randomly lodged in the solid. Inorganic nonmetallic glasses are important for optical fiber communication and for packaging solid-state circuits.

MATERIALS FOR EXTREME CONDITIONS

Performance in many areas of modern technology is limited by the materials available for construction. Jet engines, automobile engines, nuclear reactors, magnetohydrodynamic generators, and spacecraft heat shields are contemporary examples. The hoped-for fusion reactor lies ahead. Engine performance provides a convincing case. Any thermal engine, be it steam, internal combustion, jet, or rocket engine, becomes more powerful and more efficient if the working temperature can be increased. Hence, new materials that extend working temperatures to higher ranges have real economic importance.

New Synthetic Techniques

There are a number of promising synthetic techniques for producing new heat-resistant materials. Among these are ion implantation, combustion synthesis, levitation melting, molecular-beam deposition on crystalline surfaces (epitaxy), and chemical vapor deposition from glow discharges (plasmas). Most recently, laser technology has provided unusual synthetic approaches. A high-power, pulsed laser beam focused on a solid surface can locally create a very high temperature (up to 10,000K) for a very short time (less than 100 nanoseconds). Such a short-lived, high-temperature pulse can cause significant chemical and physical changes, modifying the surface, forming surface alloys, and promoting specific chemical reactions when coupled with vapor deposition. All of these techniques share the ability to form thermodynamically unstable compounds with special properties "frozen in." (Diamond is an example. This expensive gemstone is valued for its sparkling beauty and its extreme hardness even though it is thermodynamically unstable with respect to graphite under normal conditions.)

Some Examples—Real and Projected

Two examples of "exotic" high-temperature materials recently developed are silicon nitride, Si_3N_4, and tungsten silicide, WSi_2, both of technological importance in the semiconductor industry. The first, Si_3N_4, can be an effective insulating layer even at thicknesses below 0.2 microns. The second, WSi_2, is a low-resistance connecting link in microcircuits. Plasma deposition synthetic techniques allow sufficient control to permit these high-temperature materials to be deposited upon a less-heat-resistant substrate held at much lower temperatures (usually below 700K). Thus, the temperature-resistant material can be deposited without damage to the desired electrical properties of the substrate.

Polymers offer another promising route to new, "high-tech" ceramics. Silicon-containing polymers can be molded into any desired shape and then, on heating, converted to silicon carbide or silicon nitride solids that hold the desired shapes. These and other recent advances in the synthesis and fabrication of ceramics make it reasonable to anticipate the future construction of an all-ceramic internal combustion engine.

CONCLUSION

The next two decades will bring many changes in the materials we use; the materials in which we are clothed, housed, transported; the materials of our daily lives. New industries will be founded—just as polymers led to synthetic fabrics, as phosphors led to television, as semiconductors led to computers. Metals will be used less often as deliberately designed materials outperform them in their traditional functions. Chemists' ability to carry out this design and, hence, to control the properties of new materials is leading to their increasing role in these fields. Ultimately, that control depends upon understanding the composition, bonding, and geometry of materials at the atomic/molecular level—the chemist's home territory.

What we can do with this understanding then depends on what we can make—and synthesis is again the chemist's bag. That is why industries dependent upon use of new materials are looking for bright young chemists to add to their scientific staffs. That is why more chemists are being attracted to research in the materials sciences.

SUPPLEMENTARY READING

Chemical & Engineering News

"The Organic Solid State" by D.O. Cowan and F.M. Wiygul, vol. 64, pp. 28-45, July 21, 1986.

"Solid Ionic Conductors" by D.F. Shriver and G.C. Farrington, vol. 63, pp. 42-53, May 20, 1985.

"Liquid Crystals, A Colorful State of Matter" by G.H. Brown and P.P. Crooker, vol. 61, pp. 24-37, Jan. 31, 1983.

"Conducting Polymers R & D Continues to Grow" (C.& E.N. staff), vol. 60, pp. 29-33, Apr. 19, 1982.

Science

"A Chemical Route to Advanced Ceramics" Science staff article, vol. 233, pp. 1-132, July 4, 1986.

"Optical Activity and Ferroelectricity in Liquid Crystals" by J.W. Goodby, vol. 231, pp. 350-355, Jan. 24, 1986.

"Electroactive Polymers and Macromolecular Electronics" by C.E.D. Chidsey and R.W. Murray, vol. 231, pp. 25-31, Jan. 3, 1986.

Scientific American

"Materials for Information and Communication" by J.S. Mayo, vol. 255, pp. 58-65,

Oct. 1986.

"Materials for Aerospace" by M.A. Steinberg, vol. 255, pp. 66-91, October 1986.

"Materials for Ground Transportation" by W.D. Compton and N.A. Gjostein, vol. 255, pp. 92-101, October 1986.

"Advanced Metals" by B.H. Kear, vol. 255, pp. 158-167, October 1986.

"Advanced Polymers" by E. Baer, vol. 255, pp. 178-91, October 1986.

Chem Matters

"Polymers" pp. 4-7, April 1986.

"Polysaccharides" pp. 12-14, April 1986.

"Silly Putty" pp. 15-17, April 1986.

"Liquid Crystal Displays" pp. 10-11, April 1984.

"Liquid Crystals" pp. 8-11, December 1983.

R_x-Snake Bite

High blood pressure anyone? Maybe you'd like a dose of snake venom? Yes, it's true! Hypertension sufferers may find their future treatment coming from this unlikely source—and from sustained research in chemistry and physiology.

This story began 30 years ago when scientists discovered the chemical mechanisms by which blood pressure is elevated in humans. Chemical techniques isolated two closely related substances, angiotensin I and angiotensin II. In the human body chemistry, II is produced from I with the help of a specific enzyme, "angiotensin-converting enzyme" (ACE). Though I has no physiologic effect, its reaction produced angiotensin II, the most potent blood pressure-elevating substance known. Thus I provides a reservoir from which II can be made as needed to maintain a normal blood pressure level, a conversion controlled by the enzyme ACE.

It is no surprise that there is also a substance provided by Nature to lower blood pressure—this substance is called bradykinin, which, along with angiotensin II, seems to complete the control mechanism. To raise pressure when it is too low, make some angiotensin II using ACE. To lower blood pressure when it is too high, a dash of bradykinin will do the trick.

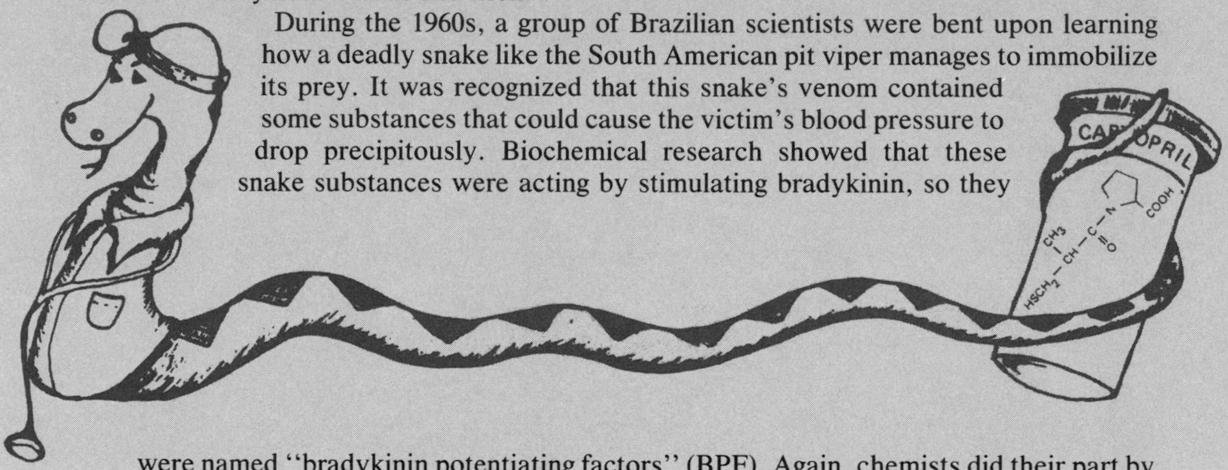

During the 1960s, a group of Brazilian scientists were bent upon learning how a deadly snake like the South American pit viper manages to immobilize its prey. It was recognized that this snake's venom contained some substances that could cause the victim's blood pressure to drop precipitously. Biochemical research showed that these snake substances were acting by stimulating bradykinin, so they were named "bradykinin potentiating factors" (BPF). Again, chemists did their part by purifying BPF from the pit viper venom and identifying several compounds that carried the activity. Chemical analysis showed them to be specific peptides.

The next chapter in this story began when ACE had been purified and characterized. That opened the door to understanding how the snake venom BPF did its work. Some of the peptides in BPF block ACE to interfere with the production of angiotensin II. Then, as a bit of a surprise, it was discovered that ACE derived part of its control function from an ability to inactivate bradykinin. Realizing this, the canny pit viper provides some peptides in its venom to protect bradykinin from inactivation! Thus these BPF peptides deprive the body of its ability to use ACE, either to raise blood pressure by producing angiotensin II or to moderate the lowering action of its own control substance, bradykinin.

With this understanding, teams of biologists and chemists recently began a systematic attack on hypertension, one of the most insidious causes of death in our stressful world. They synthesized a series of peptides modeled on those found in snake venom but designed for therapeutic use. Success came with the synthesis of the compound captopril. It acts as an ACE inhibitor, and clinical trials have amply demonstrated its ability to lower abnormally high blood pressure. No wonder that the medical profession has great expectations for ACE enzyme inhibitors in the treatment of our hypertensive population.

III-E. Better Health

In the next decade chemistry will contribute to the solution of some of contemporary biology's most important problems. All life processes are regulated by chemical interactions between macromolecules and smaller molecules of diverse structural types. Ultimately, our ability to control complex biological events will depend upon our understanding at the molecular level, so chemistry is in a position to make far-reaching contributions to physiology and medicine.

The following discussions illustrate how advances in chemical knowledge and technology have led to the discovery of new and improved medicines and therapeutic drugs in recent years, and indicate where rapid progress can be anticipated in the future.

NOTABLE SCIENTIFIC ADVANCES DURING THE LAST 15 YEARS

There have been significant changes in recent years in the methods used to discover new medicinal compounds. Remarkable progress has been made in understanding how chemical reactions control and regulate biological processes. Such understanding of the chemical mechanism of drug action permits a logical approach to the discovery of new medicines replacing the traditional trial and error screening procedures. Two important frontiers deserve special mention, those of enzyme inhibitors and of receptors.

Enzyme Inhibitors

Enzymes are powerful catalysts that work in highly specific ways. They assist in most of the chemical transformations of life, including the production of the chemical messengers that regulate body processes. These messengers are called hormones and neurotransmitters. Hormones, in animals, work in the bloodstream. Neurotransmitters work in the spaces between nerve cells. Both act to send messages throughout the body to trigger the chemistry of a multitude of bodily processes such as muscle contraction and adrenalin release. One way to affect these messengers, and hence the processes they control, is to affect the enzymes that produce them.

A substance that interferes with the action of an enzyme is called an *enzyme inhibitor*. Because our understanding of enzymes is now reaching the molecular level, we are making great strides in designing compounds that inhibit enzymes. Of particular importance have been molecular structure determinations through computer-aided high-resolution X-ray crystallography. Combining knowledge about how enzymes accelerate chemical reactions with knowledge about the coiling of proteins (tertiary structure) has been fruitful.

There are two approaches to the design of enzyme inhibitors now being pursued. One is based on the belief that enzymes act by stabilizing a transition or intermediate form of the reactant molecule. A compound is designed and synthesized to mimic this transition structure. Because the mimic compound resembles the transition structure, it can occupy the active region of the enzyme and thus block its normal action. These compounds are called "blockers." They work by

PEPTIDE

BLOCKER

ENZYME ACTIVE SITE

BLOCKER CAN NOT BE HYDROLYZED AND BLOCKS ACTIVE SITE OF ENZYME

ENZYME HYDROLYZES PEPTIDE AND RELEASES PRODUCTS

TRANSITION STATE ENZYME INHIBITOR FOR PEPTIDE HYDROLYSIS

successfully competing with the transition molecule for attachment to the enzyme active site.

A second approach again involves a compound designed to fit the enzyme active region. This time, the compound is planned so that it will react with the enzyme to inactivate it permanently. These are the so-called "suicide" or "mechanism-based" inhibitors and work by disabling the enzyme.

As evidence of their successful use in therapy, enzyme inhibitors have been designed and shown to be effective in treatment of hypertension, atherosclerosis, and asthma. Aspirin is a familiar example—it is now known to work by inhibiting the enzyme cyclooxygenase. As a result of this understanding, a whole family of cyclooxygenase inhibitors, such as indomethacin, have been synthesized and found to be medically effective in killing pain and reducing swelling.

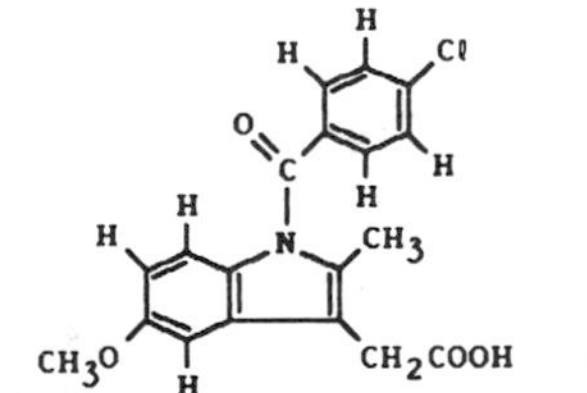

Indomethacin: An Enzyme Inhibitor That Works Like Aspirin

Receptors

A related research field concerns the so-called "receptors." These macromolecules are involved in triggering biological processes. Apparently, they cannot function until they have been activated by their appropriate hormones. They then recognize and bind biologically active molecules. This has the effect of catalyzing and controlling reactions with these molecules, as they are "held" by the receptors in a strategic manner.

Until recently, receptors were studied only indirectly. Various compounds were tested for their ability to either stimulate or block a biological process. Conclusions were then made about the structural features required by a molecule to fit a given receptor. Over the last 10 to 15 years, more powerful approaches have been developed using radioactive molecules which allow easier evaluation of the structural requirements for receptor binding. In addition, physiochemical means (NMR, spectroscopy) have been useful in isolating and characterizing receptor molecules. Two types of agents have been defined which bind to receptors—they are called agonists and antagonists. *Agonists* are compounds which trigger a biological response and include naturally occurring hormones and neurotransmitters as well as drugs generated by chemists. *Antagonists,* on the other hand, are compounds which block biological responses by binding to a receptor, thus preventing the agonist from binding and fulfilling its job.

Some chemical messengers can bind to more than a single receptor type, and thus take part in more than one type of biological action. For example, histamine triggers allergic reactions by binding to a receptor designated H_1, but it also promotes gastric acid secretion in the stomach by activating the so-called H_2-receptor. Too much gastric acid causes severe damage to the stomach lining and results in an ulcer. But a drug has been discovered which works specifically as an H_2-receptor antagonist. This drug, called cimetidine, binds to the H_2-receptor and blocks it, resulting in less gastric acid and great relief for the patient.

Norepinephrine, the chemical messenger for the part of the nervous system that controls adrenalin flow, has been shown to bind to at least four types of receptors assisting in several types of biological responses. Compounds that act as specific antagonists have already proven their value in treating cardiovascular disease, cancer, disorders of the central nervous system, and endocrine disorders.

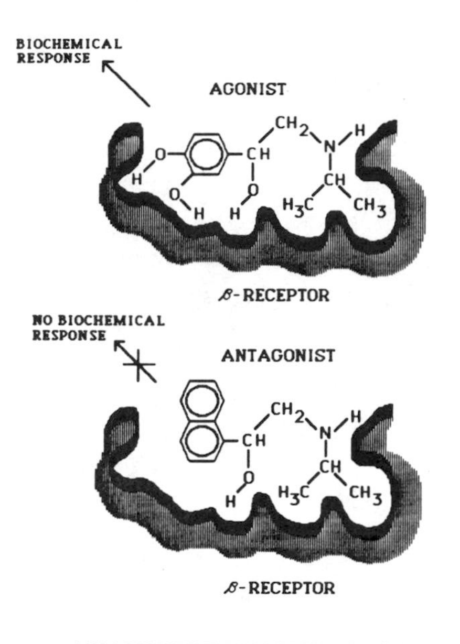

ANTAGONIST DOES NOT PRODUCE A RESPONSE BUT BINDS TO RECEPTOR AND BLOCKS ACCESS OF AGONIST.

TWO TYPES OF DRUG BINDING TO RECEPTORS

$CH_3-NH-C(=N-CN)-NHCH_2CH_2-S-CH_2-$(5-methylimidazol-4-yl)

CIMETIDINE
CONTROLS PEPTIC ULCERS

These themes, enzyme inhibition and receptor function, have wide applicability. They will come up again and again as we turn, now, to examples that display the breadth of chemical progress that has been made in recent years in the development of new therapeutic agents.

ANTIBIOTIC RESEARCH

Antibacterials

Prior to World War II, sulfonamides were the only effective antibacterial agents available. During and after World War II, antibiotic research had a major impact in decreasing disease in both humans and animals.

During the period 1945 to 1965, penicillins had come into large-scale use, and the cephalosporins (a group of antibacterial fungi) had been discovered. The tetracyclines, chloramphenicol, erythromycin, and aminoglycosides were being used to treat infectious diseases. In addition to antibiotics obtained by fermentation, man-made antibacterial agents such as nalidixic acid and nitrofurans were also being discovered. During the past 20 years, major efforts have been made to improve the range, potency, and safety of the antibiotics available. This has involved the identification of new fermentation products, and chemical alterations to improve less-than-ideal natural products (semisynthesis), as well as the introduction by synthesis of new structural types. The newer semisynthetic penicillins include agents that are not only active against common bacteria but also that are effective against the *Pseudomonas* group of bacteria, which are an increasing problem in the hospital environment. The early cephalosporins have been successfully altered to provide new compounds possessing remarkably broad usefulness and high potencies combined with increased safety.

Much of the effort in antibiotic research has concentrated on the problem of resistance development, especially in the hospital environment. Unfortunately, antibiotics may become ineffective as bacteria develop a resistance to them over time. For example, certain bacteria can gain the ability to produce enzymes that inactivate the antibiotic. Progress has been made in the design and synthesis of inhibitors to disarm these bacterial enzymes. Other bacteria can become resistant to antibiotics by preventing the antibacterial agent from entering the bacterial cell. Here again, advances have been made by both semisynthetic alterations and the discovery of new agents.

Antivirals

Viruses are the smallest of the infectious organisms. While antiviral chemotherapy is in its infancy compared with antibacterial therapy, breakthroughs are being made. Viruses do not contain much genetic information, so they exhibit only a few unique biochemical steps that are possible targets for a chemical agent. Viruses take over and control the cells of the host in order to survive and multiply. This means, unfortunately, that most of the steps in viral biology are identical, or closely similar to, those of the mammalian host. It is therefore difficult to attack the virus by chemotherapy without also endangering the host. In order to discover a safe chemotherapeutic agent, it is necessary to identify a biochemical pathway that is unique to the virus-infected cell. Viral DNA polymerase enzymes represent such a target. These enzymes are involved in the synthesis of viral nucleic acids. Examples of compounds that function as viral polymerase inhibitors are known, but often these compounds are suitable only for local application. The antiherpes drug acyclovir is effective either when locally applied or after oral or intravenous administration. Its relative safety is due to the fact that it is ignored by cellular enzymes under normal conditions. However, in the presence of certain viral enzymes, acyclovir is converted to a drug that blocks DNA synthesis by the virus.

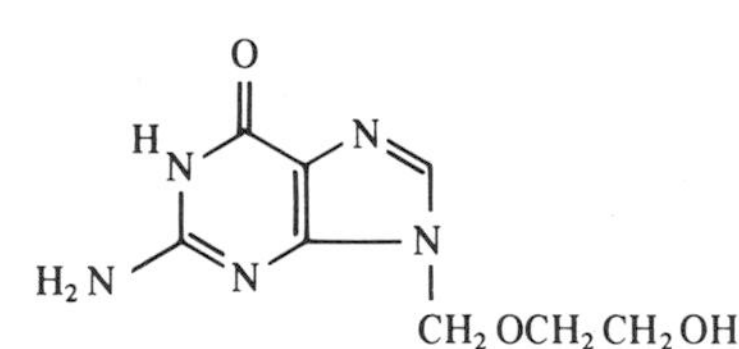

ACYCLOVIR: AN EFFECTIVE ANTIHERPES DRUG

CARDIOVASCULAR DISEASE

Diseases of the heart and blood vessels are presently the major cause of death in the United States. Therefore, high blood pressure (hypertension) and high blood cholesterol levels (hypercholesterolemia) have been the subject of extensive research.

Hypertension

Death rates for coronary heart disease in the United States fell 20.7 percent between 1968 and 1978. Improvements in the control of moderate and severe hypertension have undoubtedly contributed to this decline in coronary heart disease fatalities in the United States.

The earliest drugs used for hypertension had such serious side effects that they were used only when blood pressure reached life-threatening levels. Now several types of antihypertensive agents are used extensively for the treatment of mild and moderate hypertension, and with few negative effects.

Adrenalin is a hormone that stimulates automatic nerve action, including that which keeps the heart pumping. Its release is regulated by what is called the adrenergic nervous system. While the cause of recurring hypertension remains unknown, it has long been recognized that this adrenergic nervous system and its chemical messenger, norepinephrine, play a major role in regulating blood pressure and cardiac function. Over the years chemists have supplied clinicians with many useful antihypertensive agents which influence the activity of the adrenergic system. α-Methyldopa, tremendously valuable in the treatment of hypertension, is known to act within the central nervous system by means of an adrenergic receptor. The recognition that norepinephrine acts on several different subtypes of receptors has allowed compounds to be designed which lower blood pressure by different mechanisms. Two widely used compounds that block the action of norepinephrine are timolol and propranolol. They provide effective treatments for certain heart disorders and have also been shown to reduce the risk of death and recurrence of a heart attack. Timolol has also become the primary treatment for glaucoma, a disease that affects the eyes.

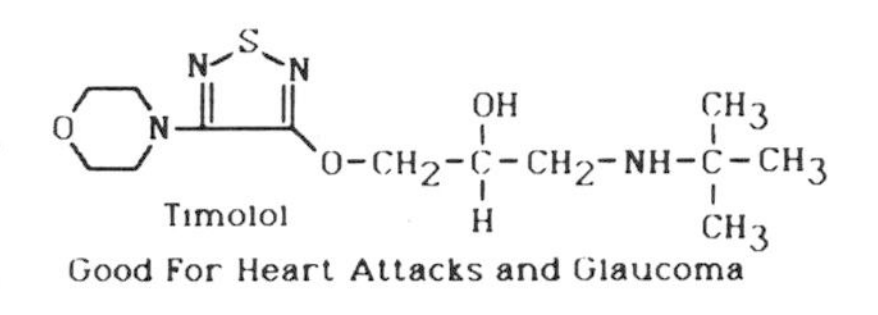

Two other classes of antihypertensive compounds include "calcium channel blockers" (also effective against angina and stroke), and the so-called angiotensin-converting enzyme inhibitors, typified by captopril and enalapril. They also show much promise for the treatment of heart failure.

Very recently, chemists working with biologists have discovered, identified, and synthesized a group of peptides released in the heart. These peptides have been named atrial natriuretic factors. Their biological properties are now being investigated to decide their possible usefulness in the creation of new therapeutic agents. We already know that these compounds tend to increase urine discharge, to relax blood vessels, and to lower blood pressure.

Atherosclerosis

The second major cardiovascular threat is an inappropriately high level of cholesterol in the blood, hypercholesterolemia. An intensive search has been under way for many years for safe and effective drugs that will lower cholesterol levels to the normal range either by blocking the synthesis of cholesterol or by encouraging its breakdown. HMGCoA reductase is a critical enzyme in the steps leading to the formation of cholesterol by the liver. Effective treatment for hypercholesterolemia

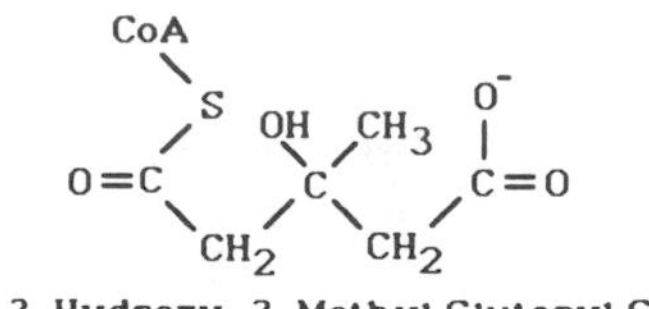

3-Hydroxy-3-Methyl Glutaryl CoA
HMGCoA: Critical Enzyme In Cholesterol Formation

may be provided for the first time by an exciting new enzyme inhibitor which works on HMGCoA reductase.

Heart Failure

For the last two centuries, digitalis has remained central in the management of heart failure in spite of its serious side effects. The search has been under way to find less toxic agents which also work to help weakened heart muscle to function. The most thoroughly investigated alternative has been through increasing levels of cyclic adenosine monophosphate (cAMP), which stimulates heart contractility. An increase in cAMP levels in the cells can be accomplished directly by agents such as prenalterol, dopamine, and dobutamine, or indirectly with caffeine or theophylline, which block the enzyme responsible for inactivating cAMP, phosphodiesterase (POE).

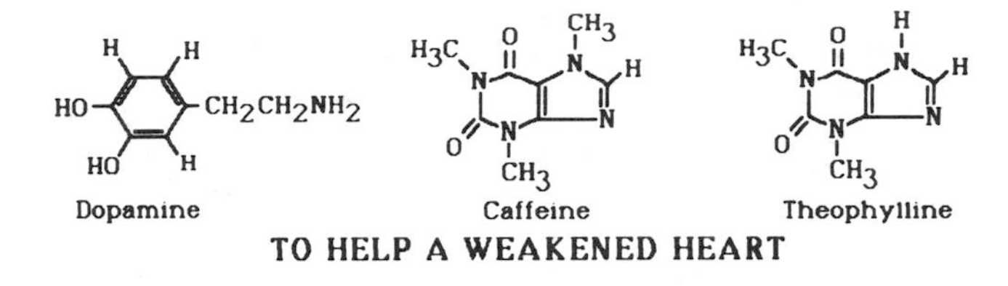

TO HELP A WEAKENED HEART

Within the last 10 years, the traditional treatment of congestive heart failure using digitalis and diuretics has been assisted or replaced by drugs that have no direct cardiac action but that increase the heart's pumping efficiency by dilating, or widening, blood vessels. These new *vasodilators* (such as the above-mentioned captopril and enalapril) can be expected to have a significant impact on the management of congestive heart failure over the next decade.

Arrhythmia

Another common heart ailment is irregularity in the force and rhythm of the heartbeat. Two of today's widely used antiarrhythmic drugs, quinidine and digitalis, trace their origins back over 200 years. Since the eighteenth century, these compounds have been used to treat this potentially deadly condition characterized by abnormal cardiac rhythm. Now we are making progress in determining how these chemical agents work. The heart's pumping cycle is regularly triggered by electrical signals involving movement of sodium and calcium ions (Na^+ and Ca^{2+}). Drugs that inactivate the sodium ion channel (quinidine, procainamide, lidocaine), depress the calcium ion channel (verapamil), inhibit sympathetic activity (propranolol, timolol), or prolong the nerve impulse (amiodarone) have been discovered.

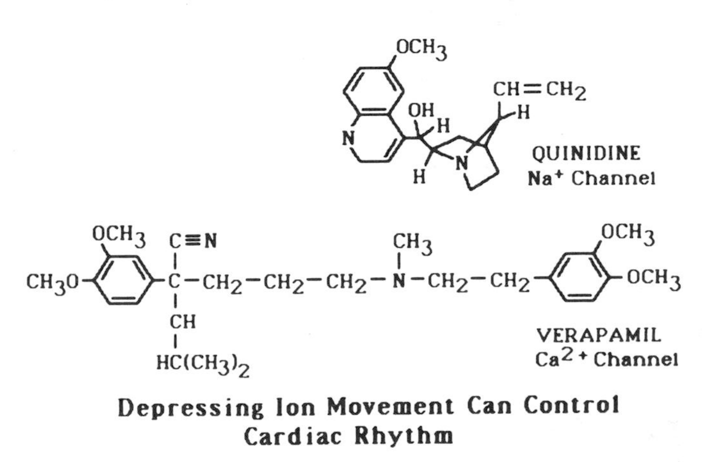

Depressing Ion Movement Can Control Cardiac Rhythm

They form the basis for current antiarrhythmic therapy and point toward a rational approach to treatment.

DRUGS AFFECTING THE CENTRAL NERVOUS SYSTEM (CNS)

The cost of direct care for mental illness is estimated to be 15 percent of our total national health care expense. Approximately 2.5 percent of our population receives treatment for mental or emotional disorders each year. Antidepressants and tranquilizers have enabled men and women to live useful lives who would not otherwise have functioned effectively.

Early therapeutic agents for treating mental illness were discovered through trial and error clinical testing. This permitted only slow advance as chemists synthesized related compounds with more desirable therapeutic effects. More recently, however, chemists working with neurobiologists have begun to learn the biochemical mechanisms by which these drugs work. As a result, alternative approaches for achieving therapeutic effects in psychosis, depression, and anxiety are now being discovered.

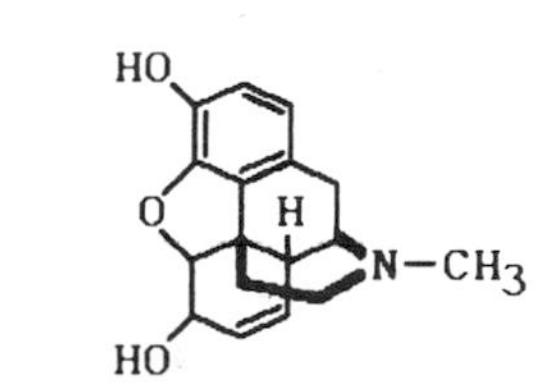

Morphine: An Addictive Pain Reliever

Among the most important pain relievers (analgesics) that act on the central nervous system are those that come from the opium poppy. Morphine, a widely used opiate pain reliever, is being replaced by synthetic drugs that are not as addicting and carry fewer side effects. In addition, drugs which are useful in treating addiction to heroin, opium, and morphine are now available. Ten years ago, two peptides were isolated from the brain and found to have actions similar to that of morphine. These compounds, called enkephalins, were then chemically characterized and synthesized. This discovery has had a profound impact on CNS research.

Try—Gly—Gly—Phe—Leu

Leucine Enkephalin Peptide Chain
A New Pain Reliever

The treatment for Parkinson's disease is a typical example of a biochemical approach to CNS therapy. Parkinson's is characterized by muscle tremors and paralysis, and is caused by a shortage of the compound dopamine. It is treated with levodopa, which gains access to the brain and is converted there to dopamine by the enzyme dopa decarboxylase. A further advance came when chemists combined carbidopa with levodopa. Carbidopa prevents levodopa from being metabolized outside the brain, thus allowing the active agent to be formed only where it is wanted, within the brain. Side effects are thus minimized.

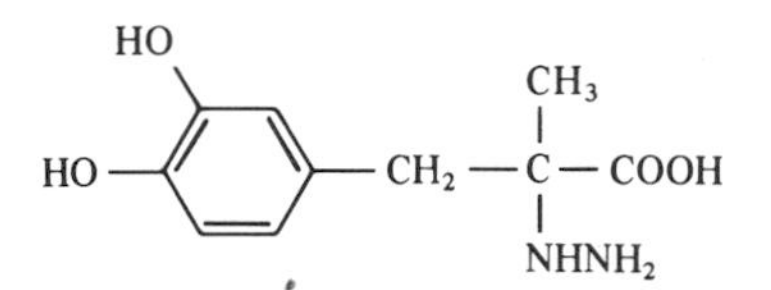

S-CARBIDOPA FACILITATES L-DOPA TREATMENT OF PARKINSON'S DISEASE

During the past decade we have made great strides in understanding the process of chemical signaling within the central nervous system of mammals. Ten years ago, only eight or nine monoamine or amino acid compounds were known that seemed to be neurotransmitters. Now over 40 more small peptides have been added to the list, each of which has a possible messenger function. The opportunities for important advances in therapy through combined chemical and biological research are quite promising.

CANCER RESEARCH

The group of diseases collectively known as cancer is second only to cardiovascular disease as a cause of death in the United States, where cancer will strike one out of four persons alive today. Cancer is characterized by uncontrolled cell growth in the body. It is gratifying that cancer research has entered a fruitful phase. New developments can be conveniently divided into those dealing with our understanding of the origin of cancer, carcinogenesis, and those relating to cancer treatment through chemotherapy.

Carcinogenesis

The discovery that organic compounds can act as carcinogens in experimental animals in the 1930s led eventually to the finding that, in high enough dosage, quite a number of compounds have the ability to cause cancer in tissues of mice, rats, and other mammals. Today, some naturally occurring and some synthetic chemicals in the environment are suspected of being capable of causing cancer in humans, so interest in the detection of these agents and how they work has increased greatly.

Several important facts about cancer-causing compounds (carcinogens) were established before 1965. Several different chemical carcinogens were found to bond covalently with cellular macromolecules (proteins, RNA, DNA), and this was found to be related to the cancer process. These findings set the stage for much further research.

The majority of known chemical carcinogens are actually "pro-carcinogens," in other words, they must be activated in the body to form chemically reactive molecules known as ultimate carcinogens. For example, benzo(*a*)pyrene, a pro-carcinogen, reacts in a series of enzyme-catalyzed reactions to produce an ultimate carcinogen which then binds to DNA. This DNA molecule with carcinogen attached is called a DNA adduct. It is the ultimate carcinogens that react with the nucleic acids and proteins in cells to disrupt their normal functions in cell growth. The major enzyme systems that transform pro-carcinogens have been identified and studied.

The chemical basis for the reactions forming carcinogen-DNA adducts is well understood, but how these adducts actually cause cancer in animals has not been demonstrated. It is known, however, that when carcinogens are chemically processed by the body, the end products can cause changes in the DNA (mutagenic effects) of bacterial and animal cells. There is a relationship between compounds which are mutagenic and those which are carcinogenic. If it can be demonstrated that a compound is a mutagen, it could possibly be a carcinogen. This evaluation is done routinely in the laboratory with the "Ames test," which uses a special strain of *Salmonella* culture. However, not all mutagens are carcinogens, and there are many natural mutagens present in a normal diet.

When cells become *malignant,* they grow abnormally and are considered life threatening. Perhaps the most promising and certainly the most dramatic recent development in cancer research is the recognition that certain genes in normal cells are closely tied to the development of malignancy. Importantly, these genes resemble or are identical to genes from certain viruses (oncogenes) that transform normal cells to malignant ones. Organic chemists can determine (1) the nucleotide sequence of the normal gene and of the oncogene and (2) the amino acid sequence of the proteins made from these genes. Changing only a single nucleotide in a gene from a bladder, colon, or lung cell can replace a particular amino acid by another in the gene product, and thereby make an otherwise normal cell malignant. The striking achievement is that we now understand *on a molecular basis* the difference between the protein of a normal cell and that of a malignant cell, at least for some transformations. That kind of understanding brings us closer to logical development of new therapies.

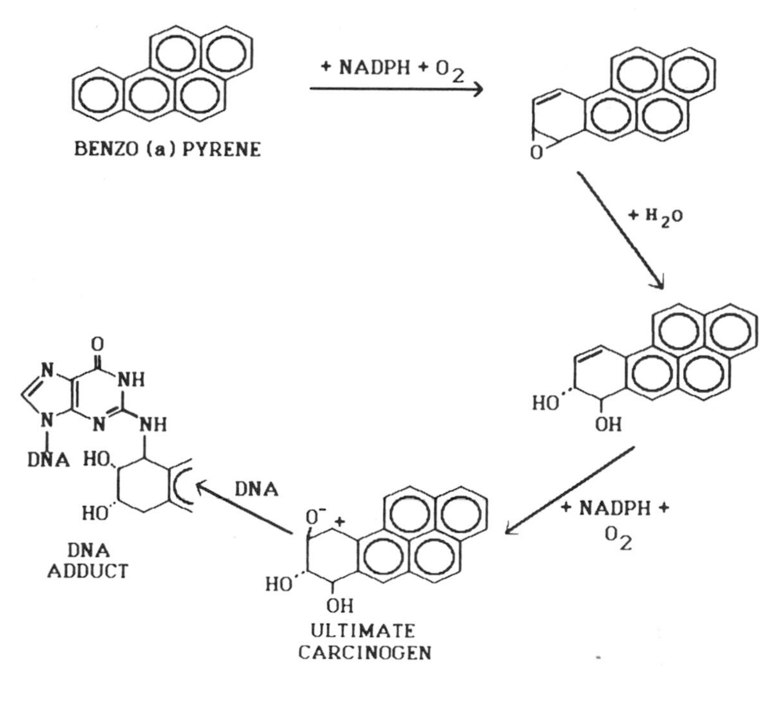

ENZYMATIC REACTIONS OF CARCINOGENESIS

Chemotherapy

Compounds used for the treatment of cancer were originally poisonous substances extracted from natural sources or of synthetic origin. The role of the medicinal chemist has been to design and synthesize new drugs with improved therapeutic value. Many new and clinically important antitumor agents have been isolated from microorganisms in the last 15 years, and their chemical structure has been determined. In a number of classes of these compounds, it has been possible to prepare semisynthetic compounds with reduced toxic side effects. Some of these antibiotics interact with DNA in the malignant cell by interleaving in the helical DNA coils. This mechanism has furnished a model for the design of new synthetic compounds now in clinical trial.

The first synthetic anticancer agent was given the name nitrogen mustard; it acts by alkylation of DNA. Similar compounds that work more selectively on only disease-affected DNA have since been synthesized, yielding more effective drugs like cytoxan. One group of widely used anticancer drugs known as the "antimetabolites" are fashioned after natural substances that upset metabolic processes. Other compounds with high electron affinity, such as misonidazole, make tumor cells more sensitive to radiation therapy.

About 40 anticancer agents have proven to be clinically useful. The most significant breakthroughs in treatment have resulted from combination therapy, using two or more drugs together. For example, in 1963 advanced Hodgkins's disease in adults was incurable, but today 81 percent of patients have their health restored with combination therapy. Complete cure can also be achieved in 97 percent of all children with acute lymphocytic leukemia. Over the last 30 years, the greatest progress in chemotherapy has been made in the treatment of cancer in children. For several tumor types, the percentage survival for children so afflicted has risen from below 20 percent to above 60 percent.

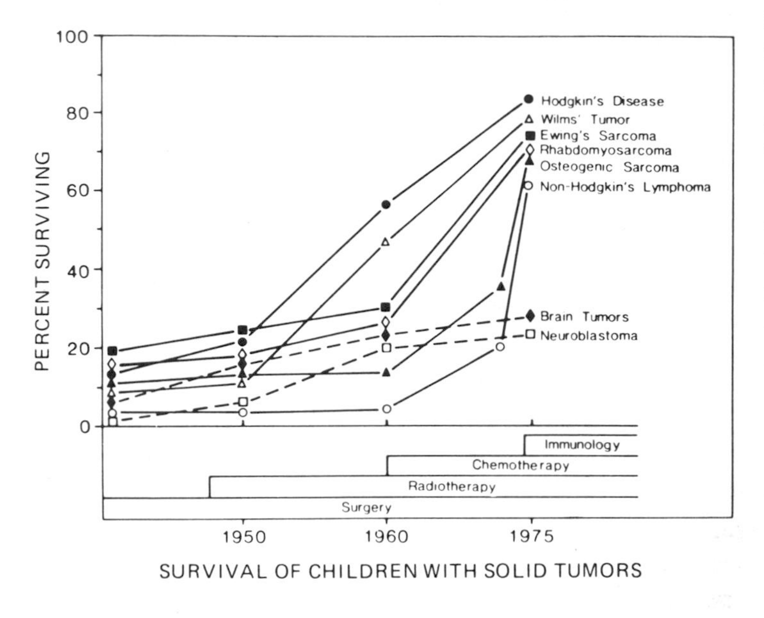

SURVIVAL OF CHILDREN WITH SOLID TUMORS

There remains a pressing need for more effective and less toxic anticancer drugs, in particular for treating slow-growing solid tumors, lung cancer, and brain tumors. Immunologists and cell biologists are discovering differences between the surfaces of normal cells and tumor cells which may provide new directions for drug design. In addition, chemists will play a critical role in the search for drugs which can stimulate the host's immune response.

INFLAMMATORY AND IMMUNOLOGICAL DISEASES AND DEFENSE SYSTEMS

Inflammatory and immunological diseases such as arthritis and rheumatism are major medical problems; they affect 7 percent of the total population. The isolation, characterization, and partial synthesis of cortisone in the 1940s enabled clinicians to make the dramatic discovery of its potent anti-inflammatory effect. This era was followed by the discovery of a family of nonsteroidal anti-inflammatory/analgesic drugs, typified by indomethacin, which are in wide use today. The biochemical mechanism of action of these compounds, like that of aspirin, has been shown to be inhibition of the enzyme cyclooxygenase. The anti-inflammatory steroids and the cyclooxygenase inhibitors have provided great medical benefits, but they do not arrest the progress of diseases such as rheumatoid arthritis. Nevertheless, the recognition that many inflammatory diseases represent disorders of the immune system has been particularly important. Chemistry provides the opportunity for us to understand the chemical basis of these events.

The immune system is that part of an organism's body that fights disease and invasion by foreign substances. In the last 20 years, much has been learned about the group of enzymes and other proteins that help our body decide when a foreign organism is present and that coordinate a response to its presence. The production of antibodies is one such response. Antibody molecules are produced in the bloodstream by plasma cells which in turn come from white blood cells. Antibodies work to neutralize foreign proteins or polysaccharides found in the blood that may cause illness. Chemists have made major contributions to understanding the nature of antibody molecules, first demonstrating that they are proteins, and then actually determining their chemical structures, as well as those of the genes that code for these proteins. From this has emerged a picture of nature's design of these molecules. They have a "variable region," in which the amino acid sequence varies according to which foreign substance the antibody is attacking, and a "constant region," which stays essentially the same for most antibodies. The variable region of the antibody molecule recognizes and binds specific intruders, whereas the constant region is concerned with the actual removal of the foreign substance. This knowledge opens promising new research avenues. The urgency of progress in this area is underscored by the need to develop effective therapies for Acquired Immune Deficiency Syndrome (AIDS).

ADVANCES IN FERTILITY CONTROL AND FERTILITY INDUCTION

Our understanding of the human reproductive cycle moved ahead rapidly when we discovered the role of the hypothalamus and pituitary glands, both deep inside the brain. These organs produce hormones and neurotransmitters to control the reproductive cycle. Other hormones are released elsewhere in the body, in response to these hypothalamic or pituitary chemical messengers. Thus, the body controls a wide range of responses, from causing an egg to be released from an ovary, to triggering the production of breast milk. When chemists determined the molecular structure of these hormones, it became possible to begin to influence fertility, the human body's ability to reproduce.

Oral contraceptives, or birth control pills, have been used with enormous impact worldwide on population control. They are made up of two groups of compounds called estrogens and progestins, including many synthetic analogs. Unfortunately, multiple side effects, including blood clots, migraine headaches, stroke, and heart disorders have been associated with their early use. In the last several years, attention has been devoted to reducing the doses of estrogen and progestin as well as balancing the ratio of the two in order to achieve oral contraception with minimal side effects.

Chemical methods have also been discovered to assist in reproduction. Clomiphene blocks estrogen receptors in the hypothalamus and in the pituitary gland. When this hormone antagonist is administered with appropriate timing in the reproductive cycle in women, it interferes with the normal feedback by estrogen to the hypothalamus and pituitary glands. Interference results in the desired hormonal surge by the hypothalamus and pituitary gland, often producing ovulation and thus fertility.

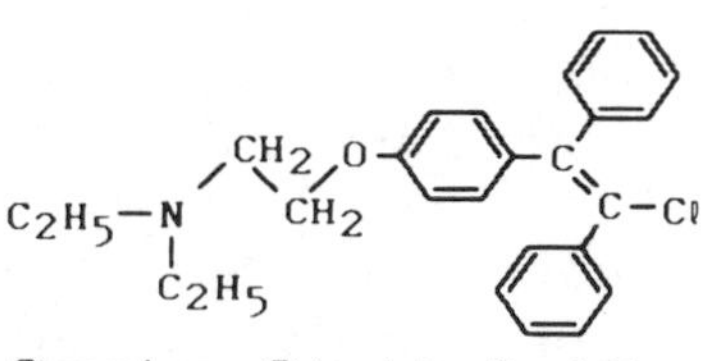

Clomiphene: Stimulates Ovulation

Gonadotropin-releasing hormone (GnRH) is secreted by the hypothalamus. It stimulates the pituitary gland to release a wide range of hormones involved in the reproductive system. Many compounds similar to GnRH (a 10-amino acid polypeptide) have been chemically synthesized and tested. Certain side effects have decreased enthusiasm for use of these analogs in contraception, but they remain of interest and are receiving attention for treating certain cancers. Dramatic medical successes have been achieved using analogs of GnRH in patients who were born without the ability to produce GnRH, a rare disorder. An analog of GnRH is administered using small, sophisticated pumps which are worn by the patient, and the drug is released in a pulsing fashion to mimic its normal secretion pattern by the hypothalamus. Patients in their 20s who have never undergone puberty can be brought through all of the successive stages of puberty and even fertility. This combination of impressive drug design with advanced drug delivery systems is an indication of future advances in the reproductive field.

Finally, there are major new directions which should also result in important therapeutic advances. Evidence from several laboratories indicates that we will soon know the molecular structure of inhibin, the key hormone involved in regulating sperm production. Synthesis of compounds similar in structure to inhibin should enable the medicinal chemist to develop a form of male birth control. It is possible that this kind of chemical influence would have fewer side effects than the use of oral contraceptives in females.

The role of the brain in regulating reproductive function has been observed for a long time. Factors such as stress, exercise, and depression are known to change or halt menstrual cycles in adult women or delay the beginning of puberty in adolescents. Women athletes and women who suffer from an eating disorder called anorexia nervosa often temporarily lose their menstrual period. Compounds found in the brain such as endorphins and enkephalins may prove useful in restoring normal menstrual cycles to these women. It is interesting to note that these compounds are naturally produced by the brain and work in the same manner as some of the drugs derived from the opium poppy. They may be involved in killing pain, causing pleasure, or changing emotions. Hopefully, the next decade will see great impact from chemical design of hormone analogs for treatment of sexual and reproductive disorders.

VITAMINS

Throughout mankind's history, vitamin deficiencies have been a major cause of death. In the eighteenth century it was found that small amounts of citrus fruit, which provides Vitamin C, could prevent a deadly disease called scurvy on long sea voyages. In 1912 the "accessory food factors," which the human body needs to function properly, were given the name "vitamins." Since that time, many vitamins have been isolated and identified. Though these compounds are not themselves enzymes, they are found to be necessary for the functioning of many enzymes. Hence, they are called "coenzymes" or "cofactors." A few of the advances and discoveries in this area are described below.

The isolation and characterization of Vitamin B_{12} as the dietary ingredient required to prevent a fatal form of blood disorder called anemia was reported in 1948. Determination of its molecular structure in 1956 by X-ray crystallographic and chemical studies showed it to be by far the most complex of any of the known vitamins. Its synthesis in 1976 was a landmark of organic chemistry. There have been major advances in our understanding of the functions and mechanisms of action of the coenzyme forms of Vitamin B_{12}.

Considerable progress has been made in the understanding of the flavins, of which riboflavin, Vitamin B_2, is an example. The flavins in various forms act as coenzymes for oxidation-reduction reactions required for normal metabolic processes. Over 100 flavoproteins are now known. It is of interest that a modified flavin has recently been discovered to be a coenzyme in methane-producing bacteria, which may be of future interest in the development of methane as an energy source.

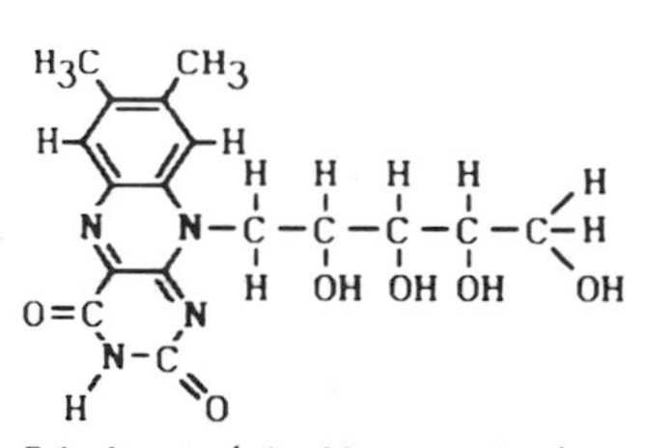

Riboflavin (aka Vitamin B_2)
Keeps Our Metabolic Fires Burning

It has long been known that Vitamin D is required for the prevention of a bone disorder called rickets. Without enough Vitamin D a child's bones may grow in a deformed manner. By the use of advanced chemical and spectroscopic techniques, it has now been shown that Vitamin D is actually a precursor to the true active substance, a hormone. Vitamin D is changed in the body to a highly powerful dihydroxyl compound that regulates absorption of calcium from the diet, its reabsorption in the kidney, and the metabolism of calcium in bone. It is not yet understood how this Vitamin D hormone carries out its functions, but research is in progress. It has been synthesized and shown to be effective in the treatment of a number of bone diseases. Trials are in progress to evaluate its usefulness in osteoporosis, a disease that causes brittle bones. New functions of Vitamin D hormones will undoubtedly be discovered, now that the compound is available for research.

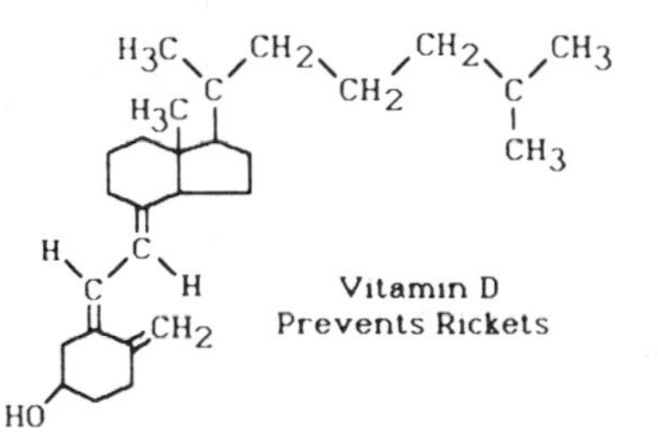

Vitamin D
Prevents Rickets

Another vitamin whose molecular structure is now known is K. Vitamin K is required as a coenzyme for the production of three or four proteins that help blood to clot. We still need to clarify how Vitamin K does this—knowing the structure is a key step toward that end.

Vitamin K: Helps Blood To Clot

For some time, we have known that a compound that comes from Vitamin A is required for the detection of light as it strikes the eye. However, Vitamin A is now recognized to play an essential role in the growth of animals as well. It also plays an important role in the development of bone, the formation of sperm in the male, and the development of the placenta in a pregnant female. Vitamin A must be

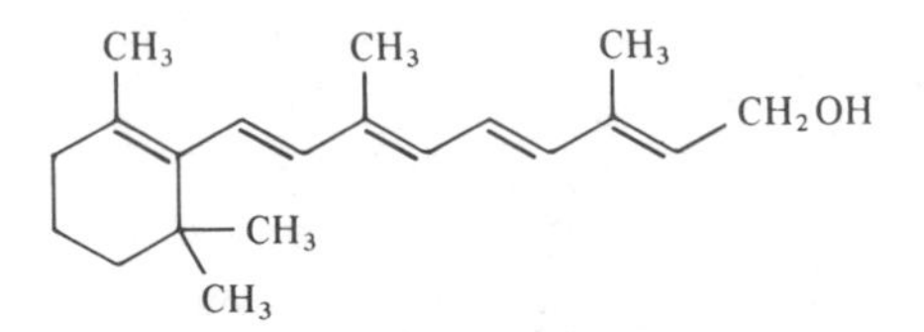

VITAMIN A: ESSENTIAL TO VISION AND GROWTH

converted into several related compounds before it can satisfy all these functions, and much progress has been made in determining the chemical changes involved. For example, it appears to be converted to retinoic acids for use in skin tissues, and some of these acids and synthetic analogs are useful in the treatment of skin disorders such as acne. Another significant development is the observation that Vitamin A compounds can retard some chemical carcinogenesis.

CONCLUSION

In this section, many examples have been given in which we are developing chemical understandings of drug action at the molecular level. Such knowledge permits us to anticipate what kinds of molecular structures are needed to deal with a particular disease, or to achieve a desired clinical result. Thus, we are entering a period in which drugs can be logically and deliberately planned—this is called "rational drug design."

It is tempting to speculate in which disease categories the most dramatic discoveries will occur during this decade. It is likely that new directions in receptor-related research will have an impact on drug discovery in cardiovascular diseases, especially atherosclerosis and hypertension, as well as on diseases of the endocrine system like diabetes. Recent research with the oncogenes of viruses has begun to provide an understanding on the molecular level of certain human cancers, opening promising new frontiers for drug discovery in cancer research. Progress in our ability to regulate the immune system should open up new approaches to the treatment of many inflammatory diseases, such as arthritis. Developments in neurobiology should lead to new drugs that act on the central nervous system. Finally, the discovery of new enzyme inhibitors and of hormone and neurotransmitter antagonists will certainly lead to the discovery of important new drugs. But this will not be the end. Science is starred with examples of unanticipated breakthroughs that prove to be more important than the advances we can foresee.

SUPPLEMENTARY READING

Chemical & Engineering News

"Synthetic Antiviral Agents" by R.K. Robbins, vol. 63, pp. 20-21, Dec. 16, 1985.

"Designer Drugs" by R.M. Baum (C.& E.N. staff), vol. 63, pp. 7-16, Sept. 9, 1985.

"Platinum Complexes of Vitamin C Show Anticancer Potential" (C.& E.N. staff), vol. 62, pp. 29-30, Sept. 17, 1984.

"New Drugs for Combatting Heart Disease" by H.J. Sanders (C.& E.N. staff), vol. 60, pp. 28-38, July 12, 1982.

Scientific American

"Materials for Medicine" by R.A. Fuller and J.H. Rosen, vol. 255, pp. 118-125, October 1986.

Chem Matters

"Penicillin" pp. 10-12, April 1987.

"Smoking" pp. 4-8, February 1986.

"Toothpaste" pp. 12-16, February 1986.

"Nuclear Diagnosis" pp. 4-7, December 1985.

"Lead Poisoning" pp. 4-7, December 1983.

A Pac-Man for Cholesterol

Since the 1960s we've known that high levels of cholesterol correlate with heart ailments, the major cause of death in the United States. What we need is a Pac-Man to chomp up the cholesterol in the blood and reduce "hardening" of the arteries that carry blood from the heart (atherosclerosis). Now a lowly fungus—not unlike the famous mold, penicllium—may have shown us one.

A normally functioning human cell uses a dual system for meeting its cholesterol needs. First, the cell has its own factory to manufacture cholesterol. In addition, the cell's exterior has a number of lipoprotein receptors that can grab onto cholesterol-containing lipoproteins as they pass by in the blood stream and pull them inside. The cell fixes the number of these Pac-Man-like receptors so that just the right amount of imported cholesterol is added to the factory-made product. If the inner cell cholesterol level falls too low, more receptors are added to extract more from the blood stream.

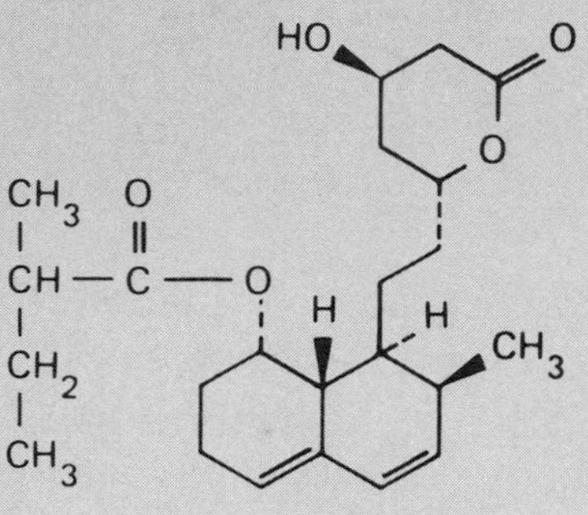

There's an idea! If the cell's cholesterol factory could be slowed down, would the cell produce more receptors to make up the difference from the blood stream supply? A chance to test this scenario came when a biochemist discovered that certain fungi produced something that inhibited cholesterol synthesis. Chemists joined in the plot, purified the effective compound, determined its structure, and named it COMPACTIN. Knowing this structure, chemists were able to synthesize close relatives of compactin that are even more potent. Chemical tests with these new chemicals indicate that the scheme works as planned. The inhibitor slows down the cellular cholesterol factory, the cell produces more lipoprotein receptors, and the blood cholesterol level drops.

The importance of this advance is shown by the fact that the average person with double the normal blood level of cholesterol can expect to live only 45 to 50 years. For the few unlucky people with triple the normal amount, life expectancy drops to 30 to 35 years. To complicate matters, 1 in 500 Americans has the genetic disease

familial hypercholesterolemia (FH). Victims of FH don't produce enough receptors at their cell surfaces, so lipoproteins accumulate in the blood and eventually cause heart attacks. Thus clinical researchers are excited to find that the new cholesterol-inhibitors work with FH patients, bringing the blood levels of cholesterol all the way down to the normal level. Much research remains, but these moldy chemicals offer immediate hope to FH sufferers and, in the future, to all people with abnormally high blood cholesterol.

III-F. Biotechnologies

A living thing is a chemical factory that takes in raw materials (foods and nutrients) and, with its chemical work force, converts those nutrients into the wide range of products needed for the maintenance and operation of the living organism. And what does this chemical factory make? Its primary function is to build other factories very much like itself. This property, called *reproduction,* means that the factory carries blueprints that tell how to assemble a new factory that can function independently. These blueprints carry all of the instructions needed to build the new factory (again from nutrients that must be available) and to fashion the new sets of chemical blueprints and a whole new work force to make it an independent, self-sufficient organism.

Today, we have a basic understanding of the chemical structures and functions of the molecules and macromolecules that are involved in these chemical factories. The blueprints are called DNA molecules (DNA means *d*eoxyribo*n*ucleic *a*cid). These DNA molecules are designed to make it easy to copy themselves to make new sets of blueprints. In addition, they carry all of the instructions needed to create the work force of the new organisms, the proteins. By far the most important members of this protein work force are the enzymes. They are the engineers who guide the construction of virtually every part of the organism. Enzymes are highly selective catalysts for reactions needed in the chemical synthesis of the many substances used in the operation of the organism. Enzymes achieve their selectivity through pattern-like or mold-like surface structures that can recognize the correct reactants among the nutrients and then shape the product to the desired structure.

Biotechnology can be described as our attempt to adapt a part of one of nature's factories to our own use, to manufacture a product we want. One way to do this is to locate and put to work a part of the factory that already does what we want. This is the kind of biotechnology that has been used for centuries when we use natural enzymes to ferment sugar to make vinegar and wine and when we ferment starch to make bread. But modern biotechnology is much more ambitious. Now scientists are learning how to alter the actual blueprints so nature's factory will make a new substance that was not in its product line before. To see how this is becoming possible, we will examine DNA and how it encodes the instructions it contains. Next, we will see how these instructions are used to construct specific proteins, including enzymes. Finally, we will see how new instructions are inserted into natural DNA to give the new set of blueprints, which will be called recombinant DNA.

DNA—WHAT IS IT?

DNA is a fascinating chain-like structure made up of long strings of sugar and phosphate molecules. Attached to the sugar molecules of these long chains are heterocyclic amines (commonly called "bases") which form cross-links between two strings. When flattened out, a double-stranded DNA molecule resembles a ladder. It is truly a macromolecule—its molecular weight may be as high as 10^9. Despite the complexity and size of a DNA molecule, it actually contains only four different amine bases: adenine, thymine, cytosine, and guanine (abbreviated, A, T, C, and G). Adenine and thymine have geometrically fixed capacities to form hydrogen bonds to each other. They match so well that an adenine base can "recognize" a thymine base and bond to it in strong preference to the other bases. Cytosine and guanine match in a similar way. Thus, A always bonds to T and C always bonds to G.

This recognition capability permits two such sugar-phosphate strings to twist into the famous double-helix structure that was experimentally discovered using X-ray crystallography. Thus, the covalent strands of two complementary DNA molecules are held together in the helix shape by the much weaker hydrogen bonds. Because the bonding between these amino bases is so specific, the helix can form only if the sequence of bases on the first string is perfectly matched to the sequence on the second string.

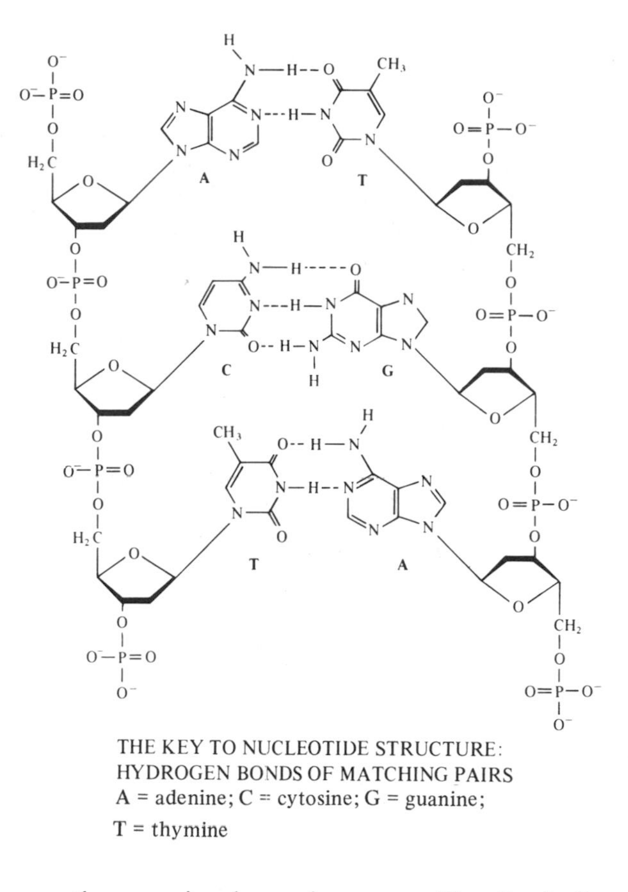

THE KEY TO NUCLEOTIDE STRUCTURE: HYDROGEN BONDS OF MATCHING PAIRS
A = adenine; C = cytosine; G = guanine; T = thymine

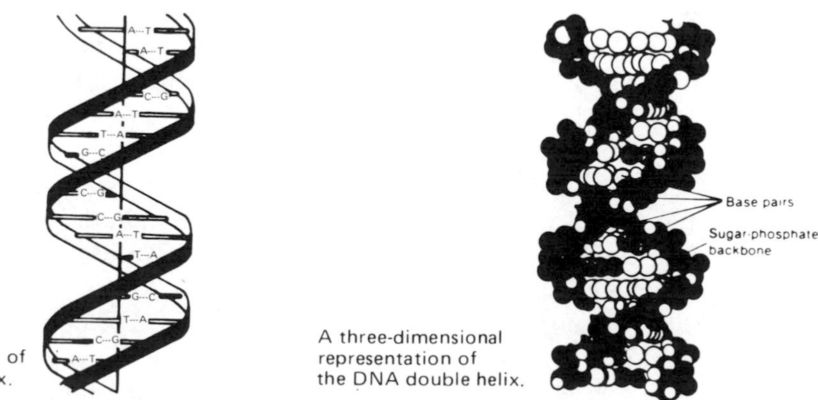

A schematic diagram of the DNA double helix.

A three-dimensional representation of the DNA double helix.

TWO VIEWS OF THE DNA DOUBLE HELIX

The sugar-phosphate units, each with an attached amine base (A,T, C, or G) can be thought of as labeled building blocks, called "nucleotides," from which a DNA macromolecule can be formed. The order in which these nucleotides line up creates an information code in the molecule. This code is how the DNA molecule carries the information to create the proteins needed by a living organism. This information can be copied to produce duplicate DNA molecules through enzymatic synthesis. The double strands unzip their hydrogen bonds to expose a single strand. This strand then serves as a sequence guide in enzymatic synthesis of an identical copy. This process involves making and breaking complementary hydrogen bonds which, because of the low bond energies, can be done without breaking the much stronger sugar-phosphate covalent bonds. Thus, the genetic coding in DNA and its duplication are accomplished through a delicate orchestration of chemical bond energies and molecular structures.

PROTEINS—WHAT DO THEY DO?

Proteins, too, are macromolecules—their molecular weights are in the range 10^4 to 10^5. This time, the macromolecular skeleton is linked together by amide or "peptide" bonds. Each amide bond is formed by elimination of water to bind two α-amino acids through a covalent chemical bond. There are 20 different amino acid building blocks that go into making proteins. Each amino acid has its own particular R group attached. Thus, these 20 amino acids make up a molecular alphabet containing 20 letters. The order in which these amino acid letters link up "spells out" and determines the molecular structure of the protein and therefore its biological function.

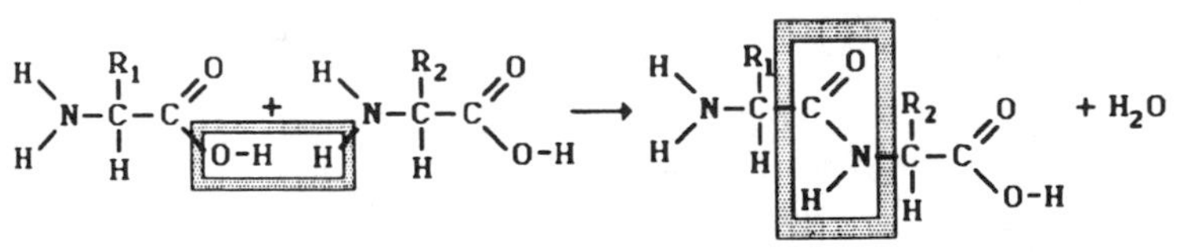

PROTEINS ARE LINKED BY AMIDE (PEPTIDE) BONDS

Proteins carry out an astonishing range of biological functions. Nearly all chemical reactions in organisms are catalyzed by a category of proteins called enzymes. The breakdown of foods to generate energy and the synthesis of new cell structures involve thousands of chemical reactions that are made possible by protein catalysis. Proteins also serve as carriers—an example is globin, which transports oxygen from the lungs to the tissues. Muscle contraction and movements within cells depend on the interplay of protein molecules designed to generate coordinated motion. Another group of protein molecules, called antibodies, protect us from foreign substances such as viruses, bacteria, and cells from other organisms. The operation of our nervous system depends on proteins that detect, transmit, and gather information from the world around us. Proteins also serve as hormones that control cell growth and coordinate cell activities.

Thus, life depends on the interplay of two classes of large molecules, nucleic acids (DNA) and proteins. The genetic inheritance of an organism is stored in its DNA, which serves both as a pattern for the formation of identical copies of itself for the next generation, and as the blueprint for the formation of proteins, the controllers of nearly all biological processes.

The arrangement of the bases in a DNA molecule is the code that tells the amino acids in what order they should link up to become a particular protein. In order to construct proteins, a third macromolecule is used to read the information coded in

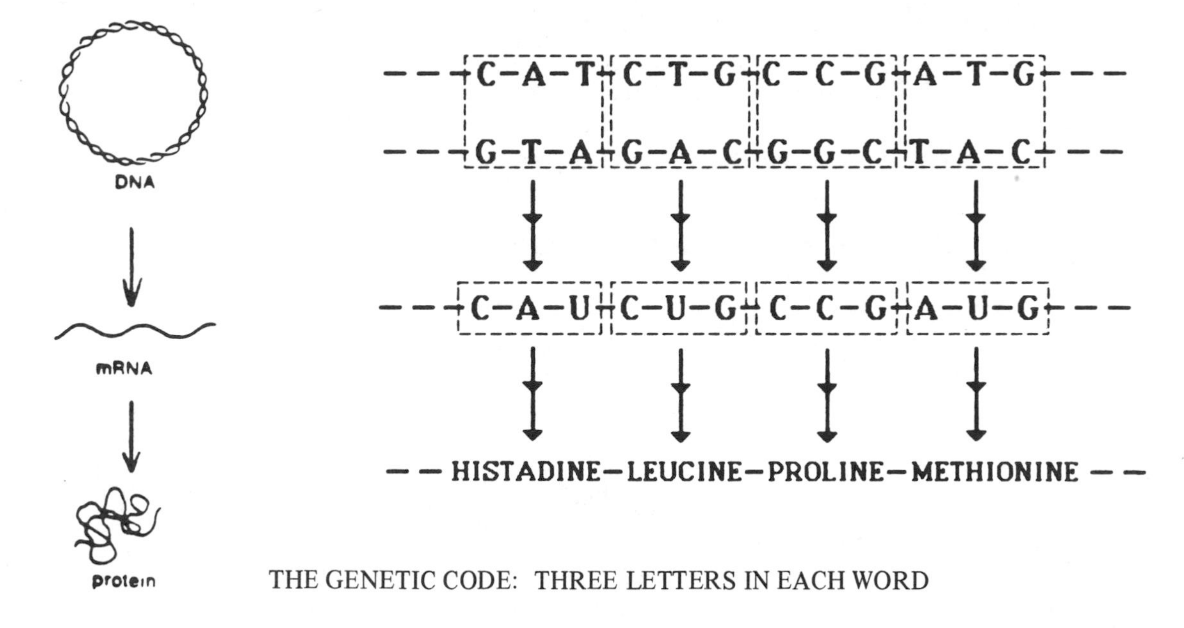

THE GENETIC CODE: THREE LETTERS IN EACH WORD

DNA. This molecule is called ribonucleic acid (RNA). It reads only one chunk of a DNA molecule at a time. An enzyme separates the two strands of DNA, and RNA begins to make a copy of the code using nucleotides. As before, adenine (A), cytosine (C), and guanine (G) are used, but now uracil (U) replaces thymine (T) in the sequence. Thus is generated the macromolecule RNA, some of which is called "messenger RNA" (mRNA). This mRNA is the information-carrying link between the gene DNA and the desired protein. It takes three nucleotides to identify a particular amino acid—thus, CCG in sequence stands for the amino acid proline, while CAU is the word for histidine in the genetic dictionary.

Figuring out this genetic code took decades of painstaking and difficult research, most of it involving chemistry. Chemistry provided the methods for determining the sequence of amino acids in the protein chains (usually called polypeptide chains). Chemists also learned how to assemble amino acids in a desired sequence so they could make polypeptides in the laboratory and even small proteins identical in structure and function to those isolated from natural sources.

More recently, chemists have developed rapid chemical means for deciding the order of nucleotides (called "sequencing") in single-stranded DNA. This breakthrough was of extreme importance because it allowed scientists to determine the primary molecular structure of a gene. Oddly enough, sequencing at the gene level is performed with less difficulty than sequencing of the encoded protein. As a result, the rapid sequencing of DNA has provided an immense expansion of our knowledge of protein structures.

Of equal importance, and at the heart of modern biotechnology, has been the development of simple, rapid, chemical strategies for gene synthesis. Two chemical methods are now in use. In the first, phosphoryl esters are formed with the alcoholic OH groups of the sugars by dehydrating agents. In the second, a preformed intermediate molecule is synthesized (a phosphoramidite) that can be used to form the desired backbone phosphate linkage. The second method has been adapted to a solid-state support, thus permitting routine synthesis of nucleotide chains (oligonucleotides) of lengths up to 50 base pairs.

All of these developments in chemistry provided tremendous leaps in our ability to understand biological molecules in chemical terms. Without these advances, biotechnology as it exists today would not be possible.

RECOMBINANT DNA TECHNOLOGIES

A recent development in biotechnology is called recombinant DNA technology or genetic engineering. It combines nucleic acid chemistry, protein chemistry, microbiology, genetics, and biochemistry. The first step in genetic engineering is to isolate and identify genetic material (DNA) from one organism. It is then modified so that it can be inserted into a new "host" organism. When this host organism reproduces itself, those that have accepted the insertion also reproduce the desired gene.

Molecular biologists have discovered two classes of proteins which make it possible to manipulate pieces of DNA in a precise manner. *Restriction enzymes* catalyze the cutting of DNA at specific nucleotide sequences. *Ligation enzymes* catalyze the joining together of two fragments of DNA in a particular nucleotide order. For example, a restriction enzyme called Bam Hl recognizes the double-stranded sequence GGATCC, and cuts between the two G nucleotides to create fragments as follows:

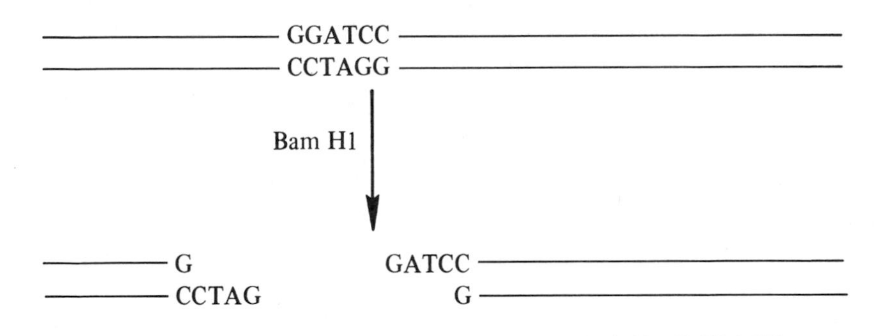

ENZYME Bam H1 CUTS DNA AT A PARTICULAR PLACE

Then the enzyme DNA ligase can take fragments like those created above and join them together to form a single continuous duplex chain as follows:

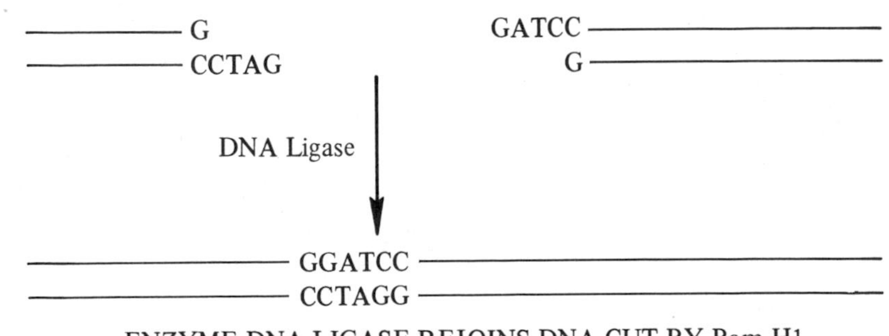

ENZYME DNA LIGASE REJOINS DNA CUT BY Bam H1

Now suppose that a foreign segment of DNA from another organism is also cut with the same matching ends. The DNA ligase will then catalyze the insertion of this foreign sequence into the host DNA. The result is called recombinant DNA.

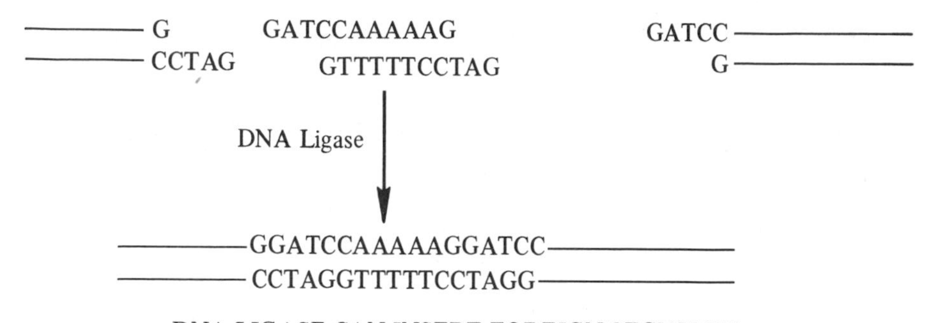

DNA LIGASE CAN INSERT FOREIGN SEGMENTS

The foreign segment is cut out of a donor DNA. The host DNA is called a *plasmid.* It is a ring of DNA that can be independently reproduced within bacterial cells. If the construction works successfully, the cells can direct the synthesis of

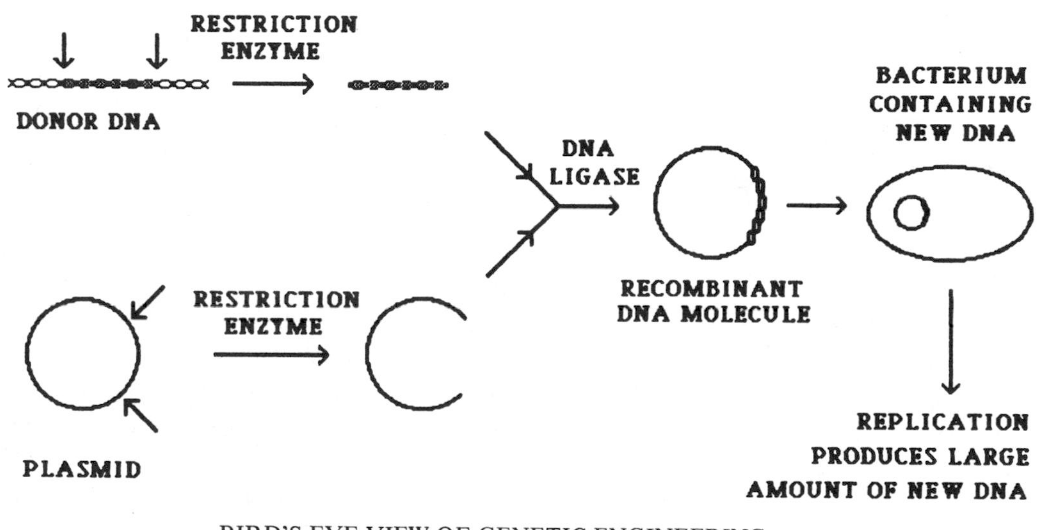

BIRD'S EYE VIEW OF GENETIC ENGINEERING

mRNA, and finally, protein. The goal is to alter the DNA to encode it for a particular desired protein. The genetically engineered bacteria can then be grown as colonies of identical bacteria (clones), all of which will then produce the particular protein for which the synthesis information was encoded by the original DNA fragment.

Specialized techniques have been developed in order to analyze and identify particular DNA fragments, including those containing specific genes. Separations technology has been developed to isolate such DNA fragments. Other analytical techniques have been developed to identify the genetically engineered cells in which the desired DNA has been introduced as well as those within which the DNA (through the intermediary mRNA) is directing the synthesis of proteins. Once again, the isolation of the protein molecules requires the application of separations technology. Thus, the application of chemical techniques to biological systems is at the very heart of recombinant DNA technology.

BIOTECHNOLOGY APPLICATION TO MEDICINE

Various genes have been chemically synthesized, cloned, and used to direct the synthesis of a desired protein through recombinant DNA technology. For example, insulin is a protein which is used to treat diabetes. The gene that led to the production of human insulin was synthesized by chemists in 1978 and was engineered into a plasmid and introduced into the common bacterium, *E. coli*. Another example is human growth hormone, a protein which is a sequence of 191 amino acids. A gene encoding this protein was created by joining together some naturally isolated DNA with some chemically synthesized DNA. This protein was produced in *E. coli* in 1979 and is being tested as a potential medication for dwarfism and similar conditions caused by a shortage of this hormone.

The production of a hormone is not the only type of protein for which recombinant DNA technology is useful. Classical vaccines developed to protect against viral infections are often isolated from natural sources. A vaccine works by stimulating the body to produce antibodies when "killed" virus cells or fragments of a virus are injected into a person. The body can then resist that particular viral infection. Of course, there is a risk associated with introducing the active disease-causing portions of a virus into someone's body. Now, using recombinant DNA technology, the DNA which codes for the protein on the outside of a virus can be produced. Thus, we can stimulate immunity to a disease by injecting just the protein coat of the virus involved and thereby create a safer vaccine which cannot accidentally cause the disease or be contaminated by other viruses.

These examples illustrate the great power of recombinant DNA technology to synthesize, on a potentially large scale, valuable protein materials which would be difficult or too expensive to produce by other means. They represent the combined efforts of chemists, biologists, and other scientists and are a prime example of the complete interdependence of the different disciplines. The potential of recombinant DNA technology, however, has barely been touched. Chemically prepared DNA sequences can be used to screen an individual in order to locate genetic defects that may indicate a special sensitivity to the appearance of disease. It is even foreseeable that genetic diseases could be corrected through the replacement of defective genes or through the addition of a genetically engineered gene. Perhaps the most important contribution that recombinant DNA technology can achieve will be the expansion of knowledge about the regulation of genes within cells.

Naturally occurring molecules are often found that are biologically active and thus medically useful. But these molecules are often not the ones chosen for a pharmaceutical product. A chemically similar molecule (an analog) or a fragment of the naturally occurring product may be used instead to lower the cost or to avoid undesirable side effects. Recombinant DNA techniques can produce these modified products. Polypeptide hormones have many types of useful biological activity, but suffer from the disadvantage of not being active when taken orally and of having a short time of effectiveness. Further progress in the chemical modification of proteins may remove these limitations. Often, a protein produced using recombinant DNA technology requires modification before its biological activity can be realized. This was true for the insulin described earlier. Chemical modification of the insulin protein produced by *E. coli* led to a biologically active compound, a new hormone.

On the other hand, the desirable pharmaceutical may be a compound that blocks or is antagonistic to the biological activity of some naturally occurring biomolecule. In this case, recombinant DNA technology can provide a good source of the biomolecule which can then be used to test chemically (or biotechnologically) synthesized compounds in order to develop a useful pharmaceutical drug.

BIOENGINEERING

An increasingly important part of modern medicine is the development of safe and effective methods for delivering drugs as well as the creation of assemblies that can replace failed human parts. This involves chemical as well as engineering development. Examples include cardiac pacemakers, heart valves (and now artificial hearts), tendon replacements, and heart-lung and kidney dialysis machines. Recent research into blood substitutes has led to some promising possibilities such as fluorocarbon chemical emulsions and serum constituents such as albumin and factor VIII (recently reported to be produced by recombinant DNA technology). Thin membranes used as artificial skin and cultured epithelial cells promise major advances in burn treatment. Materials for tooth implants and bone replacement are being developed. Insulin-releasing pumps that can be implanted in the body of someone suffering from diabetes can make insulin treatment more regular and controlled, thus reducing serious health threats. In the longer term, it may become possible to implant genetically engineered cells directly into an organism that will provide treatment for genetic and hormonal deficiencies.

BIOCATALYSIS

Enzymes, the proteins that act as catalysts in biochemical reactions, are the main focus of yet another branch of biotechnology and chemistry. The ability of recombinant DNA technology to control the synthesis of enzymes will surely extend the application of the microbe as a biocatalyst. First, it will be possible to produce almost any enzyme found in nature inexpensively. Second, and more exciting, is the possibility of perfecting present techniques for preparing biocatalysts that currently do not exist in nature through careful DNA synthesis. X-ray crystallographic techniques have provided the chemist with detailed understanding of the three-dimensional struc-

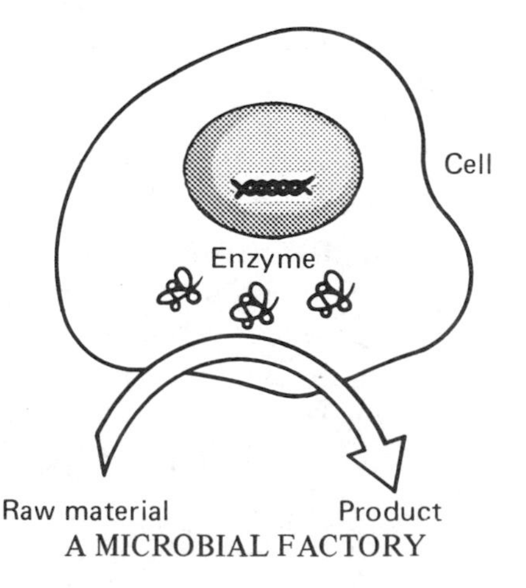

A MICROBIAL FACTORY

ture of some enzymes. Further chemical research to increase the understanding of the relationship between the chemical structure of enzymes and their catalytic activity will be needed before logical design of such biologically produced synthetic biocatalysts can be achieved.

A recent aid in biocatalysis has been the development of a technique called *enzyme immobilization*. In this technique a solid support is used to actually hold the enzyme still. This stabilizes the enzyme and increases the amount of material that it converts to the desired product. It also simplifies purification of the product because the enzyme is more easily separated from the end product. One example of this technology is the use of the immobilized enzyme penicillin acylase to convert the naturally occurring antibiotic penicillin G into 6-aminopenicillanic acid (6-APA). The acylase enzyme removes the chain of atoms that are linked at the amino nitrogen (N) atom in penicillin G. Other chains of atoms are then chemically added at this nitrogen to produce various semisynthetic penicillins for medical use.

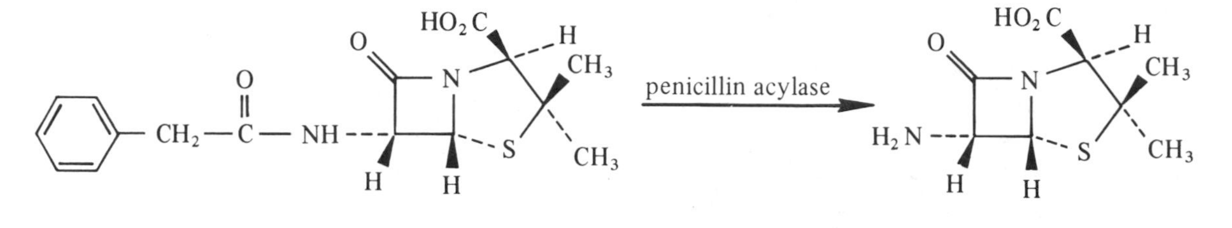

Penicillin G

6-Aminopenicillanic acid

ENZYME IMMOBILIZATION IMPROVES PENICILLIN

In another example, corn starch can be enzymatically converted to glucose. An immobilized enzyme, glucose isomerase, is then used to convert some of the glucose to the sweeter fructose. Over 2 million metric tons of this high-fructose corn syrup are produced annually in the United States.

Immobilization technology does not necessarily require the isolation of a particular enzyme. Whole cells containing the enzyme can be immobilized on a solid surface. For example, whole cells of the bacterium *E. coli* have been immobilized and used to catalyze the chemical conversion of fumaric acid and ammonia into aspartic acid, one of the amino acid building blocks of proteins. In addition, immobilized yeast cells can be used in the fermentation by which we produce alcohol (ethanol). This process has been demonstrated industrially in a large test plant facility.

No discussion of biocatalysis would be complete without talking about biomass. At this point, a relatively small amount of the total available biomass in the United States is converted into useful chemicals through biotechnology. There is increasing interest in biomass conversion, as the Earth's supply of raw materials of fossil origin (like crude oil) is limited and nonrenewable. The potential volume of cellulosic materials (plant matter) that could be converted into industrial chemicals, however, is large. Large-scale conversion of biomass into industrial chemicals requires a relatively constant, low-cost source of biomass. From a technical point of view, molasses, starch from corn or wheat, and sugar are well suited for fermentation. They are readily converted into glucose, and additionally, microorganisms are known for converting glucose into many useful chemical products. These starting materials, however, are also needed for food and are subject to wide variations in price and supply depending upon crop success and trade policies.

The potential biomass available from agricultural and forestry waste is estimated to be 10 times greater than the sources mentioned above. This biomass is less subject to the changes of both price and availability. Unfortunately, it is mostly made up of lignocellulose (lignin, cellulose, and hemicellulose). Lignin, a woody compound found in plants, resists biocatalytic breakdown and physically interferes with the fermentation of the cellulosic materials. Thus, lignocellulose biomass must be chemically pretreated to remove the lignin. Except for use as a combustible fuel, no large-scale uses for lignin have been developed and it often becomes a waste. Biocatalysis of these abundant sources of biomass is therefore waiting for further development in the chemical modification of the raw materials.

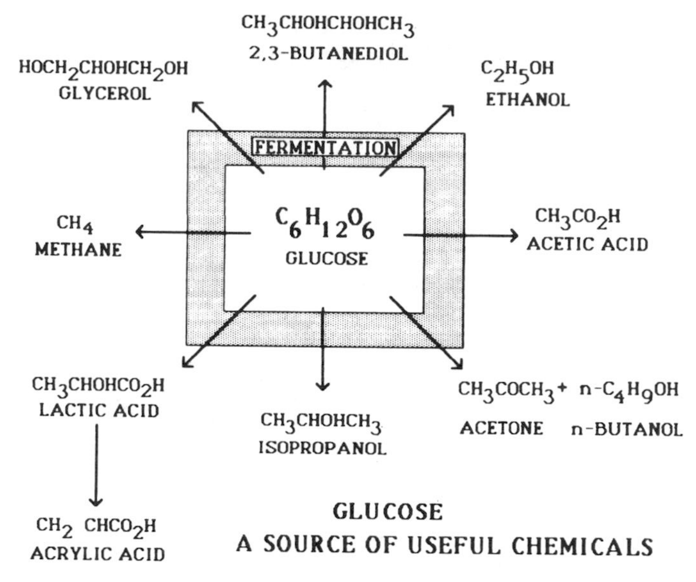

GLUCOSE
A SOURCE OF USEFUL CHEMICALS

CONCLUSION

Over the past two decades, progress in biotechnology has been dramatic. It is now possible to program living cells to generate products ranging from relatively simple molecules to complex proteins. We have only begun to realize the immense potential of recombinant DNA technology as a means of obtaining protein materials that were previously very costly or unobtainable in large quantities. Biocatalysts have already established themselves in the large-scale production of various industrial chemicals. Continued progress in biotechnology will require the cooperative efforts as well as the individual advances in several disciplines including chemistry, chemical engineering, molecular biology, microbiology, and cell biology.

SUPPLEMENTARY READING

Chemical & Engineering News

"Biomaterials in Artificial Organs" by H.E. Kambic, S. Murabayashi, and Y. Nose, vol. 64, pp. 31-48, Apr. 14, 1986.

"ACHEMA Features Biotechnology's Bigger Role in Chemical Technology" by J.H. Krieger and D.A. O'Sullivan (C.& E.N. staff), vol. 63, pp. 31-40, June 24, 1985.

"Single Cell Protein Process Targeted for Licensing" by J.H. Krieger (C.& E.N. staff), vol. 61, p. 21, Aug. 1, 1983.

"Mammalian Cell Culture Methods

Improved'' (C.& E.N. staff), vol. 61, p. 26, Jan. 10, 1983.

Science

''Solid State Synthesis'' (Nobel Address, Chemistry, 1985) by B. Merrifield, vol. 232, pp. 341-347, Apr. 18, 1986.

''Automated Chemical Synthesis of a Protein Growth Factor for Hemopoietic Cells, Interleukin-3'' by I. Clark-Lewis, et al. (5 co-authors), vol. 231, pp. 93-192, Jan. 10, 1986.

Magnetic Fluids—Attractive Possibilities

When we hear the word "magnet" most of us picture a horseshoe-shaped affair with nails and paperclips clinging to it, or one of those things we put on the refrigerator to hold up messages. But what images come to mind when we try to picture a magnetic *liquid?* The possibilities stretch the imagination.

The first question is whether or not such liquids even exist. Are there vast deposits of magnetic fluids waiting in the depths of the earth for excavation? The answer is no. Can we make a magnetic fluid by melting iron or nickel or cobalt or some other ferromagnetic material? Sorry, won't work. Every magnetic substance loses its magnetism when heated above its own characteristic temperature.

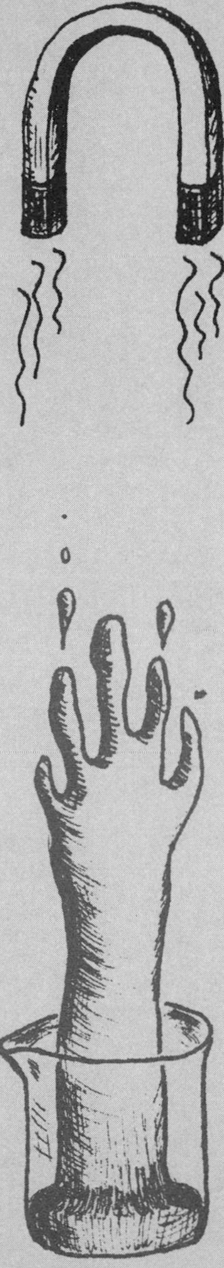

Yet magnetic fluids do exist, but the term refers to a suspension of small particles of a ferromagnetic material in a liquid. Scientists have been trying to make magnetic fluids this way since the 1770s, when one of them mixed up a batch of iron filings and water. We've come a long way since then, and magnetic fluids are now a practical reality.

One of the chief problems with a magnetic fluid is keeping the tiny particles from bunching up. Two forces are at work trying to get these particles to clump together. One is the attraction of the little magnets to each other, and the other, stronger force is the van der Waals attraction between the atoms that make up the particles. Chemists have discovered ways of dealing with these two forces. First, they use extremely small particles to prevent clumping from magnetic attraction—as small as 100 Ångstroms in diameter! Then, they coat the particles with substances called "surfactants," which blanket each particle with a layer one molecule thick. This combats the van der Waals force by keeping the particles at a distance from each other.

One of the first practical magnetic fluids was made from a mixture of iron oxide particles (FeO and Fe_2O_3), called magnetite, in kerosene. The surfactant used was oleic acid, a long molecule with a carboxyl group for a "head" and an 18-carbon chain for a "tail." Since that time many different particles, liquids, and surfactants have been used with success.

Magnetic fluids possess many unusual and useful properties. On an iron part, they can "stick" magnetically right where they are put, or they can be manipulated and moved around with a magnetic field. They can provide a very tight sealant to prevent contamination of sensitive equipment. As a lubricant, the fluid can be positioned exactly where expected wear might occur. Ferrofluids, as they are called, are used in seals and bearings around rotating shafts in machinery. Several drops can create an impenetrable seal around a shaft while still minimizing friction. Ferrofluids are used in airtight seals in the furnaces used for growing silicon crystals; in seals for gas lasers, motors, and blowers; and in computer disc drives where seals operate at extremely high rpm and where a single particle of dust on the recording head can destroy its surface. Magnetic fluids are also used in loudspeaker systems, and in magnetic inks like those found on our bank checks. There is even talk of using them in medicine, to close off arteries temporarily without damage. The future of magnetic fluids stretches more than the imagination—this "attractive" innovation stretches the magnet itself.

III-G. Economic Benefits

INTRODUCTION

The chemical industry has enormous scope. It encompasses inorganic and organic chemicals used in industry, plastics, drugs and other biomedical products, rubber, fertilizers and pesticides, paints, soaps, cosmetics, adhesives, inks, explosives, and on and on. The value of U.S. chemical sales in recent years has been in the neighborhood of $175 billion-$180 billion, with a favorable balance of exports over imports of about $8 billion-$12 billion. Employment in U.S. chemical and allied product industries is over a million people, including over 150,000 scientists and engineers. The numbers are large, and the effect on the economy is important. And even they do not adequately indicate the far-reaching presence and impact of chemistry throughout our society. Chemical products are supplied to countless other industries to be processed and resold. Additionally, chemical processes are abundant and growing in modern manufacturing. Mechanical operations, such as cutting, bending, drilling, and riveting, are being replaced by etching, plating, polymerization, cross-linking, sintering, etc. For example, electronic microcircuits are produced through a sequence of perhaps 100 chemical process steps. Finally, chemistry is the science on which our understanding of living systems is based. Heredity is now understood in terms of the chemical structure of genetic material. Disease and its treatment are chemical processes. Every medicine that a doctor prescribes is a chemical compound whose effectiveness depends upon the chemical reactions it stimulates or controls.

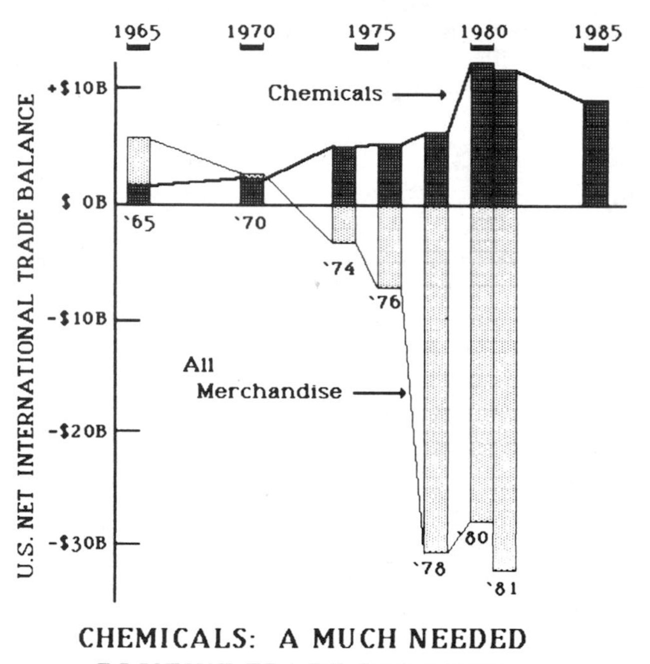

CHEMICALS: A MUCH NEEDED POSITIVE TRADE BALANCE

The business climate of the chemical industry is complex and changing. Here in the United States the situation is particularly difficult due to many diverse factors that are unique to our society. Antitrust law in the United States strongly discourages cooperative actions on the part of U.S. corporations. Abroad, cooperative partnerships between foreign corporations and the government are encouraged. Governmental policies in regard to science-based industries are frequently more favorable abroad than in the United States.

International activity in the petrochemical arena is increasing as nations controlling cheap and abundant feedstocks establish their own manufacturing complexes to refine crude oil and produce polymers and other products higher up the value scale. It seems probable that this foreign effort will be concentrated in commodities with the largest established markets (e.g., ethylene glycol, polyethylene).

The chemical industry must also respond to an active public concern for health and safety connected with possible exposure to toxic chemicals. This movement is most advanced in the United States, where concerns range from sensibly prudent to panicky. Economically, these responses must lead to higher costs to achieve the desired environmental protection, worker safety, proof of safety and effectiveness of new products, and protection against product liability. Of course, these costs are always paid for by the consumer, but they are significantly affecting the ability of U.S. industry to compete when the same industries abroad do not feel the full impact of these pressures.

It is not surprising that the viability of chemical companies in the United States has become a cause for national concern. The advanced standard of living in the United States owes a great deal to the innovations and productivity of the nation's chemical businesses. Preservation of this quality of life depends to no small extent upon whether the United States can remain a strong and leading participant in chemistry-based technologies. A key factor responsible for past success has been the strength of U.S. university research and the effective use of its new discoveries to develop new products needed by society. Vigorous support of this academic research community is a critical first requirement for maintaining the ongoing health of the U.S. chemical industry.

ENERGY AND FEEDSTOCKS

Energy and chemical feedstocks are tied together through their overwhelming dependence on petroleum. Energy uses account for most of the consumption of these organic materials. Burning of petroleum goes on at an ever-increasing pace, and the future supply crisis is directly related to this fact. Throughout much of the world, people take petroleum-derived heat and transportation for granted. Thus, the inevitable depletion of the earth's petroleum resources will strongly affect the style and standard of living of people everywhere. The effects of depletion should become evident within two decades and severe within four. Hubbert has estimated that 80 percent of the world's ultimate production of oil and gas will be consumed between 1965 and 2025. This estimate, made in 1970, seems to be consistent with current discovery and consumption rates. Its alarming implications are not fully recognized by the public.

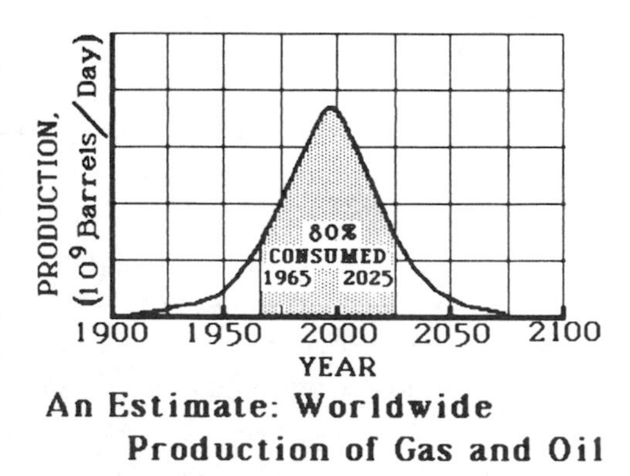

An Estimate: Worldwide Production of Gas and Oil

Petrochemical uses of petroleum account for only a few percent of the total—3 to 5 percent by most estimates. Thus, the chemical industry is not the cause of the approaching era of depletion, but the effects will be felt within the industry as feedstocks and processes change. However, petrochemical uses are characterized by higher retail prices, and they can withstand the coming price increases brought on by decreases in oil and gas reserves better than uses involving combustion. Further, processes are already known for the conversion of coal to suitable forms for use as feedstocks, and coal deposits are more abundant. Therefore, it is expected that the impact of petroleum depletion on chemical feedstocks will be much less damaging than its impact on energy production.

RENEWING OUR INDUSTRIES

International competition is a general problem for U.S. industry. Steel, automobiles, communications, textiles, and machine tools are examples of industries that have encountered significant problems. It is instructive to consider the response to these pressures in the automobile industry. It shows the central role of chemistry in maintaining and improving the U.S. position.

The U.S. automobile industry evolved into a gigantic business during the first half of this century. In the 1950s and 1960s American products enjoyed great success. The vehicles were large, heavy, and powerful. Fuel was abundant and inexpensive. There was no reason to conserve, so fuel economy of the American car was not considered by buyers. Furthermore, few foreign-built cars made their way to North America. By the mid-1960s, however, Volkswagen had entered the U.S. market with sales of more than half a million small economy cars per year. During the 1970s the market was further affected by cars manufactured in Japan. Pursuing an aggressive policy of collecting design, technology, engineering, and assembly information from other countries, the Japanese developed the most automated and efficient car-building facilities in the world. These facilities, and a commitment to quality, yielded the world's most fuel-efficient and low-cost cars at a time when gasoline prices began to soar.

At the same time, antismog legislation was passed in the United States that required better fuel economy and placed strict limits on air pollution from automobile emissions. The American car was required to change dramatically, and the investment required by the manufacturers was very high, approximately $80 billion. The antismog objectives are being achieved through many developments involving chemistry: new and lighter materials, better combustion control and engine efficiency, catalytic exhaust treatment, lowered corrosion, reduced size, transmission improvements, etc.

Polymers, aluminum, and high-strength alloy steels are used to reduce the weight of the car. New chemicals for oil additives and improved rubber formulations for tubes and hoses are solving problems of engine compartment temperature brought on by aerodynamic designs featuring sloping hoods. The ride quality of the smaller cars is being improved through the use of vibration-damping butyl rubber. Tire tread compounds are being reformulated to reduce rolling resistance. New, high-solid paints are being developed to reduce air pollution from automobile painting. Chemically based rust-proofing systems are being introduced to prolong life. Contemporary U.S. cars each contain over 500 pounds of plastics, rubbers, fluids, coatings, sealants, and lubricants, all products of the chemical industry.

Further uses of plastic materials are sure to come. Reaction injection molding is a recently introduced process for making large parts such as fenders and hoods.

High-performance composite materials, i.e., stiff fibers in a polymer matrix, have already appeared as drive shafts and leaf springs. Some advanced models have frames and bodies made of composite polymers. For automobiles, the use of composite polymers may lead to new design-fabrication methods which will greatly reduce the number of parts that must be assembled. Furthermore, new designs for light aircraft have airframes that are almost entirely composites. Advances such as these will tend to reduce the problems faced by the automobile and other basic U.S.industries, problems that arise from a complex mixture of historical preferences, social pressure, legislation, and vigorous outside competition.

NEW HORIZONS

The chemical industry is changing, and chemical science is becoming importantly intertwined with other areas of science and technology. To an increasing degree chemists must be skilled at dealing with subjects in interrelated technologies. Chemistry is critical in providing materials and processes for American industry, meeting their wide range of needs from established industries (new electrode materials for aluminum production, decaffeinated coffees, sweeteners for the food industry, etc.) to rapidly growing, high-technology areas (composites for aircraft, ceramics for electronics and engines, protein pharmaceuticals, etc.). Each of these areas requires development of chemical products that respond to markets outside of chemistry. Representative examples are given below.

Biotechnology

Biotechnology is not new. The ancients knew how to bake and brew thousands of years ago. The processes of fermentation, separation, and purification have long been familiar. But as the molecular structure and basic chemistry of genetic material became known, a new era of biotechnology has opened up. (See Section III-F.) It led to gene splicing procedures that allow biochemists to cause bacteria to produce complex molecules with biological activity. Enzymes have been found that will break chemical bonds in DNA chains at specific points and allow foreign DNA to be inserted with new chemical bonds. The altered DNA will then produce proteins according to its revised code. The protein products can be hormones, antibodies, or other desired complex chemical compounds with specific properties and functions. Interferon,

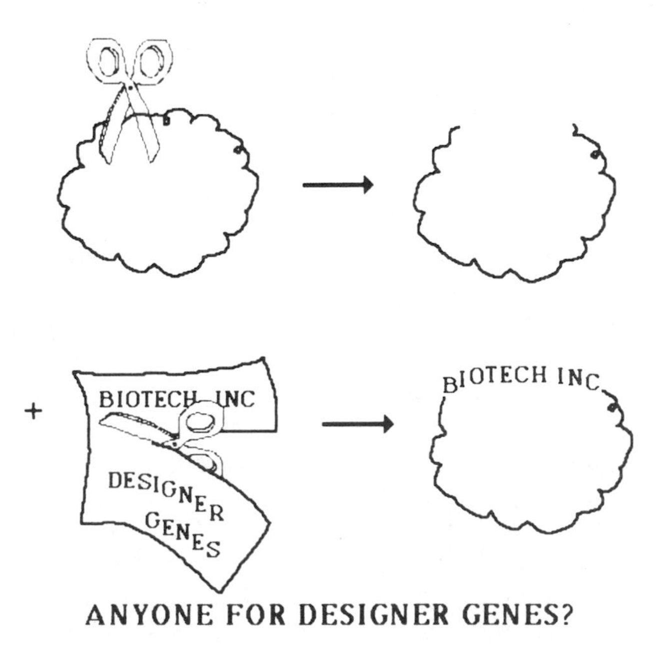

ANYONE FOR DESIGNER GENES?

produced by bacteria with a human gene spliced in place, is expected to be valuable in treating a variety of diseases. Human insulin produced through gene splicing techniques is already being marketed. Activity is intense and commercial enterprises are emerging rapidly.

The area of biotechnology is an exciting and optimistic one for scientists, engineers, and investors. Although some of the expectations may be extravagant, there can be no doubt that this area will give us many important economic developments in the coming decades. The United States is at present the world leader, with basic chemical and molecular biological research feeding an effective commercial community. Europe has strong, relevant research, and Japan has a leading position in fermentation processes. The advances that will determine the future of this field will come through a deep understanding of biology at the molecular level. Basic research on the molecular structure and chemistry of biological molecules will be a crucial ingredient as we bring biotechnology into practical use.

High-Technology Ceramics

Ceramics are materials with high-temperature stability and hardness; they tend to be brittle and therefore are difficult to shape in manufacturing. Ceramics are now of major commercial interest for components of electrical devices, engines, tools, and a wide range of other applications in which hardness, stiffness, and stability at high temperatures are essential. Major advances in their use can be anticipated because of new chemical compositions and novel fabrication techniques. For many, many centuries, ceramic pieces have been made from fine particles suspended in a liquid (a slurry) or a paste of a finely ground natural mineral. The slurry is formed or cast in the desired shape and then "fired," i.e., heated to a high enough temperature to burn away the added slurry components and to melt and join the mineral particles where they touch. We now know that the strength of the final object is critically limited by small imperfections.

A number of new chemical techniques are now being developed to synthesize new ceramic starting materials that will produce more defect-free final products. These techniques depend upon control of reaction kinetics and tailoring of molecular properties. For instance, controlled hydrolysis of organometallic compounds is used to generate highly uniform ceramic particles ("sol-gel technology"). Organometallic polymers can be spun into fibers, and then all but the polymer skeleton is burned away to produce high-temperature materials like silicon carbide. Highly uniform temperature-resistant coatings in desired shapes can be produced using high-temperature reactions of volatile compounds followed by controlled deposit of the products onto a preformed solid object. For example, jet engine parts might be made this way. Addition of suitable impurities ("doping agents") can change properties dramatically. For example, alumina ceramics can be significantly toughened by the addition of zirconia, solid ZnO_2.

Advanced Composites and Engineering Plastics

The discovery of ultra-high-strength fibers based upon graphite embedded in an organic polymer has led to development of a new class of materials now referred to as "advanced composites." A fiber, such as a graphitic carbon chain, a mineral fiber, or an extended hydrocarbon polymer, is suspended in a conventional high polymer such as epoxy. The resulting composite can exhibit tensile strength nearly equal to that of structural steel but at a much lower density. Because of this high strength-to-weight ratio, such composites are finding abundant applications in the aerospace industry. Significant weight reductions are achieved in commercial and military aircraft that use airframes and other aircraft components made of composites. Other applications include space hardware, sporting goods, automotive components (e.g., drive shafts and leaf springs), and boat hulls.

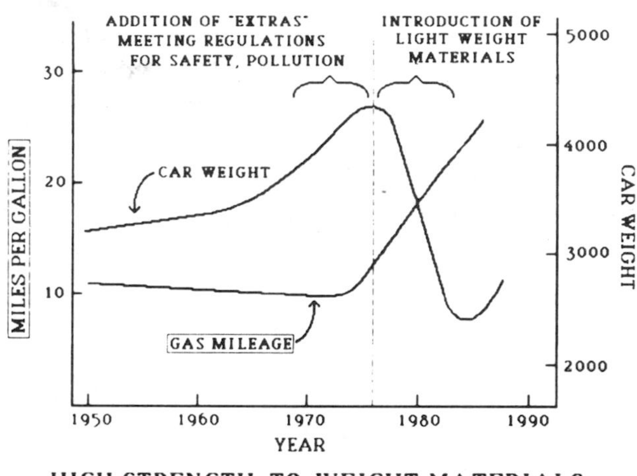

HIGH STRENGTH-TO-WEIGHT MATERIALS DOUBLED AUTO MILEAGE - 1975 TO 1985

There has also been a rapid development in designing polymer mixtures to obtain particular properties or behavior. Success with these polymer "alloys" or "blends" has required a high degree of chemical understanding of the molecular interactions at phase boundaries between two polymers that are not soluble in each other. An example is the commercial polymer blend called Zytel Y.T.®, a nylon toughened with an elastic hydrocarbon. The development of this high-performance plastic was based upon extensive studies of interactions at interfaces between different polymers.

Plastics are also being developed for high-temperature applications such as engine blocks for automobiles. A prototype "plastic engine" based on reinforced polyamide and polyimide resins has been demonstrated in an actual racing car. An engine weight reduction of 200 pounds can be achieved, with obvious benefit to fuel economy.

All of these technologies are moving forward rapidly around the world. Carbon fiber production has been well developed in Japan, while the United States is showing the way in high-strength polymeric fibers. The nature of the bonding region between the fiber and its composite environment is an important factor in structural performance but is poorly understood chemically. Research will figure importantly in the evolution of the field.

Photoimaging

The aim of photography is to produce an accurate and lasting record of the image of an object or a scene. With a history of 150 years, the silver halide process has evolved from complex procedures conducted by specialists with a working knowledge of photochemistry into a pastime expertly pursued by a large chunk of the population. The camera owner presides over remarkable feats of optics and chemistry to produce pictures on the spot, usually without having the faintest appreciation of what goes on in the camera and on the film. The result brings lifelike communication and pleasure to people throughout the world.

The chemistry of the photographic process can be usefully divided into the inorganic photochemistry of the silver halide and the organic chemistry of sensitization, development, and dye formation. When radiation strikes a microcrystal of a silver halide in the film emulsion, a faint image is formed that is believed to consist of a few atoms of metallic silver. The metallic silver functions as a catalyst for the reduction of the entire microcrystalline grain under the chemical action of an easily oxidized organic substance, the "developer." The silver halide grains in a photographic film are typically about one micron in size, and control of the size and shape of the particles is important. Although silver halides are sensitive only to light at the blue end of the spectrum, the grains can be activated at longer wavelengths with sensitizing dyes on the crystal surface. These molecules are coated onto the silver halide surface in layers less than one thousandth of a millimeter thick. Color is achieved when the oxidized form of the developer reacts with another organic compound to give a dye of the required hue. By combining the 3 color primaries, 11 colors can be achieved. Conventional color photography involves several carefully controlled chemical processes, including development, bleaching, fixing, and washing.

In instant color photography these steps must be combined in a single sheet that can be processed under existing light without temperature control. A typical instant film contains over a dozen separate layers with thicknesses of about one micron each. Physical chemical factors such as solubility and diffusion are critical, as are the chemical reactions occurring in the various layers during processing. The sophistication of the chemistry of instant color photography is difficult to comprehend considering how simple the camera is to use.

In this important area of our economy, new technological achievements continue to appear, ranging from amateur photography to such demanding and specialized uses as photoresists for semiconductor production (see below) and infrared mapping of the Earth's resources from satellites. The United States has been the world leader in photographic technology for many years in an industry in which the connection with our traditional research strength in photochemistry is clear.

Microelectronic Devices

The microelectronics revolution has already had an enormous impact on the industrialized world, and it is clear that there is a great deal more to come. The best-known device is the microprocessor, a remarkably intricate and functionally integrated electrical circuit built on a tiny bit of pure silicon, called a "chip." Some microprocessors and the latest high-capacity computer memory chips contain hundreds of thousands of individual transistors or other solid-state components squeezed onto a piece of silicon about one-quarter of an inch square.

These chips are currently made from highly purified silicon which contains impurities that have been deliberately implanted to form individual devices with desired electronic functions, such as amplification, rectification, switching, or storage of on-off logic information. These minute devices are then interconnected by metal "wires" on a microscopic scale. The fabrication of these exquisitely complex devices depends critically on thin (less than one micron thick) organic films that are sensitive to radiation. Their technology involves organic chemistry, photochemistry, and polymer chemistry.

The purpose of these films is to allow impurities or "dopants" to be added selectively to the silicon forming the pattern of a desired electrical circuit. Because steps in the process involve high temperature, a thin layer of silicon dioxide is used to mask the underlying silicon. This mask determines whether or not the silicon below is exposed for doping. Organic materials called photoresists are used to form the pattern that is transferred into this silicon dioxide layer.

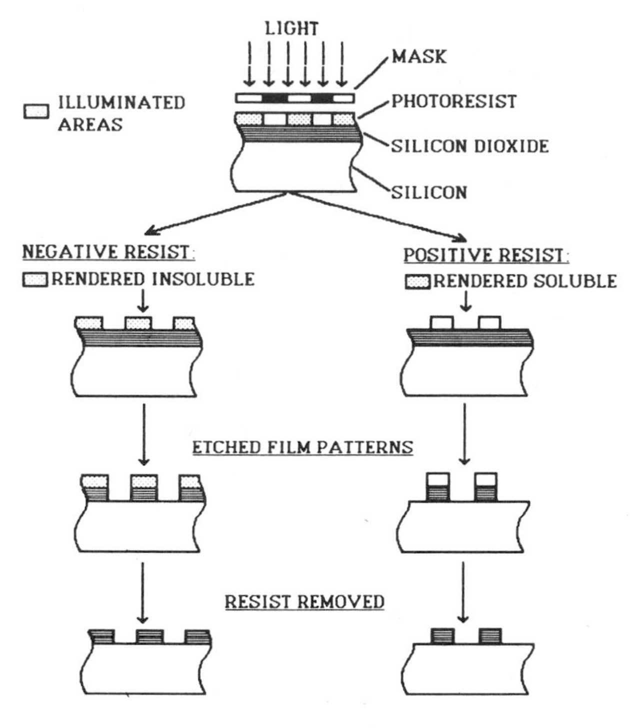

KEY STEPS IN THE FABRICATION OF SILICON INTEGRATED CIRCUITS USING PHOTORESISTS

In photolithography, chemical changes in the photoresist material are begun by exposure to light. In these changes, covalent chemical bonds are broken (or formed) at light-sensitive chemical groups attached to the polymer structure. These chemical bond changes result in a local increase (or decrease) of the photoresist solubility in a suitable solvent. Thus, after exposure through a mask, an image of the mask can be developed merely by washing in the solvent. What is not generally appreciated is that this solubility is achieved through carefully designed polymer photochemistry.

Existing organic photoresists were able to achieve the spacing between circuit elements needed in the early 1970s when individual circuit features were in the size range of 3-10 microns. However, the continued desire for smaller devices has demanded smaller and smaller features. A decade ago it became apparent that new photoresists would be needed because existing materials were not capable of defining the feature sizes (1-2 microns) soon to be required. The development of these materials has been made possible by the research in polymer chemistry, photochemistry, and radiation chemistry done in the last two decades. Because these circuit elements have dimensions close to the wavelength of light normally used for conventional photographic imaging (0.4 microns), diffraction effects caused by lines and marks on the mask become important. These effects can be reduced by using shorter-wavelength radiation. Hence, lots of effort is being spent on widespread development of resist materials that are chemically sensitive to exposure with short-wavelength ultraviolet light, X-rays, and even electron beams, instead of the near-ultraviolet light now used.

The mask itself is now made by chemically etching the desired pattern into a thin chromium film deposited on glass. The pattern is "written" into a resist film by exposure to a computer-controlled electron beam. The development of the organic resist material that is used for defining the pattern on the metal rests on relatively recent research. Many new types of chemical reactions and polymers are involved, and the advances in integrated circuit complexity could not have occurred if these new materials had not been available. Virtually none of them existed in 1970. Examples of new electron beam resists are the polymers that result from copolymerizing various alkenes and sulfur dioxide. Their synthesis and radiation sensitivity were only recently discovered.

A present trend in semiconductor fabrication is to use reactive gas plasmas from a glow discharge instead of liquid solutions to etch the material under the photoresist mask. Most organic materials are not sufficiently resistant to these vigorous conditions, and it has taken much research to provide a few useful materials. It is difficult to design materials having the necessary combination of physical and chemical properties. Their development will draw on continued research advances in polymer chemistry and photochemistry (including laser-induced chemistry).

Molecular-Scale Computers

Miniaturization of electrical devices has been one of the most significant factors in the astonishingly rapid advances that have made modern computers possible. Circuit elements in present silicon chips have dimensions near one micron, i.e., in the range of 10,000 Å. However, it may be that fabrication of microscopic devices based upon silicon and other semiconductor methods is beginning to push against natural barriers that will limit movement toward even smaller devices. Thereafter, breakthroughs will be needed. Where will we turn when existing technologies are blocked by natural limits? Irresistibly, we must contemplate molecular circuit elements that will permit us to move well inside the 10,000-Å limit. We are led to think about computer devices in which information is stored in or transformed by individual molecules or assemblies of molecules—i.e., molecular-scale computers.

In a three-dimensional architecture, use of molecular circuit elements with 100-Å spacing would provide packing a million times more dense than now possible. The materials under discussion range from entirely synthetic, electrically conducting polymers to natural proteins. Molecular switches, the basic memory elements of the proposed computer, might be based upon charge movement in polyacetylene, photochromism, or molecular orientation in solids. Ideas on connecting the molecular elements to the outside world are still vague.

As is normal, adventurous concepts generate exciting, often emotional controversy. However, the arguments of even the most sophisticated detractors are disarmed (contradicted?) by the obvious fact that their intelligent opposition is being generated in the human brain, a working "computer" using exactly the structure under challenge! In an age of machine synthesis of DNA segments and laboratory design of artificial enzymes, it would be timid to say that we will never be able to mimic the elegant circuits that each of us depends upon to read and consider these printed words. Only a few decades ago, some individuals might have classified as science fiction a proposal that someday there would be a man on the Moon, that fertility could be controlled by taking a pill, or that we could learn the structure of DNA. But since we know that molecular computers are routine accessories in all animals from Ants to Zebras, it would be prudent to change the question from *whether* there will be man-made counterparts to the questions of *when* they will come into existence and *who* will be leading in their development. The question *When?* will be answered on the basis of fundamental research in chemistry. The question *Who?* will depend on which countries commit the required resources and creativity to the search.

Molecular Computer at Work

CONCLUSION

The field of chemistry in the United States has great industrial and economic importance. The consistent and significant positive balance of payments is an indication of considerable strength. The continuing flow of innovations that benefit society is encouraging. U.S. universities are among the best in the world and year by year draw students from throughout the world for graduate study. We have a lot going for us.

The United States must work hard and be creative to maintain its leadership in view of social values that lead to antitrust regulation, environmental restrictions, health and safety requirements, and high wage rates, all of which tend to increase the cost of U.S. chemical products. Hence, we must insist upon logical and objective justification of any restraints imposed by legislation while maintaining a balanced concern for the important social values represented in current regulations. And we must continue to stimulate the academic and industrial research that maintains the impressive knowledge base that makes our progress possible. We must attract some of the finest young minds to the field of chemistry, as only a sustained and vigorous approach will be effective in keeping pace in the essential field of chemistry, so necessary for any high-technology society.

SUPPLEMENTARY READING

Chemical & Engineering News

"Engineering Plastics: More Products, More Competition" by David Webber (C.& E.N. staff), vol. 64, pp. 21-46, Aug. 18, 1986.

C_1 Chemistry: Growing Field Despite Crude Oil Drop" by J. Haggin (C.& E.N. staff), vol. 64, pp. 7-13, May 19, 1986.

"Marine Mining to Improve its Organization, Direction and Financing" by J. Haggin (C.& E.N. staff), vol. 63, pp. 63-67, Nov. 18, 1985.

"High Tech Ceramics" by H. Sanders (C.& E.N. staff), vol. 62, pp. 26-40, July 9, 1984.

Scientific American

"Advanced Materials and the Economy" by J.P. Clark and M.C. Flemings, vol. 255, pp. 50-57, October 1986.

"Composites" by T.-W. Chou, R.L. McCullough, and R.B. Pipes, vol. 255, pp. 192-203, October 1986.

"Electronic and Magnetic Materials" by P. Chaudhari, vol. 255, pp. 136-145, October 1986.

"Advanced Ceramics" by H.K. Bowen, vol. 255, pp. 168-177, October 1986.

CHAPTER IV

Intellectual Frontiers in Chemistry

A remarkable bounty of benefits has been shown to flow from chemistry. This chapter will provide abundant evidence that these benefits will increase greatly in the years to come. The basis for this optimistic expectation is that this is a time of special opportunity for intellectual advances in chemistry. The opportunity comes from our developing ability to investigate the elemental steps of chemical change and the ability to deal with extreme molecular complexity.

The Time It Takes to Wag a Tail

When your pet dog sniffs a bone, instantly his tail begins to wag. But it must take some time for the northernmost canine extremity to send the news all the way south where enthusiasm can be registered! How long does it take for that delicious aroma to lead to the happy response at the other end? Chemists are now asking questions much like this about their pet molecules! If one end of a molecule is excited, how long does it take for the other end to share in the excitement? That time may determine whether the excitation will result in a chemical reaction in the part of the molecule where the energy was injected, somewhere else, or nowhere at all.

For the canine experiment, we need a hungry dog, a quick hand with the bone, and a quick eye to read the stopwatch. For molecules, it's much harder. Only within the last few years has it been possible to measure the rate of energy movement within a molecule. But chemists now have pulsed lasers giving bursts of light with durations as short as a millionth of a millionth of a second (a "picosecond"). Comparing a chemical change that takes place in one picosecond to a one-second tail-wag delay involves the same speed-up as a 10-second instant replay of all historical events since the pyramids were built.

The alkyl benzenes provide an example. Each of these molecules has a rigid benzene ring at one end and a flexible alkyl group at the other. At room temperature, this flexible "tail" vibrates and bends under thermal excitation. But to act like our hungry dog, the molecules must be cooled to cryogenic temperatures, while avoiding condensation. Supersonic jet expansion makes this possible. When a gas mixture flows through a jet nozzle into a high vacuum, the molecules can be cooled almost to absolute zero. An alkyl

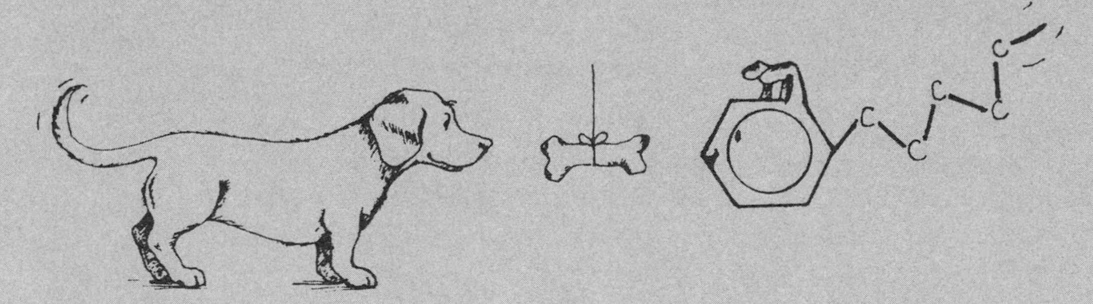

benzene molecule carried along in such a stream loses all its vibrational energy, thus relaxing the molecular tail. Then, the cold molecules intersect a brief pulse of light with color that is absorbed by the benzene ring. With careful "color-tuning," extra vibrational energy can be placed in the head without any vibrational excitation in the tail. Then we must watch the molecule to see how long it takes for the tail to wag. Fluorescence lets us do this.

When a molecule in a vacuum absorbs light, the only way it can get rid of the energy is to reemit light. Such fluorescence can be recorded with a fast-response detection system to give a spectrum that carries a tell-tale pattern showing where the extra energy was at the instant the light was emitted. Those molecules that happen to emit right away after excitation show the molecular head vibrating and the tail still cold. Those that emit later have an emission spectrum that shows that the tail is wagging. In this way, we have learned that the time it takes for the alkyl benzene tail to begin to wag depends on how long the tail is. Surprisingly, the longer the alkyl, the faster the movement out of the ring. The result shows what determines energy flow within molecules (the "density of states"). Such information might one day clarify combustion and help us make fine chemicals out of coal.

IV-A. Control of Chemical Reactions

Ultimately, success in responding to society's needs depends upon the ability to control chemical change, a control made possible by our understanding of chemical reactivity. Today, this understanding is being broadened and deepened at an astonishing pace because of an array of powerful new instrumental techniques. These instruments permit us to pose and answer fundamental questions about how reactions take place, questions that were beyond reach only a decade ago. They account for the recent acceleration of progress in the most basic aspects of chemical change.

MOLECULAR DYNAMICS

Chemistry is the science concerned with the changes that occur around us when one set of chemicals turns into another set of chemicals. Such a change, a chemical reaction, is understood at the atomic level in terms of one set of molecules rearranging into another set of molecules. The study of these rearrangements is called *molecular dynamics* and it encompasses:

- *molecular structure,* the stable geometries of the reactant and product molecules;
- *chemical thermodynamics,* the energy effects that accompany the change; and
- *chemical kinetics,* the time it takes for the reaction to occur.

The theory behind all chemical behavior rests in quantum mechanics. Quantum mechanics is the mathematical description of atoms and molecules devised by Erwin Schroedinger in 1926. It is based upon a wave-picture of the atom that has the potential for explaining all of the chemistry of that atom. Though this has been known for over 50 years, most of the predictive power of quantum mechanics has been out of reach because the mathematics has been too difficult to solve. In contrast, experimental progress on stable molecules has been extremely rapid. This is evident in the fact that chemists have prepared more than 8 million compounds, 95 percent of them since 1965. On the other hand, our understanding of the speed aspects of chemical change has been limited by reaction steps too fast to be observed.

Now a new era has begun. Chemical theory, supported by the power of modern computers, has emerged from empirical modeling. At the same time, we have experimental techniques that open the way to understanding the time dimension of chemical change. *Over the next three decades we will see advances in our understandings of chemical kinetics that will match the advances in molecular structures over the last three decades.*

Fast Chemical Processes

A chemical reaction begins with mixing reactants and ends with formation of final products. In between, there may be a succession of steps, some extremely rapid. To understand the reaction completely, we must clarify all the steps between beginning and end, including identification of all of the intermediate molecules that are involved in the steps.

Fifteen years ago, we could track intermediate molecules only if they hung around at least as long as a millionth of a second. The many interesting studies on this time scale only increased the chemist's curiosities because it became clear that a whole world of processes took place too rapidly to be detected at that limit. Nowhere was that more apparent than in the centuries-old desire to understand combustion, perhaps the most important type of reaction known.

Laser light sources have spectacularly expanded these experimental horizons over the last decade. One of their unique capabilities is to provide short-duration light pulses with which to investigate chemical processes that occur in less than a millionth of a second all the way down to a millionth of a millionth of a second (i.e., down to a picosecond, 10^{-12} sec). At the state of the art, physicists are learning how to shorten these pulses even more; pulses as short as 0.01 picoseconds (10 femtoseconds) have been measured, and kinetic studies are beginning in the 0.1-picosecond range. At one-tenth of a picosecond, frequency accuracy is limited to about 50 cm^{-1} by a fundamental physical principle—the Uncertainty Principle (See Section V-A). These developments imply that chemists can now investigate a reacting mixture on a time scale that is short compared with the lifetime of any intermediate molecular species involved. The exploitation of this remarkable capability has only just begun.

The absorption of visible or ultraviolet light by a molecule adds enough energy to redistribute the bonding electrons, to weaken chemical bonds, and to produce new molecular geometries. The outcome might be a high-energy molecular structure difficult to reach by chemical reactions stimulated by heat. So the excited electronic states reached by absorption of light furnish a new chemical world that we have only begun to understand and put to practical use.

When a molecule absorbs light, it gains energy. One of the ways it can dispose of the energy is to reemit light, generally of a different color than the absorbed light. If this emission occurs quickly, it is called *fluorescence*. "Quickly" can mean anywhere from within a microsecond to a picosecond. The blue light emitted by a Bunsen burner flame and the spectacular display of the Northern Lights are examples of fluorescence. If the light emission occurs more slowly, it is called *phosphorescence*. "Slowly" can mean anywhere from a millisecond to several seconds or even minutes. Some clock dials that glow in the dark and the blue glow of evening ocean tides are examples of phosphorescence.

We have some basic understandings about the differences that cause these two behaviors. When two electrons are shared in a chemical bond, they must have opposite magnetic spins (as expressed in the Pauli Principle). But if absorption of light adds enough energy to move one of these electrons to another part of the molecule, the Pauli Principle no longer limits the electron spins. Then they can be oriented opposed to each other, like two magnets whose fields cancel each other, to give a "singlet" state. But they can also be oriented parallel so that the two magnetic fields add together. This is called a "triplet" state. We have learned to associate fluorescence with light emission processes that begin and end in singlet states. Phosphorescence, however, requires moving from a triplet to a singlet state (or the reverse). Apparently, the need to change the electron spin makes the emission much more difficult, so it occurs more slowly.

There has been a spectacular increase in our ability to clarify what is going on in these excited states since lasers have come into the chemistry laboratory. We can now excite particular states (by control of the laser color, or wavelength), and we can measure the time it takes for reemission to occur (by use of laser pulses of very short duration). Even for the fastest fluorescent processes, we can measure the radiative lifetimes, and by measuring the wavelength of the light emitted (spectral analysis) we can see how rapidly energy moves within the molecule and where it goes. Thus, we are beginning to map and understand the high-energy electronic states of molecules so that they can be used to open new reaction pathways.

Benzophenone is a substance that demonstrates how lasers are being used to probe these high energy states. When benzophenone in ethanol solution absorbs ultraviolet light at a 316-nm wavelength, it reemits light at two different colors, at wavelengths of 410 and 450 nm. If the exciting light (316 nm) is delivered in a laser pulse of 10 picoseconds duration, "prompt" emission is seen at 410 nm, with intensity that decreases with a 50-picosecond half-life. This fluorescence is followed, however, by weaker emission, still at 410 nm, but with a longer half-life (a microsecond). This slower fluorescence disappears at lowered temperatures and is replaced by longer-wavelength phosphorescence at 450 nm with an even longer lifetime (a millisecond).

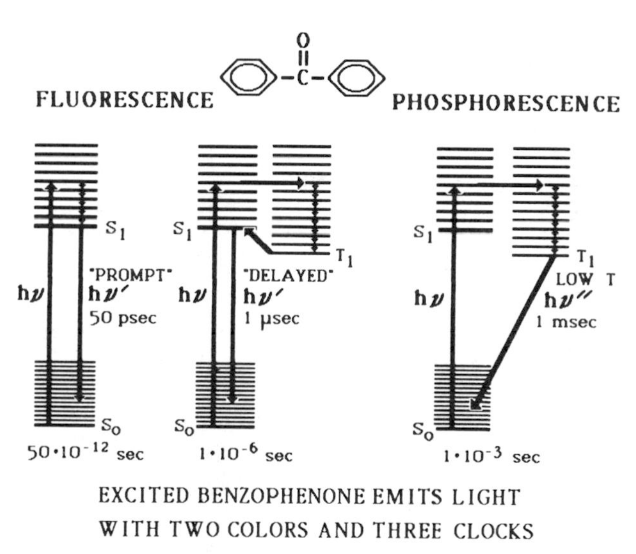

EXCITED BENZOPHENONE EMITS LIGHT WITH TWO COLORS AND THREE CLOCKS

Photochemists have been able to interpret these clues about the excited states of benzophenone. Absorption at 316 nm reaches a singlet state (S_1) but with extra energy placed in the vibrational motions of the benzophenone. This vibrational excitation is lost so quickly in liquids (warming the solvent) that even the "prompt" fluorescence back to the ground state (S_0) occurs at longer wavelengths (410 nm). On the other hand, the low-temperature behavior shows that benzophenone also has an excited triplet state (T_1),that can be reached via S_1. Once occupied, T_1 emits phosphorescent light with the characteristic long lifetime of a triplet-singlet transition ($T_1 \rightarrow S_0$). The temperature dependence of the delayed fluorescence shows that T_1 is lower in energy than S_1 and by how much.

The set of processes clarified here have lifetimes that range from 50 picoseconds to a millisecond, a difference of 20 million. The observations reveal the excited states of benzophenone and the rates of movement between them. These understandings are of extreme significance because they can all be applied to natural photosynthesis, a process scientists would like very much to master. There are many other types of laser-based, real-time studies of rapid chemical reactions now being made, including chemical isomerizations, proton transfers, and photodissociations. Some of the phenomena to follow also depend upon use of short-pulse laser excitation instrumentation.

Energy Transfer and Movement

In all chemical changes, the pathways for energy movement are determining factors. Competition among these pathways determines the product yields, the product state distributions, and the rate at which reaction proceeds. This competition is highly important in stable flame fronts (as in Bunsen burners, jet engines, and rocket engines) and in explosions, shock waves, and photochemical processes.

When two gas phase molecules collide, vibrational energy can be transferred from one molecule to another. Thus, a vibrationally "cold" molecule might be heated up and caused to react or a vibrationally "hot" molecule might be cooled off so it cannot react. These transfers of vibrational energy between and within molecules as a result of collisions between them have long been recognized as central to determining reaction behavior in flames. But progress has been slow because the processes have been too fast to measure. Now a variety of techniques—almost all based on laser methods—has opened the way to providing critical data related to the pathways and rates of energy flow. These data, in turn, furnish a basis for the development of useful theory. As much has been learned about vibrational energy movement in the last 15 years as was learned in the preceding half-century.

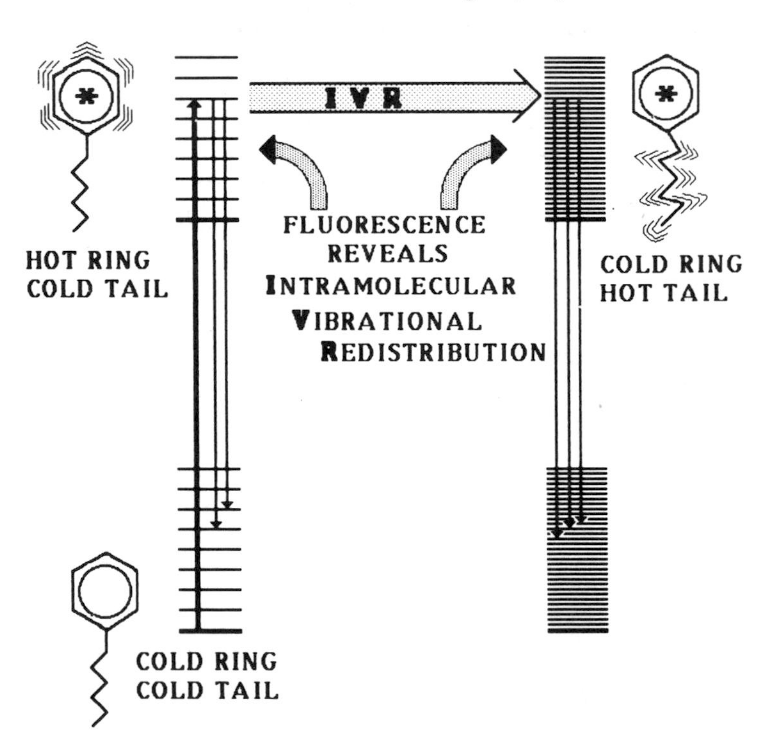

As tuned lasers became available they were used to excite particular vibrations in a molecule. Then, experiments were devised to permit us to watch this carefully placed energy move into other parts of the molecule or into another molecule if collisions occurred. Fluorescence provides one way to follow this energy movement. The light reemitted during fluorescence carries a spectral signature that shows what part of the molecule is vibrating at the moment of emission.

A clear-cut example is provided by recent studies of the alkyl benzenes, $C_6H_5\text{-}(CH_2)_nCH_3$ with n from 1 to 6. This molecule has a structure like that of a tadpole, where n determines the length of its tail. Tuned-laser excitation allows us to deposit prescribed amounts of vibrational energy in the benzene end of the cold molecule (in the head of the tadpole). When this energy is reradiated, its spectral signature displays its vibrational excitation at the instant of radiation. Since this light emission is a time-dependent process, we can monitor the movement of energy from the original location of excitation into the rest of the molecule. This movement in absence of collisions is called Intramolecular Vibrational Redistribution (IVR). Light emitted in the first picoseconds shows that the energy has not yet left the benzene unit where it was absorbed. The time scale for appearance of vibrational excitation in the alkyl tail depends upon the tail length. For $n = 4$, vibrational energy moves out into the tail in 2 to 100 picoseconds. In contrast, for $n = 1$ (ethylbenzene), it is a thousand times slower; it takes 100 nanoseconds or more. Thus, we have direct evidence about the factors that determine IVR energy movement in an isolated molecule.

State-to-State Chemistry

When two gaseous reactants A and B are mixed and react to form products C and D, the outcome is determined by statistical probabilities. The different encounters that may happen between A and B include all the possible energy contents, specific different types of excitation, and all the ways molecules may be oriented in space at the moment of collision. Not all of these collisions are favorable for reaction—most collisions have too little energy, or the energy is in the wrong place, or the collisions are at an awkward geometry. If we are to understand fully the factors that permit chemical reactions to occur, we should control the energy content of each reactant, i.e., control the "state" of each reactant. Then we could systematically vary the amount and type of energy available for reaction. Finally, we would like to see how the available energy is lodged in the products. Such an experiment is called a "state-to-state" study of reaction dynamics, and 20 years ago it was beyond all reach. Now, with modern instrumentation, chemists are realizing this goal.

The earliest efforts, based upon chemiluminescence, revealed a part of the picture: the energy distribution among the products. For example, when a gaseous hydrogen atom and a chlorine molecule react they form hydrogen chloride and a chlorine atom. These reaction products emit infrared light. Analysis of the spectra from that light shows that the energy released in the reaction is not randomly distributed between the final products. Instead, a large fraction of it (39 percent) is initially located in the vibration of the hydrogen chloride product. Discoveries like this won John Polanyi (University of Toronto) a share of the 1986 Nobel Prize in Chemistry. This measurement led directly to the demonstration of the first chemical laser—a laser that derived its energy from the hydrogen/chlorine explosion. Chemical lasers differ from conventional lasers in that the energy to produce their light comes from a chemical reaction instead of an electrical source. These beginnings led to the discovery of dozens of chemical lasers, including two sufficiently powerful to be considered for possible initiation of nuclear fusion (the iodine laser) and for possible military use in the "Star Wars" program (the hydrogen fluoride laser).

"Molecular beams" move even closer toward "state-to-state" investigations. A molecular beam is a stream of molecules produced by a suitably hot oven. A

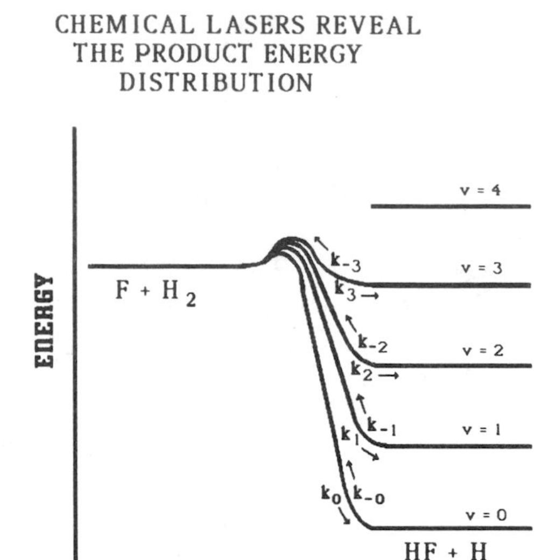

substance is placed in this oven, and when it melts and vaporizes the vapor is directed out a tiny hole to form a unidirectional beam of molecules. Outside the oven the pressure is kept extremely low—so low that no molecular collisions occur. The molecular beam can then be directed toward reactants. In such experiments, the reactants collide at such low pressures—10^{-10} atmospheres—that each reactant molecule has at most one collisional opportunity to react, and the products have none. These sophisticated instruments depend upon ultra-high vacuum equipment, high-intensity supersonic beam sources, sensitive mass spectrometers for detectors, and electronic timing circuitry for time-of-flight measurements. With this incredible control it has become possible to predetermine the energy state of each reactant molecule and then to measure both the probability of a certain reaction and the energy distribution in the products. For bringing such elegant experiments into chemistry, Yuan-Tseh Lee (University of California, Berkeley) and Dudley Herschbach (Harvard University) shared in the 1986 Nobel Prize in Chemistry.

For example, a current study has explained a key reaction in the combustion of ethylene. These molecular beam experiments show that the initial reaction of oxygen atoms with ethylene produces the unexpected short-lived molecule CH_2CHO. With this starting point, calculations have confirmed that a hydrogen atom can be knocked out of an ethylene molecule by a reacting oxygen atom more easily than that atom can be moved about within the molecule. This combustion example illustrates the intimate detail with which we can now hope to understand chemical reactions.

Multiphoton and Multiple Photon Excitation

Photochemistry has traditionally been concerned with what happens when a single photon is absorbed by an atom or a molecule. This productive field accounts for the energy storage in photosynthesis, the ultimate source of all life on this planet. Photochemistry also provides us with new ways to synthesize organic compounds and, through photodissociation, to produce a variety of short-lived molecules that play critical roles in flames and as intermediates in reactions.

Now lasers give us optical powers 10,000 times higher at a given frequency than even the largest flashlamps ever built. Clearly, these devices do not simply extend the boundaries of conventional light sources, they open doors to new processes as molecules interact with such intense photon fields. For example, at normal light intensities, the simultaneous absorption of two photons by a single molecule takes place so rarely that it cannot be detected. However, the probability of this happening increases with the square of the light intensity. Thus, if a laser increases light intensity by a factor of 10,000, then the chance of two-photon absorption increases by four orders of magnitude over the chance of one-photon absorption. This lets us do experiments in which we can prepare molecular states that cannot be reached with a single photon. Furthermore, the total energy absorbed can be enough to produce ions. This opens a new avenue to the chemistry of ions, a field of rapidly rising interest because of the discovery of interstellar ion-molecule reactions and because ions are major species in the plasmas (glow discharges) of nuclear fusion. Two-photon ionization has been used to detect specific molecules in difficult environments, like those found in explosions and in flames. Thus, nitric oxide, NO, which is an ingredient in smog, can be easily measured in a flame by counting the ions produced by a finely tuned laser probe. The probe is tuned so carefully that only the desired molecule, NO, can absorb light energy.

However, the most spectacular instance of multiphoton excitation came with the development of extremely high-power CO_2 infrared lasers. One of the most surprising scientific discoveries of the 1970s was that an isolated molecule whose vibrational absorptions are in close vibrational harmony (near resonance) with the laser frequency could absorb not two or three but dozens and dozens of photons. In a time short compared with collision times, so many photons can be absorbed that chemical bonds can be broken entirely with vibrational excitation. This unpredicted behavior is commonly called *multiple photon excitation* to distinguish it from two-photon (multiphoton) excitation.

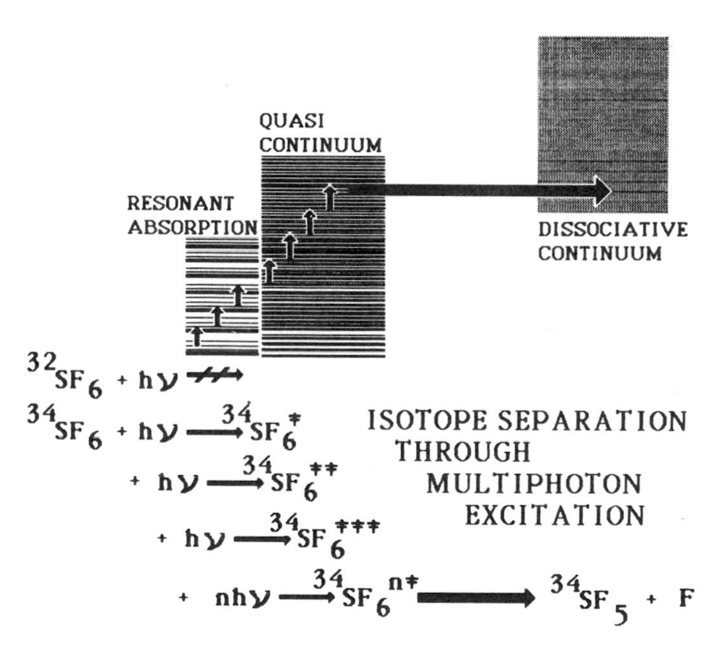

This behavior stimulated a large group of studies on energy flow within excited polyatomic molecules. Many unimolecular breakdowns and rearrangements have been triggered using multiple photon excitation. Yet, the understanding gained from this phenomena may be overshadowed by the importance of its practical uses. Infrared absorption depends upon vibrational movements whose frequencies are quite sensitive to atomic mass. As a result, the tuned laser can be used to break up just those molecules containing particular isotopes, leaving behind the others—a new method for isotope separation. For example, deuterium is present at 0.02 percent in natural hydrogen. Yet, by multiple photon excitation, this tiny percentage can be extracted using trifluoromethane, CF_3H. The process has been shown to have a 10,000-fold preference for exciting CF_3D over CF_3H. This could be of considerable importance as a source of deuterium since "heavy water," D_2O, is used in large quantities in some nuclear reactors.

Even more significant is sulfur isotope separation through excitation of sulfur hexafluoride, SF_6. This gaseous compound gave the first convincing evidence that multiple photon excitation really occurred so rapidly that collisional energy transfer could be avoided. The successful use of SF_6 for sulfur isotope separation could have heavy significance in human history. The gaseous substance that has always been used in the difficult processes used to separate uranium isotopes is uranium hexafluoride, UF_6. Because SF_6 and UF_6 have identical molecular structures, they have similar vibrational patterns. Thus, multiple photon excitation might offer a new and simpler approach to isolation of the uranium isotopes that undergo nuclear fission. It depends, of course, upon finding a sufficiently powerful and efficient laser at the lower frequencies absorbed by UF_6. It will bring more general access to the critical ingredients of nuclear energy and, unfortunately, nuclear bombs as well. The dangers of increased proliferation of nuclear weaponry can only be increased by such access.

Mode-Selective Chemistry

When two molecules collide with each other, the violence of the collision may cause their atoms to rearrange to form two new molecules (i.e., a reaction may occur). Such an outcome almost always requires that the molecular collision involve some minimum energy—enough to break some of the bonds in the reactants in order to form the new bonds in the products. This minimum energy, the *activation energy,* determines the rate of the reaction and it accounts for the dramatic effect of temperature on reaction rates.

However, the question of whether a reaction will result from a molecular collision turns out to involve more than just whether there was enough energy. There is also a question of whether the collisional energy is in the right form. To understand what this means, consider a bedspring thrown against a wall. As it bounces off, it has energy of several types. It will be moving through space, which is energy of the old-fashioned kinetic en-

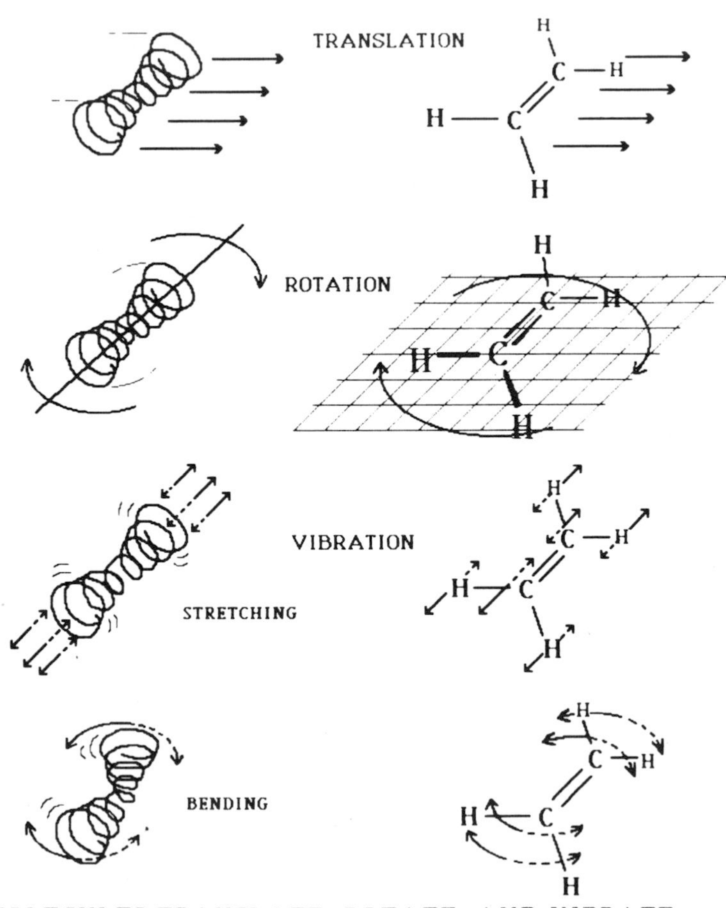

MOLECULES TRANSLATE, ROTATE, AND VIBRATE LIKE A BEDSPRING THROWN AGAINST A WALL ergy type. This is called *translational* energy. In addition, the bedspring will be tumbling in space. This, too, is a form of kinetic energy called *rotational* energy. Then, the spring will be twisting and vibrating to and fro. This *vibrational* energy consists of both potential and kinetic energy. Molecules carry energy in exactly the same ways. Whether we are talking of bedsprings or molecules, the directions of translational motion, the axes of rotation, and the spring connections (in molecules, the bonds) are called degrees of freedom. The total energy in a collision is the sum of all of these forms of energy—translational, rotational, and vibrational—from both molecules.

Chemists have long wondered whether it matters which degree of freedom carries the energy in a reactive collision. If all of the energy is in translational energy, the molecules are near each other only a short time. If the same amount of energy is brought to the collision mostly as vibration, the molecules move toward each other slowly, but now the bonds that must be broken are vibrating rapidly. Is this more or less effective?

Only since chemists have acquired lasers has it been possible to seek an answer to this fundamental question. With high-power, sharply tunable lasers, we can excite one particular degree of freedom for many molecules in a bulk sample. As long as this situation persists, such molecules react as if this particular degree of freedom is at a very high temperature while all the rest of the molecular degrees of freedom are cold. The chemistry of such molecules has the potential to show us the importance of that particular degree of freedom in causing reaction. This is called *mode-selective* chemistry.

Both unimolecular reactions and molecular beam studies of bimolecular (two-molecule) reactions escape this problem. Unimolecular reactions involve only one molecule, so collisions are not required. At sufficiently low pressures, the effects of selective excitation on reactivity can be studied. The beam experiments sidestep the problem by giving each excited molecule only one chance for collision and by noticing only those collisions which result in a reaction. Nevertheless, mode-selective reactions are not readily coming from such experiments. Apparently the problem is that vibrational redistribution takes place within molecules even without collisions (IVR). This problem is of such basic importance to molecular dynamics that it will be one of the most important study topics for the next decade.

There is, however, evidence for two-molecule mode-selective chemistry in certain solid inert-gas environments. In this situation the environment is so cold (10K) that the reactive molecules are held immobilized. They are "frozen" in a prolonged, cold collision and rotational movement has been halted. For example, fluorine, F_2, and ethylene, C_2H_4, suspended in solid argon at 10K do not react until one of the vibrational motions of ethylene is excited with a resonantly tuned laser. Then it is found that the most efficient vibrational motions are those that distort the molecular planarity. This is plausible because this type of distortion changes the molecular shape "toward" the nonplanar, ethane-like structure of the final product.

Theoretical Calculations of Reaction Surfaces

Schroedinger's wave equations of quantum mechanics have long been known to describe all chemical events. Yet quantum mechanics has been used in chemistry

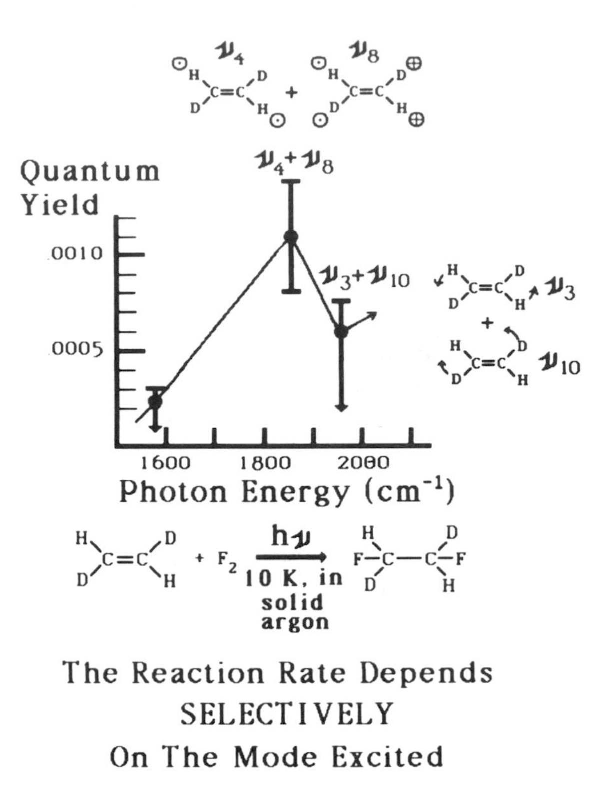

The Reaction Rate Depends
SELECTIVELY
On The Mode Excited

only qualitatively, or with severe approximations, because the equations have been too difficult to solve except for the simplest molecules (like H_2^+ and H_2). Modern computers are changing this. With today's computers, the structure and stability of any molecular compound with up to three first-row atoms (carbon, nitrogen, oxygen, fluorine) plus various numbers of hydrogen atoms can be calculated without approximations. This capability opens to the chemist many situations not readily available to experimental measurement. Short-lived reaction intermediates, excited states, and even energy barriers to reaction can now be understood, at least for small polyatomic molecules.

NEW REACTION PATHWAYS

Our increasing understanding and control of chemical reactivity is providing us with new reaction pathways in synthetic chemistry that are sure to lead to new products and new processes. Again, powerful instrumental techniques play a central role. Synthetic chemists are now able to identify rapidly and accurately the composition and structure of reaction products. This greatly speeds up the development of new synthetic approaches.

Organic Chemistry

Organic chemistry today involves three areas of emphasis. The first area concerns the isolation, characterization, and structural determination of substances from nature. New natural products are thus identified—alkaloids and terpenes from plants, antibiotics from microorganisms and fungi, peptides and polynucleotides from animal and human sources. Chromatography permits purification and characterization of substances present in only trace amounts from complex mixtures. Thus, workers in pheromone chemistry regularly separate out microgram amounts of these biologically potent molecules. The next challenge lies in determination of their composition, overall structure, and three-dimensional stereostructure. Here nuclear magnetic resonance, mass spectroscopy, and X-ray crystallography fill essential roles. Using proton NMR, only 100 nanograms (10^{-7} grams) of a substance can provide crucial information about the number and types of molecular linkages. With only 100-picogram (10^{-10} grams) amounts, mass spectrometry contributes by furnishing precise molecular weights up to 13,000 and, through the fragmentation patterns, providing revealing clues to substructures. Then, if 10 micrograms or more of a crystalline material become available, every stereochemical detail of structure is displayed through X-ray spectroscopy, such as interatomic distances, bond angles, and any mirror-image relationships present.

Physical organic chemistry is the second major area of emphasis; it seeks to relate changes in physical, chemical, and spectroscopic behavior of organic compounds to changes in molecular structure. It deals with the detailed pathways by which reactants become products—it predicts what intermediate species or structures are present and determines how the reaction pathway is influenced by solvent environment, catalysts, temperature, and pH. It provides a theoretical framework with which to predict behavior and useful synthetic routes toward materials not yet known.

Synthesis, the third area, is a process of inventive strategy. Two contemporary challenges it faces are to add to the availability of useful natural products and to synthesize new and useful substances not found in nature. Thus, thousands of pounds of ascorbic acid (Vitamin C) are synthesized annually at purities suitable for human consumption so that society can have an abundant supply of this healthful substance. Smaller amounts of 5-fluorouracil, an artificial drug that is extremely effective in treating certain cancers, are synthesized for prescription use.

Meeting such challenges has required creative evolution of the philosophy of organic synthesis. Only a few decades ago, synthetic strategies were based on clever choices from a set of already known reactions. Like moves in a chess match, the range of logical reactions was defined in advance. With the development of mechanistic reasoning, in which reactions are classified according to the mechanism by which they work, it has now become possible to invent new reactions for specific synthetic goals. Organic synthesis has had powerful success using this reasoning process.

At the same time, there has been an imaginative and fruitful expansion in the settings in which reactions are conducted. An example is solid-phase peptide synthesis in which amino acids are added in sequence to produce a desired peptide. This is all carried out under covalent attachment to an insoluble polymer support. Such polymer-bound peptide synthesis is already being applied to synthesis of important hormones and bioregulatory peptide substances. A quite different dimension now being explored is pressure. Equilibrium can be shifted to favor products with specially compact structures, and activation barriers can sometimes be affected to selectively speed up a desired process. A step in the synthesis of alkavinone, used in the synthesis of certain drugs, provides an example. At 15,000 atmospheres and room temperature, quinone will react with a correctly structured butadiene ester to form the desired bicycle ester. This process completely avoids undesired alternative structures that would be obtained if high temperature were used instead of high pressure as the control variable.

Nowhere has progress had more far-reaching significance than in our growing ability to control molecular complexities in the third dimension. This frontier, stereochemistry, can be divided into issues of surface shape (topology) and

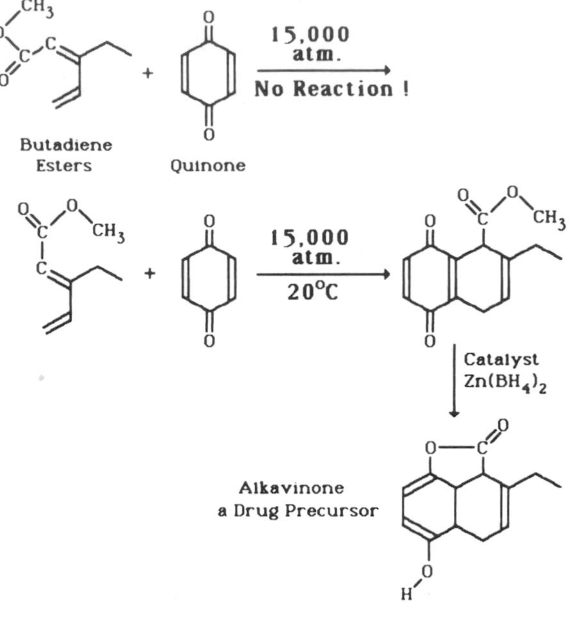

RAISING PRESSURE CAN SELECTIVELY SPEED UP A DESIRED REACTION

"handedness," the first being called "relative" stereochemistry and the second "absolute" stereochemistry. The production of a particular molecular topology already requires artful control of molecular relationships in space during reactions. However, this spatial control does not usually extend to relationships that differ only in a mirror-image sense (in handedness or chirality). When right- and left-handed molecular structures are possible, most chemical reactions will produce a mixture of the two.

Of course, a left-handed glove will not fit a right hand, so it cannot serve the function of a right-handed glove. It is the same in nature, where this "handedness" aspect of molecular structure assumes critical importance. Biological molecules must have proper topological conformation (relative stereochemistry), but for them to be functional, nature also insists upon a particular handedness (absolute stereochemistry). A molecular "right-handed" glove can play a crucial role in a biological reaction while its "left-handed" counterpart will be totally ineffective or, worse, may introduce undesired chemistry.

Though stereochemistry has been recognized for almost a century, major advances have been made within the last decade. In one technique, an extra molecular fragment of defined handedness is attached to a reactant. This "chiral auxiliary," properly placed, can govern the handedness of products that come from that reactant. The auxiliary is then removed from the product and reused in another cycle. Certain stereospecific propionates are synthesized in this manner and later used as precursors for making other biological molecules. Even more exciting is the use of asymmetric (chiral) catalysts to direct the handedness of the products. Asymmetric reduction is now a key step in the industrial synthesis of the important anti-Parkinson's disease agent L-dopa. A more general application has been the development of asymmetric epoxidation through asymmetric catalysis. When an oxygen atom is inserted equally into either face of a carbon-carbon double bond to produce an epoxide, two mirror-image-related products result. With inexpensive and recyclable chiral catalysts, it is now possible to prepare whichever one of these two stereoisomers is needed. The resulting stereospecific epoxide can be used in many synthetic pathways, carrying along and preserving the left-right character. In a major application of this method, all of the six-carbon sugars that occur naturally have been synthesized perfectly with nature's preferred handedness.

The significance of these new frontiers of organic synthesis can be seen in health applications. For example, the prostaglandins are a family of fatty acids that contain 20 carbons and include a five-membered ring. They seem to affect the action of hormones and, thereby, have important effects on the body, ranging from regulation of blood flow to stimulating childbirth. We now know the structures of several of them, and we understand both their biosynthesis and their laboratory synthesis. Their synthesis in nature begins with polyunsaturated fatty acids that are a natural requirement in the diets of mammals.

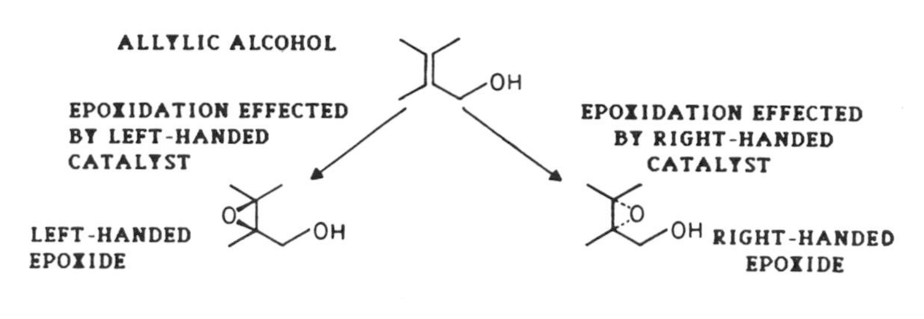

NOW THE CATALYST CAN FIX THE DESIRED "HANDEDNESS" OF THE PRODUCT

Surprisingly, these same polyunsaturated fatty acids are useful starting points for synthesis of another family of molecules, the leukotrienes, that have great potential for a variety of drug uses, including control of asthma. The ability of chemists to synthesize chemically modified prostaglandins and leukotrienes for biological testing is a triumph of synthetic organic chemistry. Equally far-reaching accomplishments are connected with the synthesis of safe compounds for birth control (e.g., 19-norsteroids and 18-homosteroids), new antibiotics (e.g., modified cephalosporins and thienamycins), and drugs for hypertension (Aldomet®) and ulcers (cymetidine, Tagamet®).

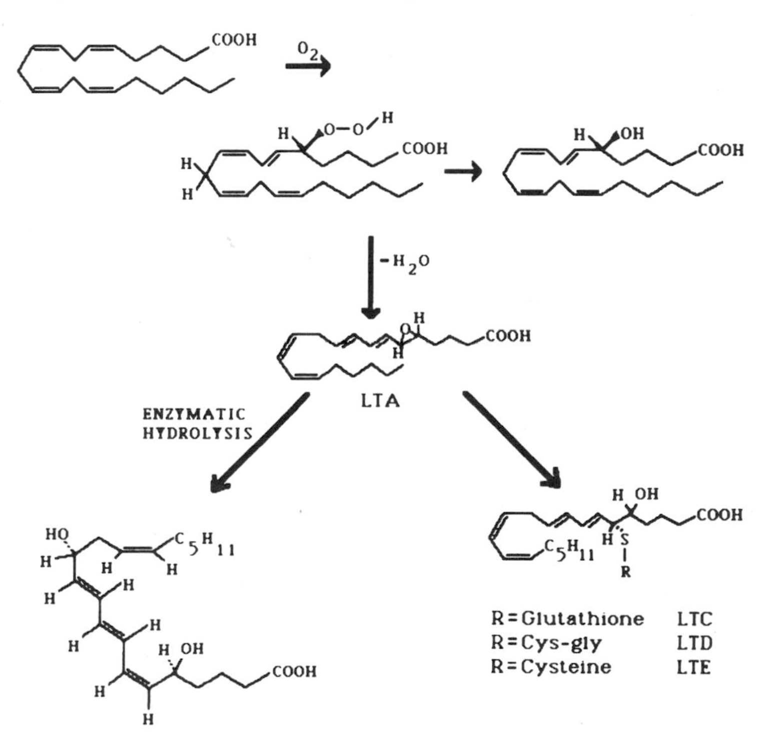

BIOSYNTHESIS OF LEUKOTRIENES

Inorganic Chemistry

There is great intellectual excitement now in inorganic chemistry, much of it at the interfaces with sister disciplines: organometallic chemistry, bioinorganic chemistry, solid-state chemistry, biogeochemistry, and other overlapping fields. For example, there is growing awareness of the crucial roles played by inorganic elements in biological systems. Living things, far from being totally organic, depend sensitively on metal ions drawn from throughout the Periodic Table. Particular metal ions have critical roles in such essential life processes as transport and consumption of oxygen (iron in hemoglobin), absorption and conversion of solar energy (magnesium in chlorophyll, manganese in photosystem II, iron in ferrodoxin, and copper in plastocyanine), communication through electrical signals between cells (calcium, potassium in nerve cells), muscle contraction (calcium), enzyme catalysis (cobalt in Vitamin B_{12}). This has led to a burst of research activity in the inorganic chemistry of biological systems. We are beginning to learn about the structures that surround the metal atoms and how these structures enable the metal atoms to react with such sensitivity to changes in pH, oxygen pressure, and to electron donors and acceptors.

The answers to many of these crucial questions will come from the active field of organometallic chemistry. The molecule makers of the field use the latest spectroscopic and X-ray diffraction techniques to unravel unexpected bonding patterns and structures. An example is provided by the large family of "sandwich" compounds that stemmed from the discovery of ferrocene, a compound in which an iron atom is placed between two flat C_5H_5 rings.

Metal atoms can bond through conventional electron pair sharing, as in the gaseous molecule $TiCl_4$ which is used in the manufacture of pure titanium metal for aircraft construction. In addition, because of the many vacant *d* orbitals in the elements from the middle of the Periodic Table (the "transition" elements), metals can act as electron acceptors (Lewis acids). Thus, in the compound $Fe(CO)_5$, iron pentacarbonyl, each carbon monoxide molecule donates a pair of electrons to a vacant valence orbital of the iron atom to form a stable structure shaped like two pyramids. The carbon monoxide molecule, and any groups that take their places, are called "ligands." Some or all of them can be replaced by other electron donors (Lewis bases) such as nitric oxide, NO; ammonia, NH_3; halide ions, F^-, Cl^-, Br^-; water, H_2O; cyanide ion, CN^-; and many more. A wide range of compounds results. For some metal atoms, even nitrogen, N_2, can be convinced to take a ligand position, thus making it more susceptible to reaction ("activation"). That is one of the ways in which organometallic chemists are attempting to discover new catalysts to "fix" nitrogen (convert N_2 into NH_3 for fertilizer use).

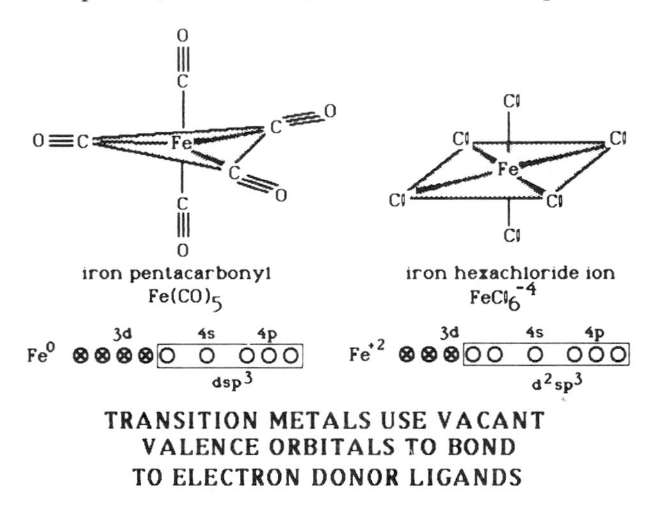

TRANSITION METALS USE VACANT VALENCE ORBITALS TO BOND TO ELECTRON DONOR LIGANDS

The key to further progress is to understand the reaction mechanisms of these molecules. Through clever choice of attached groups (ligands) and control of metal atom oxidation states, organometallic chemists have prepared remarkable compounds that show selective reactivity toward molecules previously thought too inert to participate in useful chemical transformations. For example, a saturated hydrocarbon is one with no carbon-carbon double or triple bonds, so it is relatively unreactive. Now researchers have discovered rhodium and iridium compounds with phosphine (PR_3) or carbonyl and pentamethyl cyclopentadienyl ligands that can attack the CH bonds of methane and cyclopropane. The challenge is to couple this important new reaction with other well-known transformations so that saturated hydrocarbons can be used as feedstocks. The direct conversion of methane to methanol by such a process could have a tremendous impact on the world energy situation.

In quite another direction, recent experimental developments permit us to study in the gas phase the weakly bound cluster complexes called "van der Waals" molecules. These are clusters made up of two or more molecules, all of which have completely satisfied bonding situations. The remaining interactions that such molecules exert toward one another are much weaker than normal chemical bonds. Nevertheless, these interactions are extremely important; such "van der Waals" forces account for deviations from ideal gas behavior, condensation of gases to liquids, and solubility.

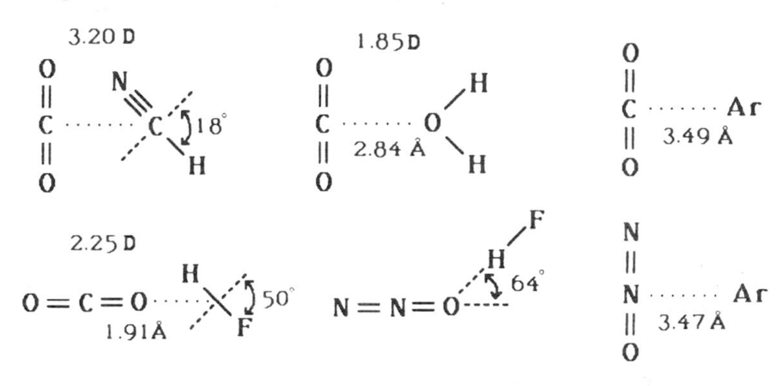

—VAN DER WAALS MOLECULES—
THE WEAK INTERACTIONS THAT GOVERN SOLUBILITY, GAS IMPERFECTION, LIQUEFACTION

Such complexes can now be prepared and studied spectroscopically in cryogenic (low-temperature) matrices and under molecular beam conditions using supersonic jet cooling. These techniques have provided a wealth of information, including molecular geometry, vibrational amplitudes, dipole moments, and ease of energy movement from one part of the complex to another. Information like this is important to the development of detailed theories of reaction rates and the prediction of reaction pathways. Further studies should help to explain such phenomena as condensation, solubility, and adsorption.

At the solid-state/inorganic chemistry intersection is the opening field of composite structures. A composite is made up of two or more materials used together to take advantage of some of the properties of each. Multilayered ceramics for interconnections between semiconductor chips are now being fabricated as well as nonmetallic electrical conducting substances composed of alternating layers. Another new class of materials of considerable interest is the ultrafine filamentary composites. Filaments smaller than a human hair (500-1,000 Å thick) are uniformly distributed throughout another material, which leads to dramatic changes in material properties. The challenge for the future will be to obtain a full understanding of such material interactions so we can design and synthesize new materials with properties to order.

Selective Pathways in Organic Synthesis

Selectivity is the key challenge to the organic chemist—to make a precise structural change in a single desired product molecule. The different reactivity in each bond type must be recognized (*chemoselectivity*), reactants must be brought

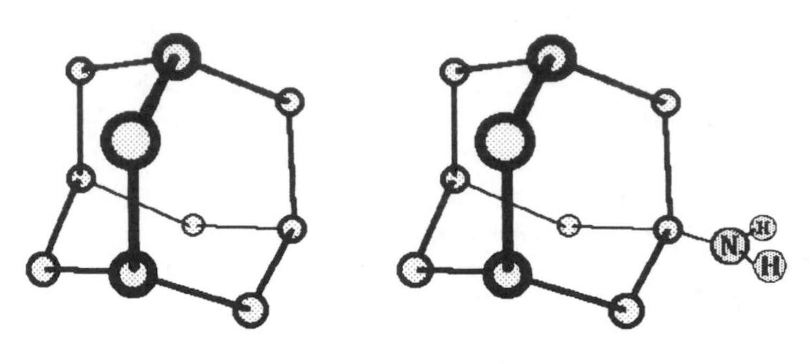

ADAMANTANE
LAB CURIOSITY

ADAMANTADINE
ANTIVIRAL AGENT

together in proper orientation (*regioselectivity*), and the desired three-dimensional spatial relations must be obtained (*stereoselectivity*). The degree to which this type of control can be achieved is shown in the synthesis of the substance adamantane, $C_{10}H_{14}$. This unique molecule resembles in structure a 10-atom "chip" off a diamond crystal. In a laborious synthesis, it was finally produced in a many-step process, but in only 2.4 percent yield. Recent research in the synthesis of polycyclic hydrocarbons now allows production of adamantane in one step in 75 percent yield. Then a surprise practical payoff came when it was discovered that adding a single amine group to adamantane gives adamantadine (1-amino-adamantane), which is an antiviral agent, a preventive drug for influenza, and a drug to combat Parkinson's disease.

Cycloaddition to make five-membered rings becomes important for a wide range of applications ranging from novel electrical conductors to pharmaceuticals (e.g., antibiotics and anticancer compounds). An example is the ring closed by a rhodium catalyst to form a critical precursor to thienamycin. In this case, the five-membered ring contains a nitrogen atom. The final product proves to be a relative to penicillin and an important drug in the battle against infectious diseases.

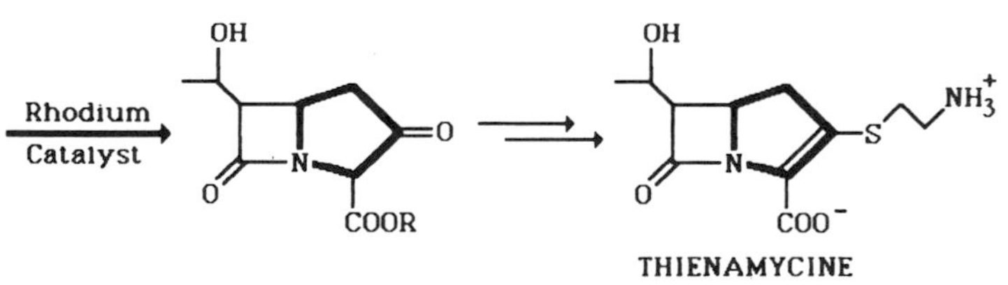

CATALYTIC CLOSURE OF FIVE-MEMBERED RINGS

At another extreme, large ring compounds have been exceptionally difficult to synthesize. Their structures are complicated by functionally important left/right-handed structural geometries (chiral centers). Their wide-ranging biological properties—from pleasant fragrances for perfumes to antifungal, antitumor, and antibiotic activities—make large ring synthesis a useful and interesting challenge. An example is erythromycin, $C_{37}H_{68}O_{12}N$, which can be shaped into 262,144 different structures derived from the many possible ways to couple the right- and left-handedness at chiral centers ($2^{18} = 262{,}144$). Twenty-five years

ERYTHROMYCIN
ONCE CONSIDERED "HOPELESSLY COMPLEX"

ago, this compound was judged to be "hopelessly complex" by R. B. Woodward, who won the Nobel Prize for synthesizing molecules as complex as quinine and Vitamin B_{12}. Today we can hope to work to such a goal, in part because of the development of specially designed templates that bring together the end atoms of a 14-atom chain to form a 14-membered ring. This provides the structural framework of erythromycin, and it has already resulted in the synthesis of a number of ingredients of musk, a scented compound used by animals to communicate and by humans to make perfume.

Crossing Inorganic/Organic Boundaries

As indicated earlier, the traditional line between organic and inorganic chemists is disappearing as the list of fascinating metal-organic compounds continues to grow. Furthermore, research in developing new inorganic substances has provided a surprising reward in their frequent applicability in organic synthesis. The borohydrides provide an example. These are reactive compounds of boron and hydrogen that are electron-deficient from a bonding point of view. But these borohydrides have proved to be valuable as selective, mild reducing agents in organic synthesis. Silicon and transition metal organometallic compounds give other examples. Silicon compounds, for example, are used to fold a long molecular reactant precisely as needed to synthesize the molecule cortisone. Now this valuable medicine can be made in fewer than 20 steps, at a yield 1,000 times higher than was achieved in the earlier, 50-step process.

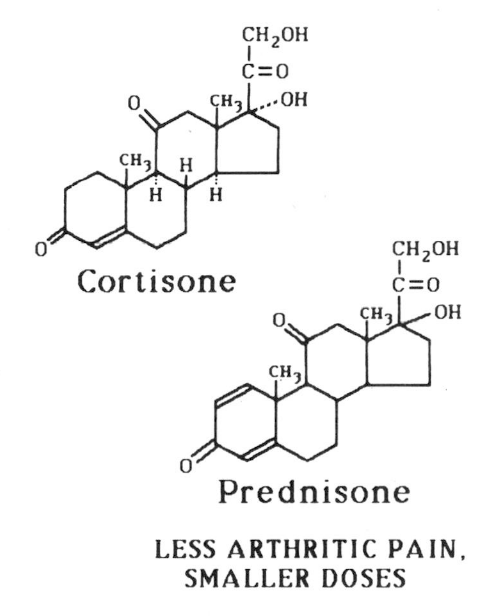

Cortisone is well known in the treatment of arthritis. Unfortunately, experience showed that relief could be temporary and that continued use of cortisone had undesired side effects. These developments made the new silicon-assisted synthetic routes all the more valuable. Several cortisone analogs were prepared and tested for their medical effectiveness. One such product, prednisone, is more effective than cortisone, even when used in much smaller doses, with the result that side effects are much reduced.

Organometallic compounds furnish important intermediate steps in many organic reactions. Organometallics are electron-rich, and because of this, nature accomplishes a lot of its electron transfer through these compounds. Organometallics are easily oxidized by both inorganic oxidants and organic electron acceptors in solution and at electrode surfaces. It has been important to establish how these compounds make and break carbon-to-metal bonds rapidly, selectively, and with stereospecificity. Recent theoretical advances have been based upon the closeness of approach of the reactants at the time of electron transfer. In this picture, each reactant is considered to consist of an "inner sphere," occupied by the electron donor or acceptor (the metal atom), and an "outer sphere," occupied by the ligands. Electron transfer reactions are classified according to the extent of penetration of these inner and outer spheres.

Pathways Using Light as a Reagent

Another promising chemical pathway is connected with the use of photons in chemical synthesis. Many natural products and complex molecules of medical importance involve high-energy ("strained") molecular structures. "Strained" molecules are those with uncomfortable or unusual bond angles. In traditional synthetic procedures, the aggressive reagents needed to force molecular reagents into uncomfortable geometry tend to threaten the fragile product. Photochemistry has been remarkably successful in avoiding this difficulty.

The reason for this success is that absorption of light can change the chemistry of a molecule dramatically. After excitation, familiar atoms can have unexpected ideas about what constitutes a comfortable bond angle; functional groups can have drastically different reactivities; acid dissociation constants can change by 5 to 10 orders of magnitude; ease of oxidation-reduction can be drastically altered; and stable structures can be made reactive. The energy absorbed by the molecule puts its chemistry on a high-energy "hypersurface" whose reactive terrain can be totally unlike the ground state surface below, the one that chemists know so well.

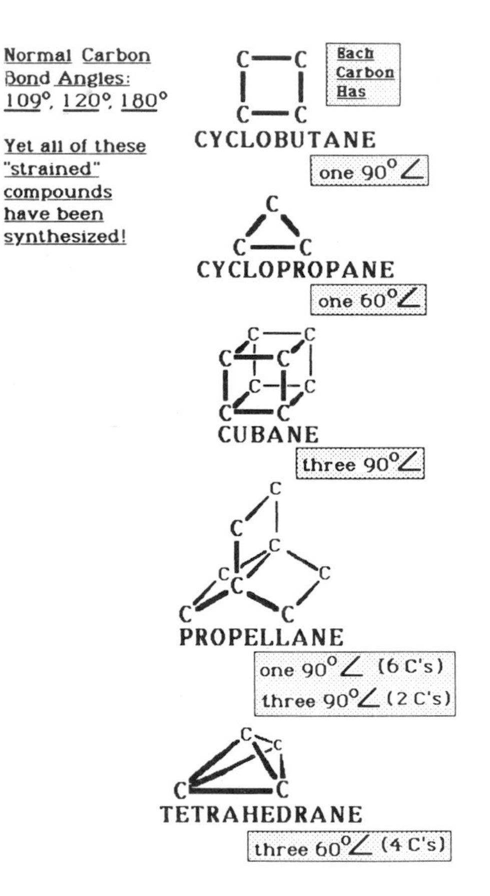

Many examples can be given to illustrate the possibilities. Most dramatic are those that involve cyclic structures that require unusual ("strained") bond angles around carbon. Thus, rings containing three or four carbon atoms are relatively unstable and, hence, difficult to synthesize. At first, they were looked for just because they were chemical oddities. Now we know that many biologically active molecules or their synthetic precursors contain such strained rings as essential structural elements, so their synthesis has assumed great practical importance. These unusual, energy-rich structures are natural targets for photon-assisted synthesis. The photon provides extra energy, and it places the reaction on a hypersurface where uncommon bond angles can be the preferred geometry. Using these principles, chemists have made many molecules of bizarre structure. Appropriately named *cubane* is an example: eight carbon atoms are placed symmetrically at the corners of a perfect cube. Once formed, the molecule is surprisingly unreactive. *Propellane* also involves eight carbon atoms, now in a structure made up of three squares sharing a side. Even more amazing is the family of *tetrahedranes* whose central structural element looks like a three-sided pyramid. Each corner carbon atom is simultaneously bound to three others at 60° angles to form four interlinked, three-membered rings.

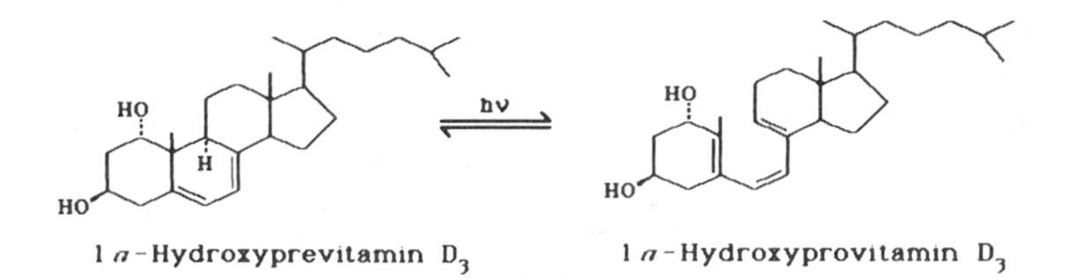

TUNED LASER IRRADIATION DOUBLES THE EFFICIENCY OF THIS STEP TOWARD VITAMIN-D_3

As mentioned above, these photochemical syntheses have proved to be much more than an intellectual chemical chess game. All these syntheses store energy in chemical bonds (the reactions are endothermic). The energy can be recovered later for its own use or to energize subsequent synthetic steps to form other desired, energy-rich molecules. Among the important biological molecules already prepared photochemically are the alkaloid atisine, several mycine antibiotics, and precursors of Vitamin D_3.

To take advantage of the benefits offered by light-assisted pathways, chemists need to become as familiar with the energy geography of the multidimensional reaction hypersurfaces as they are with the ground reaction surfaces upon which stable molecules react. Lasers will be a powerful aid in this exploration. Already it is known that a 1 percent change in the wavelength for the exciting light (from 3,025 to 3,000 Å) can double the yield in synthesis of provitamin D_3 (a precursor to Vitamin D_3). In the formation of the hormone mentioned above, the combination of wavelength control (3,000 Å) and low temperature (−21°C) can quadruple the product yield.

SUPPLEMENTARY READING

Chemical & Engineering News

"Laser Vaporization of Graphite Gives Stable 60 Carbon Molecule" by R.M. Baum (C.&E.N. staff), vol. 63, pp. 20-22, Dec. 23, 1985.

"Chiral Boranes Could Launch Third Generation of Organic Synthesis" by S. Stinson (C.&E.N. staff), vol. 63, pp. 22-23, Aug. 5, 1985.

"Work on Polymer Models of Enzymes Forges On" (C.&E.N. staff), vol. 63, May 27, 1985.

"Inorganic Macromolecules" by H.R. Allcock, vol. 63, pp. 22-37, Mar. 18, 1985.

"Method Synthesizes Chiral Boranes in 100% Optical Purity" by S. Stinson (C.&E.N. staff), vol. 62, pp. 28-29, Mar. 26, 1984.

"Technique Allows High Resolution Spectroscopy of Molecular Ions" by R.M. Baum (C.&E.N. staff), vol. 62, pp. 34-35, Feb. 20, 1984.

"Selective Laser Excitation Promotes Reaction" (C.&E.N. staff), vol. 61, pp. 25-26, April 11, 1983.

"C_1 Chemistry Spurs Cluster Catalyst Work" by J. Haggin (C.&E.N. staff), vol. 60, pp. 13-21, Feb. 9, 1982.

Science

"Molecular Beam Studies of Elementary Chemical Processes" (Nobel Prize Address) by Y.-T. Lee, vol. 236, pp. 793-798, May 15, 1987.

Metals and DNA: Molecular Left-Handed Complements" by J.K. Barton, vol. 233, pp. 727-734, Aug. 15, 1986.

"Methylene: A Paradigm for Computational Quantum Chemistry" by H.F. Schaeffer III, vol. 231, pp. 1100-1107, Mar. 7, 1986.

"Theory and Modeling of Stereo-selective Organic Reactions" by K.N. Houk et al., vol. 231, pp. 1108-1115, Mar. 7, 1986.

"Selenium in Organic Synthesis" by D. Liotta and R. Monahan III, vol. 231, pp. 356-361, Jan. 24, 1986.

Scientific American

"Predicting Chemistry from (Molecular) Topology" by D.H. Ronway, vol. 255, pp. 40-47, September 1986.

"Quasicrystals" by D.R. Nelson, vol. 255, pp. 43-51, August 1986.

Jack and the Soybean Stalk

Perhaps a modern explanation for the amazing size of Jack's fairytale beanstalk can be found in *brassinolide*. This remarkable chemical is an extremely effective plant hormone that can double the growth of food plants, by both cell elongation and cell division. Only recently have chemists been able to isolate, identify, and then synthesize this valuable substance so that it can be used to increase the world's food supply.

Plant hormones have already revolutionized agriculture. They allow us to coerce cotton plants to release their cotton balls at harvest time, command fruit trees to cling to their fruit, induce Christmas trees to keep their needles, and order stored potatoes not to sprout. *Brassinolide* now can add to this list, and it is active in quantities of less than one-billionth of an ounce!

Chemists play a crucial role along the long and arduous research road from discovery to use of a new plant hormone. For example, *brassinolide* is found in minute quantities in the pollen of the rape plant (*Brassica rapus* L.). To isolate enough chemical to study, researchers laboriously collected pollen brushed off the legs of bees who had been cavorting in the rape plants. From 500 pounds of pollen so gathered, chemists were able to extract only 15 milligrams of *brassinolide,* an amount as small as a grain of sand. From this they were able to grow a single tiny crystal, so that a chemical crystallographer could analyze the molecular structure with X-ray diffraction. Just as X-rays penetrate an arm to reveal broken bones, they penetrate a crystal and reveal the geometrical arrangement of the atoms in *brassinolide*. The chemists were surprised to discover an unprecedented seven-atom ring within the molecule, a feature that must be essential to the function of this beneficial compound. With this key information, synthetic chemists have now made several close relatives of *brassinolide,* and agricultural scientists are evaluating them in greenhouse production of potatoes, soybeans, and other vegetables.

This advance involved the knowhow and interaction of plant and insect physiologists, organic chemists, and chemical crystallographers from many different laboratories. It shows that mental effort is as good as magic beans. Maybe better, Jack!

IV-B. Dealing with Molecular Complexity

As detailed in earlier sections, natural products are enormously useful in meeting society's needs. These chemical substances include regulators of plants and insect growth, agents for communication among insects, pesticides, antibiotics, vitamins, drugs for cardiovascular and central nervous system diseases, and anticancer agents. As we develop these products, chemistry becomes a key science at every stage: natural products must be detected, chemically isolated, structurally characterized, and then synthesized as a final proof of structure. Chemical synthesis also provides sufficient amounts of important natural substances needed for biological testing.

Chemical synthesis can also improve upon what nature has provided. Many natural products have useful biological properties, even though they are not ideal for our needs. For example, natural thienamycin has excellent antibiotic properties, but the molecule is unstable and therefore unsuitable for use in human medicine. A synthetic chemical substitute has provided a stable molecule which promises much as an agent for fighting infectious disease. Thus, synthetic chemists have been able to follow a lead provided by a natural product to design and prepare a new molecule with even better biological and chemical properties.

As was emphasized in the discussion of biotechnology, our understanding of macromolecules has provided new insights into their function in biological systems. These new insights have come from structural studies, synthetic alterations, increased understanding of the relationship between molecular structure and function, and the techniques of molecular genetics.

SYNTHESIS AND BIOSYNTHESIS

Modern synthetic techniques now provide access to molecules of complexity and structural specificity that were completely out of reach two decades ago. Synthesis of tailored peptides and nucleic acids of substantial size—molecules widely useful in molecular biology and biotechnology—has become routine. At the same time, our ability to understand and influence the synthetic processes of living organisms is advancing rapidly. All of this is based upon our impressive and growing power in synthetic and biosynthetic chemistry.

Synthesis of Natural Products

Over the last two decades, the synthesis of natural products has consistently advanced to new levels of molecular complexity. Chemists are now addressing a major challenge of organic chemistry, the synthesis of only one desired form of a mirror-image pair. In nature, many biological molecules can take different geometric forms that are mirror images of each other. Each form is called a stereoisomer, and often only one form is biologically useful. Every carbon atom that has four different groups bonded to it gives rise to mirror-image pairs. Such a carbon is called a chiral atom or a chiral center. The synthesis of the polyether antibiotics is a prime example of the way in which chemists have met the challenge of stereoisomers. Monensin, a compound produced by a strain of bacteria called

Streptomyces cinnamonensis, is perhaps the best-known example from among a group of about 50 naturally occurring polyether antibiotics. Three polyether antibiotics (monensin, lasalocid, and salinomycin) are currently in use for control of infectious parasitic disease in the poultry industry (coccidiosis). Monensin has an American market of about 50 million dollars annually.

Monensin presents a huge challenge to synthetic chemists: 17 chiral centers are present on the backbone of 26 carbon atoms, which means that, in principle, 2^{17} or 131,072 different stereoisomers exist for the antibiotic. Thus, to achieve the synthesis of monensin, it is essential to have a high degree of stereoselectivity.

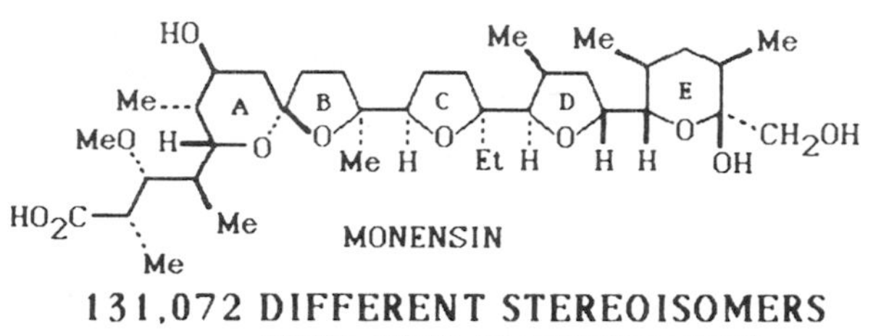

The successful total synthesis of monensin and its structural relatives (lasalocid, salinomycin, and narasin) involved revolutionary breakthroughs. Until these achievements, it was uncertain that a stereo-controlled reaction could effectively be realized in flexible noncyclic molecules. Encouraged by these results, chemists have now extended this approach to the synthesis of another group of antibiotics known as ansamycins. However, the most dramatic developments have been made in the chemistry of *palytoxin*.

Palytoxin, a toxic substance isolated from marine soft corals of the genus *Palythoa,* is one of the most poisonous substances known; intravenous injection of only 0.025 micrograms into a rabbit can cause death. Pioneering investigations by organic chemists in Japan and Hawaii led to suggestions for the overall structure of palytoxin that indicated the uniqueness of its structural complexity and molecular size. When synthetic chemists set their sights on the synthesis of palytoxin, they were opening a new page in the history of organic chemistry.

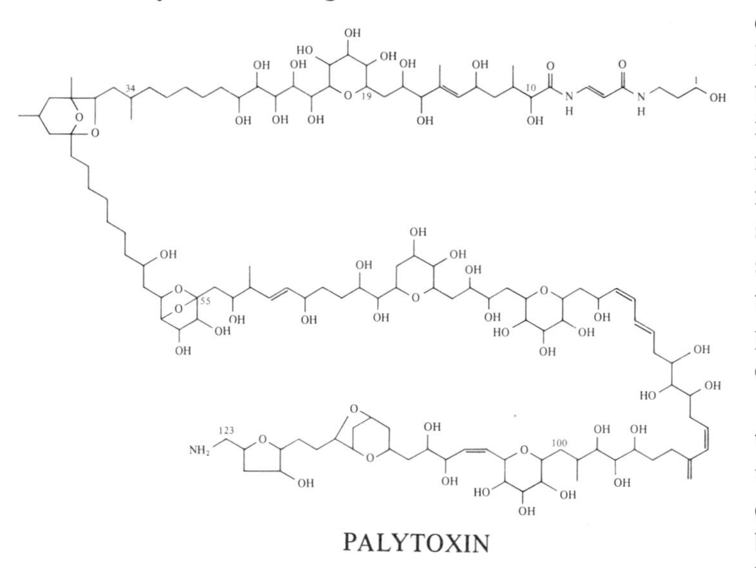

PALYTOXIN

This monster molecule contains 128 carbon atoms, 64 of which are asymmetric (chiral) centers. These centers, combined with seven skeletal double bonds, give palytoxin over two sextillion stereoisomers (2,000,000,000,000,000,000,000 = 2×10^{21})! The basic structure had established the stereogeometry of 13 of the centers—leaving 51 yet to be learned. Hence, the first step toward synthesis was to establish the stereochemistry of palytoxin.

Enormous barriers stood in the way of the determination of the stereo structure of palytoxin. Chemists had only tiny amounts of the final product and, because it was not in crystalline form, X-ray analysis was of no use. Furthermore, nuclear magnetic resonance could not be conclusive because palytoxin is structurally too complex. However, organic synthesis was ready for the challenge, based on the experience gained with the polyether antibiotics.

The researchers began with careful degradation of palytoxin to break it, chemically, into more manageable fragments. This degradation had to be gentle, so that each fragment would retain the stereochemistry it has in the parent molecule. Then, each fragment was synthesized in all of its isomeric forms to discover the shape that matched the natural product fragment. The process required that 20 different degradation fragments be synthesized, each in its various stereoisomeric forms, to identify the natural structure. The success of this endeavor has raised the sights of synthetic organic chemists everywhere.

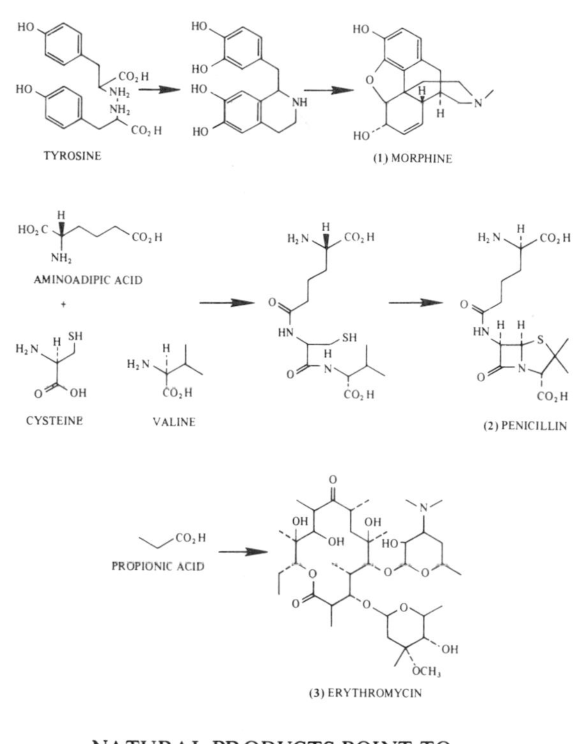

NATURAL PRODUCTS POINT TO NEW SYNTHETIC PATHWAYS

Biosynthesis of Natural Products

Natural products have for many years played a central role in the development of organic chemistry. The fact that many human medicines for the relief of pain and treatment of disease come from nature places a great deal of emphasis on this area of research. Morphine, for pain, and penicillin and erythromycin, with antibiotic properties, are all found in nature. One approach is to study the ways in which natural compounds are actually formed in nature. Only recently has it become possible to test experimentally these possible biosynthetic pathways so that many of them are now reasonably well understood.

The major experimental tool for biosynthetic investigations has been the use of isotopic tracers of the common elements, carbon (^{13}C, ^{14}C), hydrogen (^{2}H, ^{3}H), nitrogen (^{15}N), and oxygen (^{17}O). In this method, the isotope of an element is substituted for the natural isotope at a particular place in a reactant molecule. Then the isotope is looked for after reaction has taken place, using instrumental techniques. In this manner scientists can map reaction pathways. Extensive chemical degradations are no longer needed to locate sites of isotopic labeling because this task has been revolutionized by the development of stable isotope

NMR and the availability of high- resolution NMR spectrometers. Such NMR techniques have permitted the determination of the biosynthetic pathways that lead to certain powerful poisons that are produced by fungi and contaminate grain and other foodstuffs. These toxins, such as aflatoxins and trichothecin derivatives, pose major economic and public health threats.

Recombinant DNA technology provides another set of potentially powerful new tools for the study of biosynthetic pathways. The antibiotics monensin and erythromycin, both discussed above, provide excellent examples. These two substances are structurally and stereochemically among the most complex natural products. Beyond the basic building blocks for each antibiotic (the simple substances acetate, propionate, and butyrate), little is known about the details of the pathways by which these polyoxygenated, branched-chain fatty acids are assembled. Recent advances in the understanding of the genetics of the *Streptomyces* bacteria, along with the development of promising approaches to cloning for these organisms, have now made it more possible to unravel and perhaps control biosynthetic pathways at the genetic level.

The Chemical Synthesis of DNA

Section III-F on biotechnology described how nature encodes in the molecular polymers of DNA the information needed to generate a living organism. A repeating skeletal chain of sugar-phosphate ester linkages provides a stable backbone upon which a message can be written using an alphabet of the four amines: adenine, thymine, cytosine, and guanine (A, T, C, and G). These nitrogen-rich cyclic amines are chemically bonded to the sugar groups in a sequence that carries the information. Although these amines are called "bases," in fact, each one couples the ability to form hydrogen bonds acting as an electron donor (a "base") with the ability to form hydrogen bonds acting as an electron acceptor (a "proton donor" or "acid"). This hydrogen bonding capability furnishes the mechanism for replication. The DNA double helix structure is held together by hydrogen bonds between each amine "acid/base" of the first strand with a matching or complementary "base/acid" amine on the second strand. Then reading the message of the DNA molecule can be accomplished merely by making and breaking these relatively weak hydrogen bonds without danger of breaking the stronger (sugar-phosphate) bonds of the template strand.

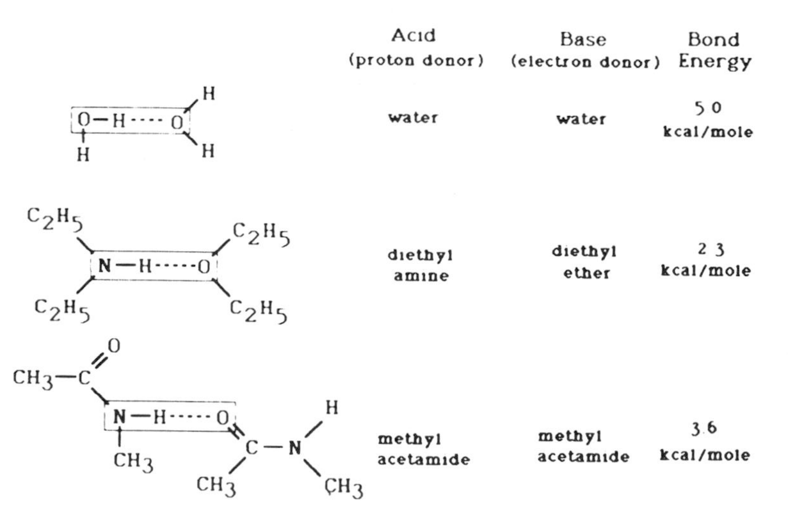

HYDROGEN BONDS - WEAK BUT IMPORTANT

The first chemical synthesis of a gene, about 15 years ago, required many person-years of effort. The remarkable (and continuing) progress since then permits synthesis of a similar size gene by a single researcher in 2 weeks. There have been several syntheses of the gene for insulin in industrial laboratories, and a noteworthy synthesis of the gene for interferon in the United Kingdom. Each of these products shows promise for major medical and commercial value. The recent synthesis of the gene for the enzyme ribonuclease was designed to permit later alterations of the gene, making it possible to deliberately change the physical and chemical properties of this protein.

Much progress is still needed. The yields of individual steps in DNA synthesis are still too low to permit routine synthesis of long molecules of DNA. State-of-the-art methods now can prepare gene fragments over 100 base pairs long, but we would like to deal with fragments 10 or 100 times longer yet. Costs for commercial custom synthesis of DNA molecules are coming down, but they can still exceed $200 per nucleotide. These synthetic nucleotide polymers of limited length are called *oligonucleotides* (from the Greek *oligos,* meaning few). Commercial machines for synthesizing DNA have only begun to meet the needed requirements for durability and dependability.

Meanwhile, there is great excitement about the examples that are appearing. Synthetic oligonucleotides have been used to clone medically valuable proteins such as Factor 8 (a blood fraction used in the treatment of hemophilia) and commercially important proteins such as renin (used in the manufacture of cheese). The next decade will see continued efforts to alter the structure of enzymes to make them more useful in industry, to alter the structure of proteins and peptides to make new pharmaceuticals, and to uncover new knowledge concerning genetic regulation and human disease.

STRUCTURES OF MACROMOLECULES

The structures of the giant molecules of living systems—the proteins and nucleic acids—offer challenges just like those encountered for smaller natural products. We must first know which atoms are bonded to which in order to describe the covalent molecular structure. Then we must learn how the chains of these large polymers are oriented in space, because the biological properties of the proteins and nucleic acids are intimately connected to their three-dimensional structures. This is especially true for proteins, whose remarkable range of biological functions has been described in Section III-F. Following are some of the characteristics of proteins that allow them to be effective in areas ranging from digestion of food to blood transport of oxygen, and from contraction of muscles to antibody protection from viruses and bacteria. These characteristics define some of the biological frontiers in which chemistry will play a central role.

Proteins Have Complex Three-Dimensional Shapes That Relate to [Determine?] Biological Function

Chemical research of the past two decades has revealed that proteins have highly intricate three-dimensional forms and that these forms are critically related to the specific biological functions of each protein. A protein chain consisting of hundreds of linked amino acids takes on a three-dimensional architecture (called a conformation) that is determined by its particular amino acid sequence. For example, collagen, a protein that gives strength to skin and bone, has the shape of a rod. Antibodies are Y-shaped molecules with cavities that recognize foreign substances and trigger subsequent reactions for their efficient disposal. X-ray crystallographic studies have given valuable information about their architecture. Enzymes have clefts called "active sites" that bring reactants together and permit the formation of new chemical bonds between them. Thus, proteins have definite conformations that are at the heart of their biological roles. Major advances have been made in viewing these conformations using X-rays, neutron and electron beams, and other probes that enable us to "see" proteins magnified more than one million times. Clarifying these protein conformations shows us how biological functions are accomplished.

We need to know much more about how proteins recognize specific sites on DNA and how they influence them. Additionally, we want to learn how peptides interact with receptor proteins to produce physiological changes in organisms. For example, the body produces a series of peptides called endorphins, compounds that have painkilling and tranquilizing effects. Understanding of how the binding of these peptides to proteins on the surfaces of cells can lead to powerful changes in mood and consciousness will be a step toward unraveling the mysteries of brain function.

Proteins Are Highly Dynamic

Chemical studies of the past decade have also shown that proteins are highly dynamic molecules. Proteins change their shape while performing their functions. For example, light changes the conformation of rhodopsin, a protein in the retina, as the first step in vision. This structural change occurs in less than a billionth of a second. Such rapid changes in protein molecules can now be detected by using pulsed lasers. Another useful approach in the analysis of protein dynamics involves cooling a protein to very low temperatures so that individual steps in its action are slowed down to permit more leisurely study.

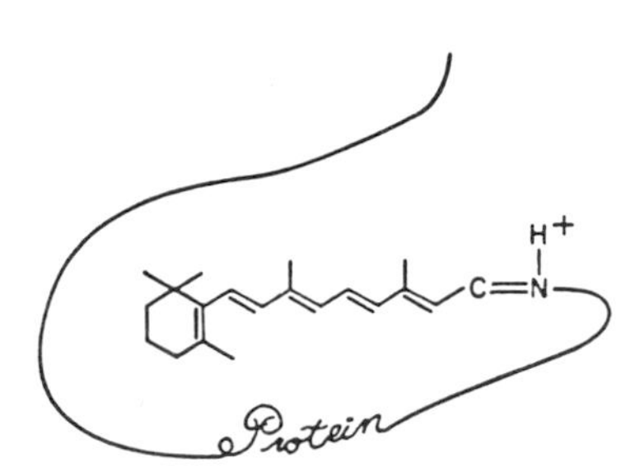

BACTERIORHODOPSIN

Proteins Display Recurring Structural and Mechanistic Themes

Even the simplest cells contain more than 5,000 kinds of proteins. Yet we are finding that structural and mechanistic themes seen in one protein frequently recur in others. For example, there is a close relationship between the enzymes thrombin (for blood clotting) and chymotrypsin (for digestion). Moreover, the structures of many proteins have stayed the same over long evolutionary periods. There is surprisingly little difference, for example, between human and mouse hemoglobins. Enzymes work in complex organisms in much the same way that they do in simple ones. This knowledge is now being used to unravel disease mechanisms, devise new diagnostic tests, and develop novel drugs and therapeutic strategies.

Structural Studies on Dihydrofolate Reductases and Their Inhibitors

Dihydrofolate reductase (DHFR) is an enzyme present in all living creatures, from bacteria to mammals. DHFR acts on dihydrofolate, which is a necessary ingredient in the complex chemistry of DNA synthesis in cells.

Quite some time ago, it was noticed that feeding of folic acid, a source of dihydrofolate, actually encouraged the growth of tumors present in laboratory animals. Rapidly dividing cells, like those found in a tumor, require equally rapid DNA synthesis and ample supplies of compounds like dihydrofolate. Hoping to find an antagonist that would block and reverse this effect (an "antifolate"), investigators set about synthesizing and testing many chemical analogs of folic acid. This approach paid off with the discovery of aminopterin and later of methotrexate. Amazingly, the essential difference between these compounds and folate itself was simply the substitution of folate's 4-hydroxyl group by a 4-amino group.

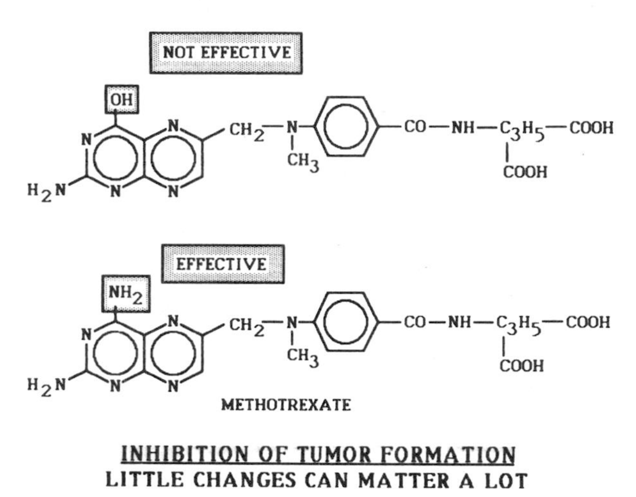

INHIBITION OF TUMOR FORMATION
LITTLE CHANGES CAN MATTER A LOT

Thereafter, it was determined that methotrexate acts by inhibiting the enzyme DHFR. In fact, DHFR binds methotrexate so strongly that inhibition is essentially irreversible. This slows tumor growth by interrupting the action of DHFR and thus interfering with DNA synthesis and cell division. Today, methotrexate is in widespread and effective clinical use for treatment of childhood leukemia, choriocarcinoma, osteogenic sarcoma, and Hodgkin's disease.

Meanwhile, other, more distant analogs of folic acid were synthesized and tested in great numbers, including the substituted 2,4-diaminopyrimidines. This program led to the discovery of an agent useful against bacteria, called trimethoprim, and one effective against protozoa, premithamine, among others. All of these antifolates act by inhibiting DHFR. In some cases they are highly specific about which species of organism they work on. For example, trimethoprim has about 100,000 times greater affinity for binding the DHFR of *E. coli (Escherichia coli)* bacteria than for vertebrate DHFR. This fact makes trimethoprim safe for use as an antibiotic since it strongly prefers the bacterial enzyme. A decade ago, study of several DHFRs by X-ray crystallographic methods was begun to determine the molecular-structural basis for their action and to point the way toward a logical, structurally based approach to drug design.

This X-ray crystallographic approach has begun to pay off. So far, the structures of DHFRs from three widely differing species, namely, the two bacteria *E. coli* and

L. casei (Lactobacillus casei) and the chicken (representative of vertebrates), have been determined. In addition, these enzyme structures have been examined as they appear when various molecules are bound to them.

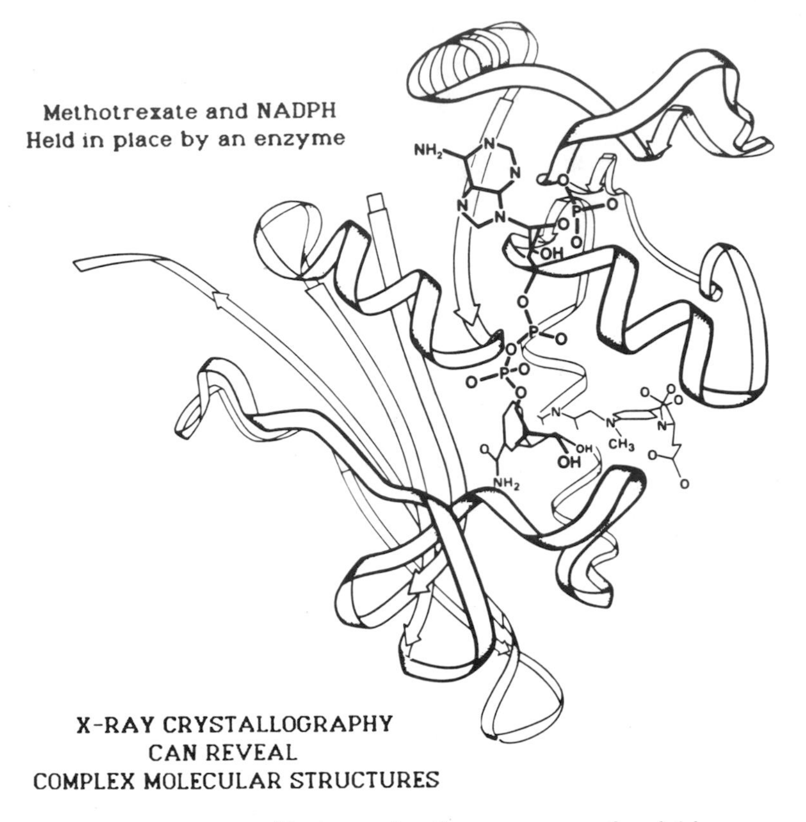

The most striking feature seen on comparing DHFR molecules from the different organisms is the close similarity in their overall foldings. Clearly, the general molecular structure of the enzyme was highly conserved during millions of years of evolution, even though only about 25 percent of the amino acid sequence remained unchanged (80 percent among the vertebrates, however).

DHFR provides an excellent model for studying how similar enzymes stimulate the rather unreactive nicotinamide nucleotides NADH and NADPH. Biochemists who study metabolic pathways have long recognized that the nicotinamide nucleotides serve as a kind of all-purpose oxidation-reduction currency, furnishing a way to exchange electrons in biological reactions. Now we are finding that the stereochemical aspects of its placement in DHFR facilitate hydride transfer through hydrogen bonds.

Frontiers in the Chemistry of Genetic Material

In higher organisms (including humans) the percentage of nucleotides in a strand of DNA which actually code for the sequence of amino acids in proteins is estimated to be only about 5 percent. What is the role of the remaining 95 percent? Recently, it was discovered that another type of information is coded in the sequence of DNA nucleotides. Apparently, information concerning the different conformations or shapes that DNA can take is stored there as well.

How are such conformational changes brought about? They happen around single bonds where relatively free rotational movements are possible. In ring structures, such rotations tend to pucker or wrinkle the ring into shapes that are not flat (nonplanar conformations). There is usually an energy barrier between the two (or three) energetically comfortable structures, called conformers, that result from such rotation. But the barriers can be small enough so that transfer between these structures can be relatively easy at room temperature. In sharp contrast to the stereoisomers of a molecule, the conformation of a molecule can be determined by secondary interactions, they may change in response to their environment, and two or more conformers can be present at once in dynamic equilibrium.

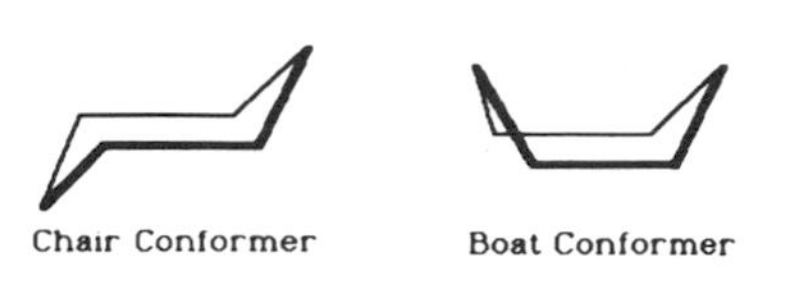

Many conformational characteristics have been discovered for the nucleic acids. For example, puckering of the furanose ring, which is common to both DNA and RNA, leads to flexibility in their backbones. Furanose is the 5-carbon cyclic sugar found in the backbone of nucleic acids. A number of different conformations can be assumed by this ring, but the most prominent is called the C2′ *endo* conformation. This conformation has been considered to be characteristic of DNA nucleotides, while another one, the C3′ *endo* conformation, was more frequently found in RNA nucleotides. We must learn more about the energy barriers between these two conformations. It is now thought that the energy barriers separating different conformations are lower for the deoxynucleotides than for ribonucleotides. In the three-dimensional structure of a transfer RNA found in yeast, which has 76 nucleotides, the majority were found to adopt the C3′ *endo* conformations. This has a significant effect on the spacing of certain phosphate groups. The phosphate-phosphate distance is close to 6.7 Å in the C2′ *endo* conformation and less than 5.6 Å in the C3′ *endo* conformation. Thus, changes in sugar pucker make the polynucleotide backbone elastic, so it can accommodate different conformations. We need to know these conformations more precisely, how easily they can interchange, and how they affect biological function.

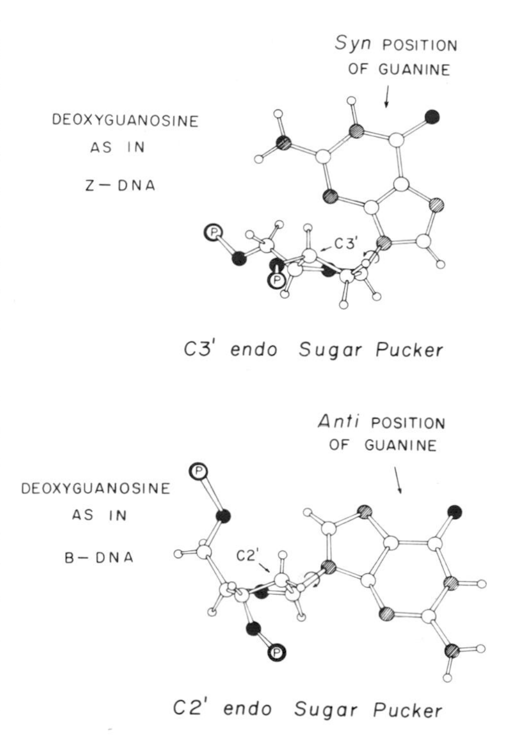

SUBTLE DIFFERENCES MATTER

For almost 30 years, DNA has been known to adopt two different right-handed conformations, A-DNA and B-DNA. They are called right-handed because the DNA spiral twists to the right. The A conformation is one in which all the deoxynucleotides have the C3′ *endo* conformation, while in B-DNA all of the nucleotides have the C2′ *endo* conformation. However, this simple classification into two possible right-handed conformations has now been changed considerably as a result of single-crystal diffraction analyses. Surprisingly, some of these analyses revealed the presence of alternating C3′ *endo* and C2′ *endo* conformations with alternating distances between phosphates.

That led to the discovery of left-handed DNA conformations in the laboratory. Polynucleotides were deliberately linked together so that purine bases and pyrimidine bases alternated with each other. Such a molecule adopts a conformation in which the purines take the C3′ *endo* conformation, while the alternating pyrimidines take the C2′ *endo* conformation. This structure is called Z-DNA; it twists to the left and exhibits an irregular tertiary structure.

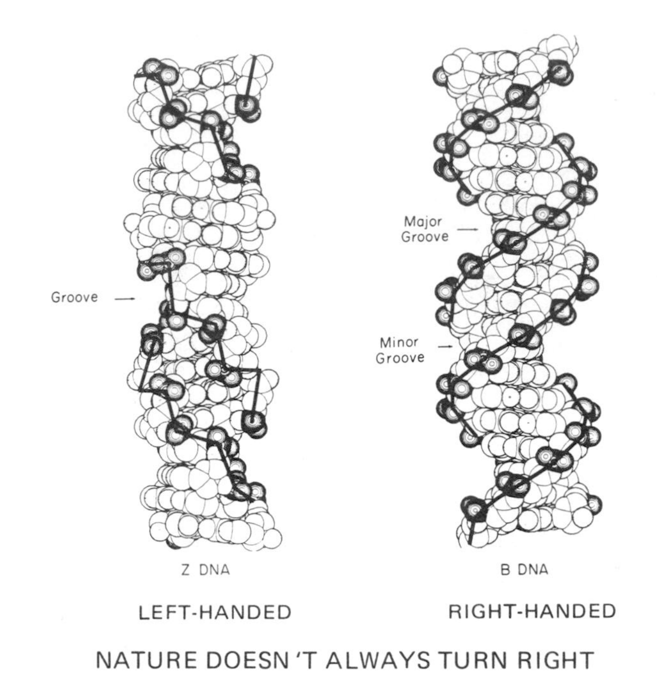

NATURE DOESN'T ALWAYS TURN RIGHT

At present, the overall view of the nucleic acids is that they are conformationally active. It is also now thought that the well-known right-handed B-DNA structure is likely to be in equilibrium with a number of other structures, including left-handed Z-DNA. The focus of much chemical and biological research will be on the nature of these conformational changes. We need to know more about how these conformational changes are affected by their environment, by modifications in the molecule, or by alterations in the nucleotide sequence.

Structure and Function in Biochemistry

Structure determines properties and properties determine function. Thus, from the simplest molecules like ethyl alcohol, to molecules with the exquisite and varied architectures of proteins, their molecular structure is inextricably related to their function as drugs, antibodies, biological catalysts, hormones, transport agents, cell surface receptors, structural elements, or muscles that convert chemical energy into work.

A prime question we wish to answer is how the structure of a protein might determine its function. One approach is to generate many structural variations of a protein in a controlled manner by precisely altering the order of its amino acids. With this method, the exact three-dimensional structure of a protein can be fixed to permit a logical analysis of the structure/function relationship.

Today we have procedures that allow us to move toward this objective. Modern molecular biology has taught us how to place almost any piece of DNA into a microorganism and thereby cause it to synthesize the protein that this DNA encodes. At the same time, modern organic chemistry has enabled us to rapidly and easily synthesize sequences of nucleotides that constitute pieces of genes. These pieces of genes can then be used to change the prescribed sequence of bases in the gene for the parent protein. Thereby, a modified protein with an altered sequence of amino acids can be generated, and a structure and function never before available can be produced.

This method for creating specific mutations of normal proteins is formally termed *oligonucleotide-directed mutagenesis*. This can lead to proteins with any structure we may desire. In addition, once a single molecule of the gene for that protein has been prepared, the protein itself can be produced forever after in microorganisms in whatever quantities may be desired.

These techniques focus on the creation of a mutant protein with a predetermined amino acid sequence. Such approaches are useful for learning the properties and functions of a protein altered in a specific manner. An alternate approach is to create a large number of structural variants, decide which ones show desired properties, and then go back and determine the structures of those desirable proteins. Such random mutagenesis can be allowed to take place anywhere in the gene of interest or, in order to better control the possible properties of a protein, can be restricted to a particular region of the gene.

At present, oligonucleotides can be synthesized in 98 percent yield at the rate of one base every 5 minutes. Improvements here could make the rapid synthesis of entire genes (rather than just oligonucleotides) a routine procedure, and thereby greatly speed up the creation of new proteins.

Great improvements can be foreseen in chemical and biochemical techniques for determining base sequences in nucleic acids and amino acid sequences in proteins. Presently, an automated instrument called a gas phase sequenator can reliably determine about 60 consecutive amino acids (called residues) from the amino end of a protein. Use of tandem mass spectrometry or other novel approaches might allow the complete sequence to be established for a protein of several hundred residues by automated techniques.

Gene Structure and RNA Splicing

The combination of a number of recent advances has yielded startling insights into the gene structure of man and other complex organisms. These advances include the ability to combine DNAs from different organisms, the ability to discover which DNA segments encode specific proteins and to isolate them, and the ability to determine the nucleotide sequence of long pieces of DNA. This new knowledge has raised many questions and opened new areas of research.

To find the DNA segment that contains a single gene from the total genetic material of a human cell is like finding the legendary needle in the haystack. The sequences that specify any one particular gene are about one-millionth of the total genetic material. The solution to the problem was to use recombinant DNA techniques to distribute pieces of human DNA into well over a million rapidly dividing bacteria, and then to grow each bacterium separately to give an entire colony of offspring of the single bacterium. Then the colony of bacteria containing the gene of interest is identified by some diagnostic technique that tests for the desired gene function. Each rapidly growing bacterial colony produces billions of identical copies of each gene that can then be isolated as a chemically pure substance. This process is called cloning. So far, DNA segments for well over 100 different human genes have been purified by this method. A similar number have been isolated from a few other vertebrate species, such as the mouse, and a greater number have been isolated from simpler organisms such as yeast.

Globin is a protein found in the blood ingredient hemoglobin. The DNA sequence that codes for the globin protein is interrupted in places by sequences that do not code for the protein. This is typical in the genes of eukaryotic cells (cells containing a nucleus)—the coding region is interrupted by one or more stretches of noncoding DNA, called intervening sequences or *introns*. Introns have also been called

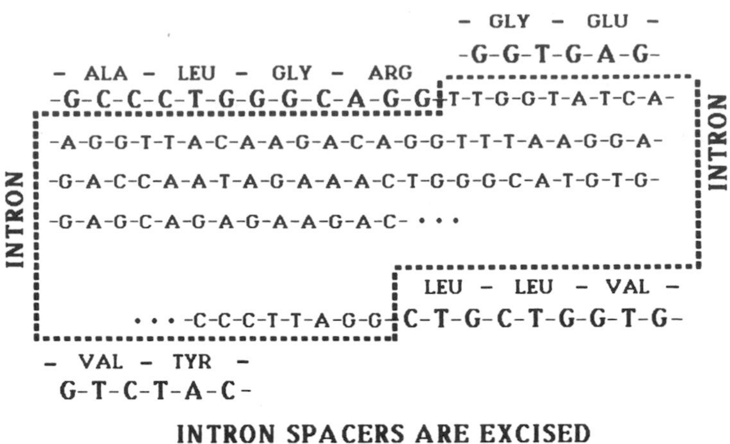

INTRON SPACERS ARE EXCISED TO GIVE MESSENGER RNA

"nonsense codes," but it has been discovered that they may have many important functions. Introns are found in most genes that code for messenger RNA and in some genes that code for transfer and ribosomal RNAs. In all cases that have been studied, the introns are copied along with the neighboring coding sequences as part of a large precursor RNA. The introns are then removed by a cleavage process called RNA splicing, which results in a functional RNA molecule with a continuous coding region. For example, there are two introns in the human globin gene. After they have been removed, the resulting messenger RNA is transported from the nucleus to the cytoplasm for translation into globin protein.

The phenomenon of RNA splicing is common in cells with nuclei, *eukaryotes,* but is thought to be absent in cells without well-defined nuclei, *prokaryotes*. It is the only major step in gene expression in which eukaryotes and prokaryotes differ significantly. Because of this it is interesting to examine just how RNA splicing regulates and affects the expression of genes. In addition, the possibility that introns in the genetic code might be responsible for the evolution of eukaryotic genes is being explored.

The impact on society of future research on gene structure and gene expression will be enormously beneficial. Many human diseases are the result of defects in gene expression. Information about the nature of the genetic changes in cancer cells may yield new avenues for pharmacological treatment of cancer. The process of aging is poorly understood; it is possible that some of the more destructive aspects of this process are controlled by the activity of a few gene products so that identification of the functions of these genes may lead to better treatments for aging patients.

SUPPLEMENTARY READING

Chemical & Engineering News

"Experts Probe Issues, Chemistry of Light-Activated Pesticides" by R.L. Rawls, (C.& E.N. staff), vol. 64, pp. 21-24, Sept. 22, 1986.

"Anticancer Drug Cisplatin's Mode of Action Becomes Clearer" by R. Dagani (C.& E.N. staff), vol. 63, pp. 20-21, Dec. 16, 1985.

"Electrochemical Techniques Benefit Bioanalysis" (C.& E.N. staff), vol. 63, pp. 32-33, Jan. 14, 1985.

"Penn Chemists Synthesize Complex Natural Antibiotics" by R. Dagani (C.&E.N. staff), vol. 62, pp. 17-19, Oct. 15, 1984.

"Potentiometric Electrode Aims to Measure Antibody Levels" by R. Rawls (C.& E.N. staff), vol. 62, pp. 32-33, Apr. 2, 1984.

Science

"Long Range Electron Transfer in Heme

Proteins'' by S.L. Mayo, W.R. Ellis, R.J. Crutchley, and H.B. Gray, vol. 233, pp. 948-952, Aug. 29, 1986.

''Transformation Growth Factor—Biological Function and Chemical Structure'' by M.B. Sporn, A.B. Roberts, L.M. Wakefied, and R.K. Assoian, vol. 233, pp. 632-634, Aug. 8, 1986.

''The Intervening Sequence RNA of *Tetrahymena* is an Enzyme'' by A.G. Zang and T.R. Cech, vol. 231, pp. 470-475, Jan. 31, 1986.

Chem Matters

''Natural Dyes'' pp. 4-8, December 1986.

''Autumn Leaves'' pp. 7-10, October 1986.

''Lipstick'' pp. 8-11, December 1985.

Something for Nothing

Grandpa used to say, "There's no free lunch!" That was his way of saying that you never get something for nothing. But now, Grandpa, we're not so sure! The recent discovery of high-temperature superconductors has everyone talking about amazing visions, like trains riding on air and electrical energy transmitted from Nevada to Alaska with no losses on the way.

Kamerlingh Onnes, a Dutchman, started it all in 1911. When he cooled metals to low temperatures, the electrical resistance, which limits conductivity, smoothly dropped with temperature. Theorists explained that a current flow requires electrons to move through the metal crystal, but as they move, they keep bumping into the vibrating metal atoms, losing energy and generating heat. If the crystal is cooled, these lattice vibrations are diminished, so there are fewer collisions to slow down the electrons. The theory confidently indicated that the resistance would reach zero only when the temperature reached the unattainable "absolute zero."

But when Onnes cooled mercury to liquid helium temperatures, he got the surprise of his life. At 4.2K, the resistance suddenly dropped so low he couldn't measure it. Below that critical temperature, T_c, an electrical current, once started, kept going for weeks, months, even years. The resistance, which usually stops such a current, had truly become zero. *The metal had become a superconductor.*

As new superconductors were discovered, the highest known value of T_c slowly crept upward. The world's record reached 15K in 1941 with the discovery of the two-element (binary) superconductor niobium nitride, NbN. Another binary, NbGe, had led the field with T_c = 23K since its discovery in 1973. Here progress stopped cold (pun intended).

Then, in 1986, the lid blew off. First, a quaternary copper oxide compound was found to become superconducting at 37K. In the next few months, rumors flew, suggesting possible T_cs at 40K, 52K, 70K, 94K, and even 240K. Working around the clock, scientists in the United States, Europe, and Japan recognized finally that certain four-element copper oxides with the layered perovskite crystal structures were true superconductors with T_c near 94K. The breakthrough was an yttrium-barium compound, $YBa_2Cu_3O_x$, with a noninteger x near 7.4. Soon it became clear that yttrium could be replaced by six or seven other lanthanide elements, while strontium or calcium could take the places of some of the barium atoms.

Where do we go from here? Anyone who feels that electrical bills are too high can be cheerful—about 20 percent of the electrical energy moved around the country is wasted in the copper transmission lines. That's enough energy to light up the whole West Coast. Since liquid nitrogen, at 77K, is a cheap coolant, superconductivity is now affordable. From tiny motors to the enormous turbines in hydroelectric plants, a new era can be foreseen. In our most powerful computers, heat dissipation limits circuit size, hence computer capacity. This problem disappears, along with the resistance, if the connectors are superconducting. But the most advertised expectations concern new uses of superconducting magnets. They will surely reduce the cost of a medical NMR whole-body imager. And superconducting magnets might levitate whole trains so they ride on a practically frictionless cushion of air.

So, getting the resistance down to zero really does give us something for nothing. And by the way, Grandpa, want to get in on the free lunch?

IV-C. National Well-Being

Research across the whole range of chemistry contributes to a better environment and to sustained economic competitiveness. But certain research areas are key to progress in these realms. For example, the surface sciences, with their implications for new heterogeneous catalysts, furnish a wellspring of critical importance to economic progress. Condensed-phase chemistry and new separations techniques also can be expected to contribute fruitful new dimensions. Next, the new frontiers in analytical chemistry support and contribute to advances in all other areas of chemistry. Analytical chemistry is the cornerstone upon which our monitoring and management of the environment is built. Finally, nuclear chemistry was nurtured in the World War II Manhattan Project, and its influence continues to be of prime importance, since the world's energy needs may involve nuclear reactors (despite Chernobyl), and world peace presently is based upon a precarious balance of nuclear arms. In each of these areas there are opening frontiers and rewarding intellectual opportunities to be pursued.

CHEMISTRY AT SOLID SURFACES

The surfaces of metals and ionic solids are, by nature, chemically reactive. The reason is clear—the bulk crystal is based upon a structure that gives each interior atom the best possible chemical bonding to neighboring atoms around it in all three dimensions. At the surface, however, the atoms have unsatisfied bonding capacity since the neighboring atoms are missing in at least one direction. Hence, this is a region of special chemical behavior, and one of unusual interest to chemists. The importance of this special behavior simply cannot be exaggerated. Corrosion occurs, of course, at iron surfaces, with obvious bad effects on many useful structures, from the lofty Eiffel Tower to the lowly nail. Estimates are that corrosion costs the U.S. economy billions of dollars annually. At aluminum surfaces, the rapid reaction that takes place on exposure to air forms a protective and quite inert oxide coating. Hence, we can safely have the convenience of aluminum foil in the kitchen, despite the fact that aluminum is flammable at sufficiently high temperatures. By far the greatest importance of surface chemistry is that it bestows extremely effective catalytic activity upon some surfaces. This capacity of a solid surface to speed up chemical reactions by many orders of magnitude without being consumed is called heterogeneous catalysis. Its great value as the basis for commercial processes of immense economic value has been noted in Sections III-B and III-C. It furnishes one of the most important and active frontiers of chemistry.

Heterogeneous catalysis is not new. What is new is the array of powerful instruments, developed over the last 15 years, that at last provide experimental access to the chemistry on a surface *while that chemistry is taking place*. Without such techniques, catalysis has remained over many decades a fairly mysterious art. Now we have instruments with which to characterize precisely the nature of the catalyst surface and to study molecules while they are reacting there. Now we are accumulating the store of quantitative data needed for catalysis to move from an art to a real science. The intellectual challenge to understand the chemical behavior of molecules on a surface has propelled surface science into the mainstream of fundamental research in most departments of chemistry and chemical engineering.

The instrumentation of the surface sciences will be described in Section V-C. Some research highlights and productive frontiers will be described here.

The Structure of Solid Surfaces

We have already discussed, in Section III-C, the role of specific metallic surfaces in the catalytic restructuring of hydrocarbons to produce gasoline. As a second example, research on the catalytic production of ammonia from elemental nitrogen and hydrogen is of comparable importance. This is because NH_3 is a critical fertilizer component, so it helps determine (or limit) the world food supply. At elevated temperatures, N_2 and H_2 can react to form NH_3 on perfect crystals of an iron catalyst. The effectiveness of a catalyst depends upon how rapidly each surface site can adsorb reactants, encourage them to rearrange chemically, and then release the products so that the site can begin the process again. The iron crystal face designated (1,1,1) is about 430 times more active than the closest-packed (1,1,0) crystal face and 13 times more active than the simpler (1,0,0) face. It is now believed that the rate-limiting step is the rupture of the strong nitrogen-nitrogen bond of N_2 (225 kcal/mole) and that this occurs with an activation energy near 3 kcal/mole on the (1,0,0) face but with nearly zero activation energy on the specially active (1,1,1) surface.

Because of such influences on catalytic action, surface structures are attracting much research interest. Small particles tend to display many different surfaces, depending on how they are prepared. As the metallic particle grows, it becomes more like the bulk material and tends to favor surfaces without terraces and kinks. Interestingly, atoms in the surface layer may be located closer to adjacent atoms in the second layer than they would be if they were located deep inside the crystal. Even more drastically, because of the incomplete bonding of surface atoms, they may seek equilibrium positions different from the packing in the bulk material in order to improve their bonding. Such "surface reconstruction" has been found for platinum, gold, silicon, and germanium.

Another important question that can now be experimentally explored is the chemical composition of the surface. Even the purest samples will have some impurities, and these may noticeably affect some properties of metals and semiconductors. A crucial question is how much a given impurity prefers to concentrate at the surface. The difference in bonding between host atoms and impurity atoms explains why the bulk portion of the material tends to reject the impurity. This same difference may cause the impurity to be a welcome addition to the surface, where host atoms alone cannot satisfy their bonding capability. There are cases in which impurities at the parts per million level are so concentrated at the surface that they can cover it completely. This strongly affects the chemistry at that surface.

Of course, this issue is always present in alloys composed of two or more elements. There is excess silver at the surfaces of silver-gold alloys, excess copper at copper-nickel alloy surfaces, and excess gold at gold-tin alloy surfaces. Some metals that do not readily dissolve in each other in bulk are found to mix in any proportion on a surface. Experimental data and understandings are especially needed at this time when a variety of binary and ternary substances are under study because of their interfacial electrical properties.

In summary, determination of the atomic structure of surfaces and surface composition is basic to understanding the wide variety of surface properties now finding important practical applications. They are the starting point for advancing corrosion science, heterogeneous catalysis, lubrication, and adhesion, as well as for producing new surfaces with novel electronic properties.

Adsorbed Molecules; Chemical Bonding at the Surface

For many decades, the strength of binding of an adsorbed substance on a surface was measured by the ease of its removal on warming. Some substances are easily removed at temperatures near or below room temperature. Such a situation is traditionally called "physisorption"; the adsorbed substance keeps its molecular shape and is bound to the surface only by weak forces, such as van der Waals or hydrogen bonding interactions. Other substances are much more tightly held by the surface and can be removed only by heating to much higher temperatures—perhaps 200 to 600°C. Here, covalent bonding to the surface is involved and the molecular structure of the adsorbed substance is probably different from what it was before adsorption. This situation is called "chemisorption," and it is almost always involved at some stage in any heterogeneous catalysis. Thus, understanding of the molecular structure and chemical properties of chemisorbed molecules lies at the heart of heterogeneous catalysis.

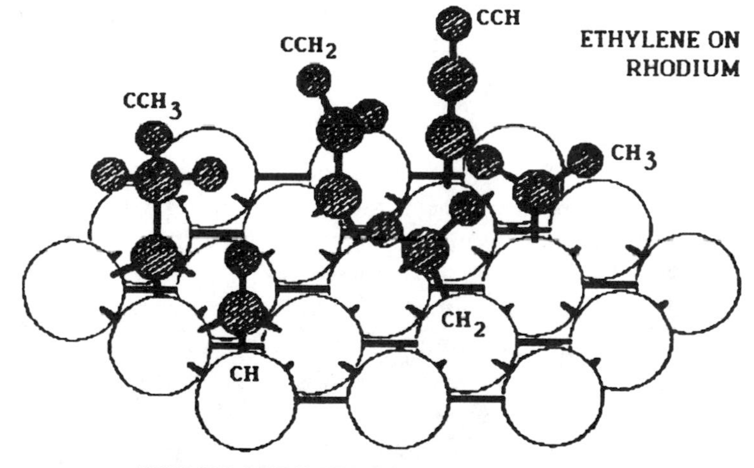

WHICH STRUCTURES ARE PRESENT?

Among small molecules, carbon monoxide on metal surfaces has historically received the most attention, largely because its spectroscopic properties allow the detection of small numbers of CO molecules on a surface. This is fortunate, for one of today's most pressing problems is the conversion of coal to useful hydrocarbon feedstocks, usually accomplished via carbon monoxide. Many catalytic schemes use carbon monoxide as an intermediate in the form of "syn gas," a mixture of CO and H_2 derived from coal (see Section III-C and Table III-C-2).

A second key system is ethylene adsorbed on catalytic metal surfaces. It has been known, from its thermal behavior, that ethylene chemisorbs on platinum and rhodium catalysts. Now, we can add information about the structures that are formed on the surface through direct observation of the vibrational frequencies of the adsorbed species. Direct observation of these frequencies through infrared absorption spectroscopy is sometimes possible, but the introduction of electron

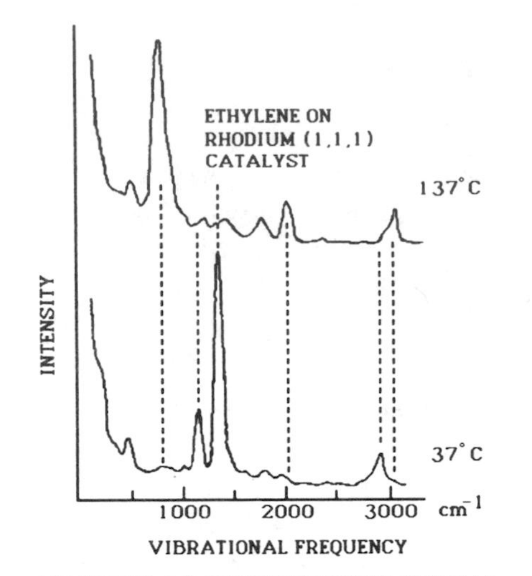

MOLECULAR FINGERPRINTS REVEAL THE REACTION PRODUCTS ON SURFACES

energy loss spectroscopy (EELS) has greatly accelerated such studies. The characteristic molecular frequencies are imprinted on the energy distribution of electrons bounced off the metal surface. These frequencies provide a fingerprint that is readily interpreted by a chemist experienced in relating infrared spectra to molecular structures (see Section V-C). For ethylene on rhodium, the EELS spectrum plainly shows that, after adsorption, the ethylene molecule has been structurally altered even at room temperature. Then, on warming 50°C or so, the spectrum begins to change still more. By the time the temperature has changed by 100°C, the spectrum shows that reactions have taken place and the hydrocarbons now present on the surface have new structures. These EELS spectra reveal, then, which of the possible surface structures (C_2H_3, C_2H_2, C_2H, CH_3, CH_2, and CH) are present at a given temperature and, hence, the sequence of their formation as the temperature is raised. Such intimate knowledge of the chemical events taking place on the catalyst surface furnishes the basis for a detailed understanding of the catalytic dehydrogenation and hydrogenation of ethylene, which is important in many chemical processes.

Coadsorption on Surfaces

Chemistry on surfaces takes on a new dimension when two substances are adsorbed on the same surface. Then attention shifts from the interaction of the adsorbate with the surface to the interaction of two different molecular species when they share the special environment provided by the surface.

The first way in which this interaction can occur is when one adsorbate changes the special environment encountered by the second adsorbate. For example, a clean molybdenum metal surface will break down the sulfur-containing molecule, thiophene, C_4H_5S. However, if elemental sulfur is coadsorbed, it chemisorbs quite strongly at the active sites needed for thiophene decomposition. Thus, sulfur "poisons" the catalyst for this particular reaction. This is of great importance because thiophene is an impurity we want to remove from gasoline.

As a second example, carbon monoxide is physisorbed on rhodium, as shown both by its ease of removal on warming and its vibrational frequency on the surface, which is close to that of gaseous carbon monoxide. If, however, the rhodium is 50 percent covered by coadsorbed potassium, CO becomes chemisorbed instead. The EELS spectrum shows a CO vibrational frequency appropriate to a bridged structure, with a frequency indicating the presence of a carbon-oxygen double bond. Under these conditions, hydrogenation of CO is encouraged, and this leads to the production of desirable higher molecular weight alkanes and alkenes (hydrocarbons possessing one or more double bonds). (See discussion of "syn gas" in Section III-C.)

Still to be mentioned, of course, is the direct reaction between the two adsorbates. In the future, this will be seen as the origin of most of the new chemistry that can take place in this special reaction domain. An obvious example has already been cited, the hydrogenation of ethylene (C_2H_4). When hydrogen adsorbs on platinum or rhodium, the H_2 molecule is split and the two atoms are separately bonded to the metal atoms. Now, when ethylene is coadsorbed, it does not encounter H_2 at all. Instead, it finds individual hydrogen atoms attached to the surface. Plainly, if coadsorbed hydrogen and ethylene react, they will follow a reaction path characteristic of the actual species on the surface and governed by activation energies that are different from those associated with a gas phase encounter between H_2 and C_2H_4.

CONDENSED-PHASE STUDIES

Many challenges facing chemistry, solid-state science, earth science, biochemistry, and biophysics involve the ability to understand and manipulate the properties of condensed phases—liquids and solids. Chemistry is central here, since these properties result directly from the interatomic and intermolecular forces between the atoms and molecules present in these phases.

Optical and Electronic Properties of Solids

Over the past 15-20 years high pressure has proved to be a powerful tool in the study of electronic phenomena in solids. Compression pushes molecules closer together, and this increases the overlap among adjacent electronic orbitals. Since different types of orbitals have different spatial characteristics, they are affected to different degrees. This ''pressure tuning'' makes pressure a powerful tool for characterizing electronic states and discovering electronic transitions to new states with different physical and chemical properties.

Many examples of electronic transitions that show pronounced response to high pressure have been found. For example, it has been possible to use high pressure to convert substances that are normally insulators into electrical conductors. This has been done for nine elements and for about 50 compounds. One application is fast electrical switches without make-and-break contacts. Also, the first organic superconductor showed its superconductivity under pressures between 6,000 and 18,000 atmospheres. Visible color changes can also be caused, as has been shown for several compounds known as anils, spiropyrans, and bianthrones (photochromic-thermochromic transitions), and for about 30 ethylene-diamine complexes (electron-transfer transitions). Such pressure studies are showing us how phosphors absorb light of one color and reradiate light of another color, and are helping us increase the efficiency of a variety of laser materials.

Liquids

Many of the fundamental processes of nature and industry take place in the liquid state. The rate of movement of molecules in solution can limit the speed with which a chemical reaction can occur, a nerve can fire, a battery can generate current, and chemicals can be purified and isolated. A properly chosen liquid solvent can accelerate a chemical reaction by a millionfold or slow it down by a similar amount.

Molecules in liquids can be highly efficient agents for storing or transferring energy. The very structure of liquid water determines our planetary environment and influences the course and nature of all biochemical processes essential to life.

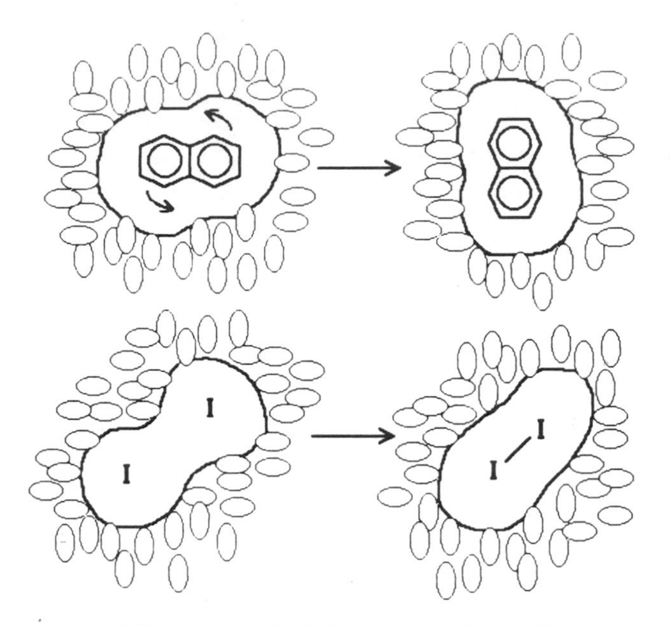

LASERS LET US MEASURE FAST CHANGES IN THE SOLVENT CAGE

The structure and dynamics of a wide range of fluids, from liquid hydrogen to molten silicates, can be investigated by a number of spectroscopic techniques, such as X-ray and neutron diffraction, nuclear magnetic resonance, and laser Raman and light scattering. Among the newer experimental approaches, pulsed laser excitation techniques are particularly powerful. On a picosecond time scale (10^{-12} seconds), we can sense the freedom of movement of a solute molecule held in its solvent cage. Now we can watch fundamental chemical events as they take place: how two iodine atoms combine in a liquid to produce an iodine molecule; how electrons released in liquid water become trapped, or solvated; how energy placed in a solute molecule like nitrogen or benzene is transferred to its solvent environment.

Quite a different opportunity area is connected with the melting of small clusters of metal atoms. We have a variety of new experimental methods for producing and studying small metal clusters, as well as the theoretical tools with which to interpret the results. We can look ahead to an understanding of how the change from the fluid liquid state to the rigid solid state emerges as cluster size increases toward bulk amounts. Furthermore, the computer can keep track of the energy and randomness associated with each arrangement, so thermodynamic data can be calculated for comparison with experiments, and then for predictions under conditions out of reach experimentally.

Critical Phenomena

For any fluid, there is a characteristic temperature and pressure above which the liquid and gaseous states are identical. Fluid behavior under these "critical conditions" can differ markedly from normal behavior and give rise to new phenomena. The past 20 years have seen a revolution in our understanding of such critical phenomena. Undoubtedly, the most important single theoretical advance in our understanding in the last 15 years has been the development of the new mathematical technique called the "renormalization group" approach. It has shown promise for quantitative description of fluid properties and their dependence upon molecular shapes and forces.

The past 15 years have seen the beneficial use of critical phenomena in a variety of applications. Critical point drying is now a standard sample preparation method in electron microscopy. Further, there are remarkable changes in the solvent power of a liquid near its critical point. These are at work, for example, in the removal of caffeine from coffee for caffeine-free instant coffee and in the extraction of perfume essences. In addition, there are valuable research applications in liquid chromatography.

Chemistry of the Terrestrial and Extraterrestrial Materials

The Earth's geochemical phenomena involve complex mixtures, frequently with a number of crystalline and glassy (amorphous) phases, and they may take place at extremely high pressures and temperatures. Recent advances in high-pressure technology have made studies possible that duplicate conditions near the earth's core. In recent years many earth scientists have studied the "geochemical cycles" of elements—that is, the changing chemical and physical environment of a given element during such natural processes as crystallization, partial dissolving, change of mineral structure (metamorphism), and weathering. These processes may lead to concentration (e.g., ore deposits) or dispersion of an element. The geochemical cycle of carbon has provided a focus for the reawakened field of organic geochemistry. Research on the stability, conformation, and decomposition reactions of fossil organic molecules has led to greater understanding of the origin and composition of coal and other organic deposits. Such knowledge has obvious value that extends from guiding our exploration for new fossil fuel deposits to helping us decide how to use the ones we have.

Meteorites are of considerable chemical interest because they include the oldest solar system materials available for research and they provide samples of a wide range of parent bodies—some primitive, some highly evolved. Meteorites carry records of certain solar and galactic events and yield data otherwise unobtainable about the genesis, evolution, and composition of the Earth and other planets, satellites, asteroids, and the Sun. Unusual isotopic percentages of many metals and gaseous elements, and compositional data—particularly trace elements—have shed light on stages of the formation, evolution, and destruction of the original parent body or asteroid where the meteorite originated.

Within the last decade, the study of meteorites has been dramatically advanced by the recognition that if these projectiles from outer space land on the Antarctic ice sheet, they are immediately entombed in an inert environment and permanently refrigerated, stopping chemical changes. The question, of course, is how does one find these meteorites in the wide and forbidding spaces of this hostile region? Nature provides an astonishingly convenient answer. The Antarctic ice sheet is a vast glacier, so it gradually flows northward, carrying the meteorites with it. Over thousands of years, snow that fell near the South Pole finally reaches the end of the glacier where the ice begins to evaporate. Here at the glacier edges, the meteorites are dropped in great numbers, essentially never having been exposed to terrestrial life forms, erosion, or weathering. Since this discovery, more meteorites have been

METEORITES: AN ANTARCTIC TREASURE TROVE

collected (in the last decade) than over all of history before. The chemical and physical analysis of this meteorite treasure trove has only just begun.

ANALYTICAL CHEMISTRY

Characterization of atomic and molecular species—their structures, compositions, etc., is called *qualitative* analytical chemistry. The measurement of the relative amount of each atomic and molecular species is called *quantitative* analytical chemistry. Both areas contribute to and benefit from the current rapid progress in science. Basic discoveries from physics, chemistry, and biology are providing new methods of analysis. In return, these new abilities are central to research progress in chemistry, other sciences, and medicine, as well as to a wide range of applications in environmental monitoring, industrial control, health, geology, agriculture, defense, and law enforcement. Further, the 10-fold growth of the analytical instrumentation industry to $3 billion in sales worldwide has been led by the United States with its nearly $1 billion positive balance of trade in this area.

A key factor in this growth has been the incorporation of computers into analytical instrumentation. The benefits here are circular; modern computers have evolved through advances in solid-state technology. In turn, these advances have critically depended upon the ability to analyze quantitatively the concentrations of trace impurities in silicon, the key element in current computer technology. Now microprobe analyzers using computer imaging techniques are answering questions critical to making microcircuitry even smaller, which will produce computers that are faster, more reliable, and cheaper.

Analytical Separations

Analyses of some complex mixtures are possible only after separation of the mixture into its components. Then, a variety of identification and quantitative measurement schemes become effective that would be confusing or impossible if applied to the unseparated mixture. Hence, devising new separations for use in an analytical context is an active field of research.

There is no single technique more effective and generally applicable than the chromatographic method. The basic principle depends upon the fact that each molecular species, whether gaseous or in solution, has its own characteristic strength of attachment to, and ease of detachment from, any surface it encounters. The differences in these attachment strengths can furnish a basis for separation. The differences can depend upon heat of adsorption, volatility, interaction with the solvent, molecular shape (including stereogeometry), charge, charge distribution, and even functional chemistry. Great ingenuity has made it possible to use the whole range of molecular properties for analytical separations that can require only tiny amounts of material.

The different instrumental methods of chromatography will be discussed in Section V-C. For this discussion, a few illustrative examples will show the potential. In liquid chromatography, a solution of the mixture of interest passes through a column loaded with a suitable particulate material. For example, if an aqueous solution of pigments (such as those contained in carrot juice) is slowly passed through a tube containing small lumps of a suitable resin, the various pigments pass through the tube at different rates. The pigments that attach most weakly to the resin wash through fastest, and the ones that attach most strongly come out last. This provides a vivid example because we can actually see the different colors of the carrot juice pigments once they are separated. Of course, the method works to separate all sorts of compounds, whether colored or not. Under the best conditions, liquid chromatography can separate and reveal the presence of as little as 10^{-12} grams of a substance in a mixture. For gaseous samples, the technique can separate literally thousands of components such as are found in flavors, insect communication chemicals (pheromones), and petroleum samples. It is even possible to separate compounds that differ only in isotopic composition (e.g., deuterium instead of hydrogen!) by this method.

Two-dimensional chromatography can give additional specificity, resolution, and sensitivity by coupling with techniques such as electrophoresis, which involves the movement of substances in the presence of a high electric field. For example, two-dimensional electrophoresis can sort 2,000 blood proteins at once by separating a mixture spot of the sample linearly under one set of conditions, and then using another set of conditions to separate further the initial line of spots at right angles. Spot locations and amounts can be measured quantitatively with computerized scanning based on National Aeronautics and Space Administration computer programs developed for satellite pictures.

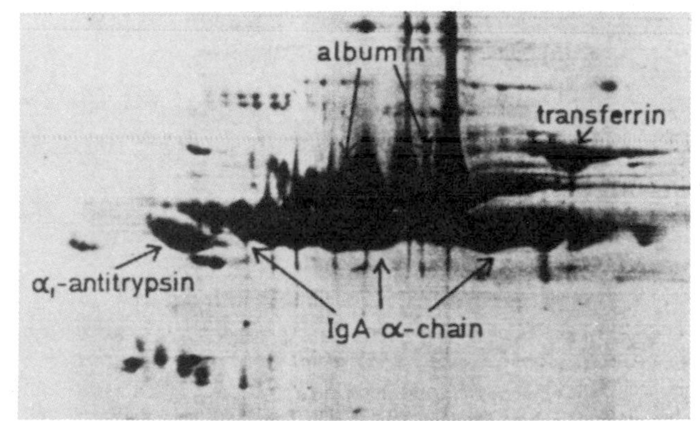

PROTEIN GEL PATTERN
HUMAN MYALOMA SERUM
5 $\mu\ell$

Optical Spectroscopy

The intellectual opportunities in this field, which introduce a variety of valuable analytical techniques, can be illustrated by two notable achievements of the last decade: the incorporation of computers as an essential part of most instrumentation, and the detection of single atoms and molecules. "Smart" commercial instruments now include microcomputers preprogrammed to carry out a wide variety of experimental procedures and sophisticated data analyses. The more powerful computers of

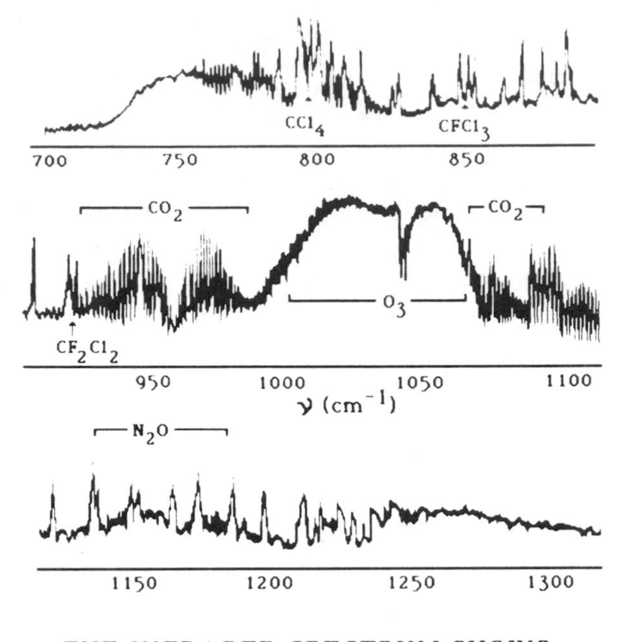

THE INFRARED SPECTRUM SHOWS ATMOSPHERIC POLLUTANTS EVEN AT NIGHT

the future will digest huge volumes of data from spectroscopic methods (especially Fourier transform and two-dimensional methods) much more efficiently. This will further improve resolution, detection limits, interpretation, spectral file searching, and immediate presentation of the results with three-dimensional color graphics to permit direct human interaction with the experiment.

Intense laser light sources are revolutionizing analytical optical spectroscopy. An immediate benefit is increased sensitivity. In special cases, resonance-enhanced two-photon ionization using tuned lasers has achieved the ultimate sensitivity: detection of a single atom (cesium) or molecule (naphthalene). Achievements in laser-induced fluorescence are approaching this same incredible limit. Laser remote sensing, such as for atmospheric pollutants, is effective at distances of over one mile; fluorescence excitation and pulsed laser Raman are particularly promising. In these latter methods, a laser pulse is emitted in the direction of the sample, which might be a smokestack plume. Then the time that it takes the fluorescence or Raman signal to return (at the speed of light) is measured to determine how far away the sample is. Thus, the signal not only tells us what substances (pollutants) are in the sample but also permits us to track them as they move away from the source.

The ability of a laser to emit a precise wavelength means there is the potentiality for the identification of one component in a mixture (without need for separation). Yet this selectivity is sometimes defeated because atomic and molecular absorptions can be much broader in wavelength than the laser line width. However, the resulting overlap can be eliminated by the wavelength narrowing that occurs at extremely cold, cryogenic temperatures. This cooling can be achieved for gaseous molecules by passing them through a nozzle to bring them to supersonic velocities. In an alternate approach, molecules can be embedded in a cryogenic solid, such as solid argon, at temperatures near that of liquid helium (a process called matrix isolation). These two complementary techniques minimize interference by rotational and vibrational absorptions and improve detection sensitivity and diagnostic capability.

Mass Spectrometry

This method involves separation of gaseous charged species according to their mass (see Section V-B), and it offers unusual analytical advantages of sensitivity, specificity, and speed (10^{-2}-second response). All of these attributes make for an ideal marriage to the computer. In the celebrated Viking Mars Probe, mass spectrometry was the basis for both the upper atmosphere analysis and the search for organic material in the planetary soil 30 million miles from home. Such sensitive soil sniffing to detect hydrocarbons might become a fast method for oil exploration. A special tandem-accelerator/mass spectrometer can detect three atoms of ^{14}C in 10^{16} atoms of ^{12}C, which corresponds to a radiocarbon age of 70,000 years. The broad applications of mass spectrometry include the analysis of elements, isotopes, and molecules for the semiconductor, metallurgical, nuclear, chemical, petroleum, and pharmaceutical industries.

In tandem mass spectrometry, one mass spectrometer (MS-I) feeds ions of a selected mass into a collisional zone where impacts cause fragmentation into a new set of fragment ions for analysis in a second mass spectrometer (MS-II). This technique, abbreviated MS/MS, offers a particularly promising frontier for analysis of mixtures of large molecules. "Soft" ionization that avoids extensive fragmentation is used first to produce a mixture of molecular ions. From this mixture, one mass at a time is selected by MS-I, and it is more vigorously fragmented to produce an MS-II spectrum that characterizes the structure of that one component. High speed and molecular specificity are important features of MS/MS. It is a powerful tool for analysis of groups of compounds sharing common structural features. It is particularly effective in removing any background signal caused by the contaminant species usually present in biological samples. It is now possible to determine the sequence of peptides with up to 20 amino acids and, in some instances, with sample sizes as small as a few micrograms.

Combined ("Hyphenated") Techniques

There is a growing appreciation for the extra benefits of using these computerized instruments in combination, such as the mass spectrometer coupled to a chromatograph (gas or liquid, GC/MS or LC/MS) or to another mass spectrometer (MS/MS), or these coupled with the Fourier transform infrared spectrometer (GC/IR, GC/IR/MS). High-resolution MS gives one part per trillion ($1/10^{12}$) analyses for the many forms of dioxin (TCDD) to see if the toxic form is present in human milk and the fatty tissue of Vietnam war veterans. GC/MS is necessary for the specific detection of 2,3,7,8-TCDD, the most toxic dioxin isomer. GC/MS is used routinely for detecting halocarbons in drinking water at concentrations far below the toxic level, polychlorobiphenyls (PCBs), vinyl chloride, nitrosamines, and for detecting most of the Environmental Protection Agency's list of other priority pollutants. MS/MS with atmospheric pressure ionization can monitor many of these contaminants continuously at the parts per billion level, even from a mobile van or helicopter. The high specificity as well as sensitivity of these methods make them especially promising for detecting nerve gases, "yellow rain," and natural toxins in foodstuffs (10^{-11} g of vomitoxin in wheat) and plants (*Astragalus* or "loco weed"). Metabolites found by GC/MS have led to the identification of more than 50 metabolic birth defects in newborn infants where early identification is critical in preventing severe mental retardation or death. One of the most exciting intellectual frontiers is the possibility that routine profiling of human body fluids can detect disease states well before external symptoms of those illnesses appear.

Electroanalytical Chemistry

Electrochemistry has a long history of analytical applications, beginning with pH meters. Today, pulse voltammetric techniques permit detection of picomole quantities (10^{-12} moles). Solid-state circuitry, microprocessors, miniaturization, and improved sensitivity have made possible continuous analysis in living single cells (with electrode areas of a few square microns). Electroanalytical methods are also useful in such difficult environments as flowing rivers, nonaqueous chemical process streams, molten salts, and nuclear reactor core fluids.

SEPARATIONS SCIENCES

Separations Chemistry

Separations chemistry is the application of chemical principles, properties, and techniques to the separation of specific elements and compounds from mixtures (including mineral ores). It takes advantage of the differences in such properties as solubility, volatility, adsorbability, extractability, stereochemistry, and ion properties of elements and molecules. As an example, the rare earth elements neodymium (Nd) and praseodymium (Pr), important in laser manufacture, must be separated from a mineral called monazite. A difficult part of this extraction is the separation from cerium, which is chemically similar. Photochemical studies show that this separation can be greatly enhanced by selective excitation to take advantage of the different chemistries of the elements under photoexcitation.

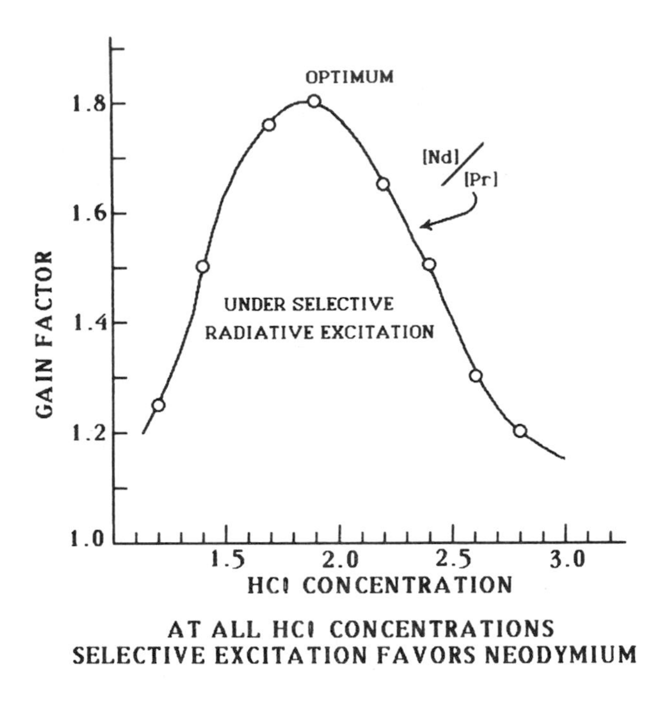

AT ALL HCl CONCENTRATIONS SELECTIVE EXCITATION FAVORS NEODYMIUM

The availability of critical and strategic materials to U.S. industry and the military is dependent in many instances on the development of practical, economical chemical separations methods. Table IV-C-1 shows our dependence on imports for some critical metals and minerals. For example, almost 90 percent of our use of platinum, in great demand as a catalyst, comes from imports. Mining of the major platinum source in the United States, in Stillwater, Montana, has not yet begun. A second important example concerns our access to uranium. About 13 percent of the nation's electrical energy is derived from nuclear energy, and a much larger percentage than that is utilized in the industrialized Northeast. Chemical separations are vitally important in the nuclear fuel cycle, beginning at the uranium mill where low-grade uranium ores (typically only 0.1 to 0.3 percent U_3O_8) are treated in selective chemical processes to produce a concentrate of more than 80 percent U_3O_8. Then, further refinement, based on transfer from one solvent to another (solvent extraction), or formation of the volatile fluoride, UF_6, produces a uranium product pure enough for use in nuclear fuel manufacture. Then, after removal from the reactor, the highly radioactive fuel is subjected to a selective chemical process to separate uranium and plutonium from the fission products for recycling or for weapons use. This step is a remarkable feat of chemistry and chemical engineering because the aim is to separate two similar elements, uranium and plutonium, from each other and also from the highly radioactive fission products, which include about half of the Periodic Table. All of this must be done in a remotely operated plant which, by robotics, handles tons of materials so radioactive that they cannot be approached by a human being.

TABLE IV-C-1 U.S. Import Dependence, Selected Elements (Imports as Percentage of Apparent Consumption)

	1950	1980
Manganese	77	97
Aluminum (bauxite)	71	94
Cobalt	92	93
Chromium	100	91
Platinum	91	87
Nickel	99	73
Zinc	37	58
Tungsten	80	54
Iron (ore)	5	22
Copper	35	14
Lead	59	<10

These are only a few examples of the many ways we depend upon separations chemistry. Future availability of many of the critical elements listed in Table IV-C-1 will depend, sooner or later, upon developing new chemical mining or separations processes that permit us to use low-grade domestic ores and the salt solutions (brines) that are found in geothermal wells. These developments will require research advances across a wide front, mainly focusing on the action of solvents and all of the properties of the liquid state that affect solvent power.

NUCLEAR CHEMISTRY

Since the days of the Curies, chemists have played a key role in the fundamental exploration of radioactivity and nuclear properties, as well as in nuclear applications to other fields. Thus, the 1944 Nobel Prize for the discovery of nuclear fission went to a chemist, Otto Hahn. Then, the 1951 Nobel Prize for the discovery of the first elements beyond uranium in the Periodic Table, neptunium and plutonium, went jointly to a chemist, Glenn Seaborg, and a physicist collaborator, Edward McMillan. Most of the advances in our understanding of the atomic nucleus have depended strongly on the complementary skills and approaches of physicists and chemists. Furthermore, the applications of nuclear techniques and nuclear phenomena to such diverse fields as biology, astronomy, geology, archaeology, and medicine, as well as various areas of chemistry, have often been, and continue to be, pioneered by people educated as nuclear chemists. Thus, the impact of nuclear chemistry is broadly interdisciplinary.

Studies of Nuclei and Their Properties

Particularly exciting advances have been made in extending our knowledge of nuclear and chemical species at the upper end of the Periodic Table. In the last 15 years, elements 104 to 109 have been synthesized and identified, often by ingenious chemical techniques geared to deal with the very short half-lives of these species (down to milliseconds). In addition to these new-element discoveries, many new isotopes of other elements beyond uranium have been found, and the study of their nuclear properties has played a vital role in advancing our understanding of alpha decay, nuclear fission, and the factors that govern nuclear stability. Fission research in particular has been quite fruitful. For example, the "nuclear Periodic Table" identifies particular stable proton-neutron combinations ("closed shells"); one of these is the tin isotope ^{132}Sn (50 protons, 82 neutrons). Changing this nucleus by only one nucleon gives a dramatic change in the nuclear fission behavior, both in the distribution of fission products obtained and in their kinetic energies. Furthermore, the study of spontaneously fissioning isomers among the heaviest elements has led to the important realization that the potential energy surfaces of these nuclei have two specially stable regions. This, in turn, opened the way to a new approach to calculating such surfaces—the so-called shell correction method.

Further exploration of the limits of nuclear stability is clearly in order, both at the upper end of the presently known nuclei and on the neutron-rich and neutron-poor sides of the region of stability defined by the stable nuclei found in nature. Newly discovered nuclear reaction mechanisms, based upon accelerating heavy nuclei as bombarding particles, promise to give access to more neutron-rich, and therefore much longer lived (minutes to hours), isotopes of elements with $Z > 100$ than have been available. This should open the way to more detailed investigations of the chemistry of these interesting elements at the upper end of the actinide series and beyond. The quest for so-called "superheavy" elements, i.e., nuclear species in or near the predicted "island of stability" around atomic number 114 and neutron number 182, has not been successful so far, but this exciting goal is still being pursued.

Space Exploration

The wide range of applicability of nuclear techniques is demonstrated in the exploration of the Moon and our companion planets during the past two decades. For example, the unmanned Surveyor missions to the Moon provided the first chemical analyses of the Moon. They employed a newly developed analytical technique that utilized the synthetic transuranium isotope ^{242}Cm. The analyses identified and determined the amounts of more than 90 percent of the atoms at three locations on the lunar surface. These analyses, verified later by work on returned samples, provided answers to fundamental questions about the composition and geochemical history of the Moon. Nuclear techniques also played an important role in the chemical analyses performed by Soviet unmanned missions to the Moon, and in experiments designed to seek life on the surface of Mars by the U.S. Viking missions. Similarly, isotopic distributions were important results in the analyses of returned lunar samples and of meteorites, making possible clarification of the history of the Moon and meteorites.

Isotopic Composition

Ever since the discovery of the isotopic composition of the chemical elements, it has been assumed that this isotopic composition is essentially constant in all samples, an assumption that provides the basis for assigning atomic weights. The only exceptions involved elements with long-lived radioactive isotopes. Since 1945, however, humans have affected the atomic weights of several elements (e.g., Li, B, U) under some circumstances. More fundamentally, it has been discovered that the solar system is not composed of an isotopically homogeneous mixture of chemical elements. Even for an element as abundant as oxygen, variations of the isotopic abundance have been noted for different parts of the solar system. Such isotopic variations have now been established for several chemical elements and provide clues to the processes that gave rise to the chemical elements, as well as to the conditions that existed at the birth of the solar system.

A startlingly large isotopic variation was discovered in the uranium of ore samples from the Oklo Mine in Gabon (West Africa) in 1972. Unusually low isotopic abundances of uranium-235 in these ores led to the astonishing conclusion that, 1.8 billion years before the first man-made nuclear reactor, nature had accidentally assembled a uranium fission reactor in Africa! This reactor was made possible by the higher ^{235}U concentration (~3 percent instead of the present-day 0.7 percent) at the time. Mass spectrometric analyses of various elements in the Oklo ore proved that isotopic compositions labeled them unmistakably as fission products. It also made it possible to deduce such characteristics of the reactor as total neutron flow (1.5×10^{21} neutrons cm^{-2}), power level (~20 kW), and duration of the self-sustaining chain reaction (~10^6 years). An important practical result of the Oklo studies is the fact that most fission products, as well as the transuranium elements produced in the reactor, did not migrate very far in 1.8 billion years. This has a clear relationship to the possibility of long-term confinement of radioactive waste products in geologic formations.

Nuclear Chemistry in Medicine

Nearly 20 million nuclear medicine procedures are performed annually in the United States (radioactive iodine thyroid treatment is one example). Advances in nuclear medicine depend crucially on research in nuclear and radiochemistry. For example, great progress in our knowledge of the chemistry of the element technetium in the past decade will clearly lead to much more effective applications of radioactive technetium, ^{99}Tc. This is the most widely used radionuclide, because the chemical properties of technetium compounds give them therapeutic activity. For example, technetium tends to concentrate in bone and particularly in cancerous bones, providing important diagnostic power.

Another important example is the development of especially rapid ways to incorporate into molecular structures short-lived isotopes that emit positrons. Two examples are the carbon isotope ^{11}C, with a 20-minute half-life, and the fluorine isotope, ^{18}F, with a 110-minute half-life. Both are produced through cyclotron bombardment. These nuclei are then placed in such compounds as ^{18}F-2-deoxy-2-fluoro-D-glucose and 1-^{11}C-palmitic acid in a time short enough to permit their use in positron emission tomography (PET), which is analagous to X-ray tomography (CAT scan). The positron technique is finding new clinical applications in studies of the nervous system and the heart, known as neurology and cardiology.

Stable isotopes, in conjunction with NMR spectroscopy, also have important applications in medicine. With ^{13}C, ^{2}H, ^{15}N, and ^{17}O tracers, NMR spectroscopy of humans will allow new insights into the molecular nature of diseases, provide a noninvasive method for their early detection, and make possible studies of metabolic processes in living subjects. This has led to one of the most exciting developments of the last few years, large object imaging. In this technique, a computer stores the NMR signals that result when an object as large as a human is slowly moved through the magnetic field of the NMR sample space. Then the computer reconstructs a three-dimensional image of the object, showing the location and local concentration of the atoms whose NMR is being measured. Thus, the presence and chemical form of key elements can be mapped in entire human organs in living patients. These powerful, noninvasive techniques were literally undreamt of 15 years ago. They have arisen in response to demands for ability to study via NMR ever larger biomolecules and working biological systems.

SUPPLEMENTARY READING

Chemical & Engineering News

"Vibrational Optical Activity Expands Bounds of Spectroscopy" by S.C. Stinson (C.& E.N. staff), vol. 63, pp. 21-33, Nov. 11, 1985.

"Progress Reported in Coupling LC and MS" (C.& E.N. staff), vol. 63, pp. 38-40, May 20, 1985.

"New Chromatography Columns Cut Need for Sample Preparation" by W. Worthy (C.& E.N. staff), vol. 63, pp. 47-48, Apr. 29, 1985.

"New Methods for Trace Analysis of Manganese" (C.& E.N. staff), vol. 63, pp. 56-57, Jan. 14, 1985.

"Microsensors Developed for Chemical Analysis" (C.& E.N. staff), vol. 63, pp. 61-62, Jan. 14, 1985.

"New Laser System Far Surpasses Mass Spec for Surface Analyses" by W. Worthy (C.& E.N. staff), vol. 62, pp. 20-22, Oct. 8, 1984.

"New Detectors for Microcolumn HPLC" (C.& E.N. staff), vol. 62, pp. 39-42, Sept. 17, 1984.

"New Methods Shed Light on Surface Chemistry" (C.& E.N. staff), vol. 61, pp. 30-32, Sept. 12, 1983.

"Archeological Chemistry" by P.S. Zurer, vol. 61, pp. 26-44, Feb. 21, 1983.

CHAPTER V

Instrumentation in Chemistry

All scientific knowledge is rooted in our abilities to observe and measure the world around us. Thus, science benefits enormously when more sensitive measuring techniques come on the scene. This is the situation in chemistry today.

The discussions to follow will identify a number of powerful instrumental methods that are now the everyday tools of research chemists. We will focus on the capabilities of today's instruments and on how much they have changed over the last decade or two.

A Laser Flashlight

A laser flashlight? Sounds like something out of Buck Rogers or Star Trek! What would that be? Well, to bring this down to Earth, let's first think about what a laser is and then how to make it into a flashlight.

Lasers are very special light sources. They put out pencil-sharp beams of pure color and so intense they can be used to cut patterns in steel. Also, they can be focused so sharply they are better than a surgeon's knife in mending the retina in your eye. Finally, they can give light pulses as short as a millionth of a millionth of a second! That's what's called a picosecond. With shutter speeds that fast, chemists can now "photograph" the fastest chemical changes known.

And how do lasers work? It all begins with a whole bunch of atoms or molecules all ready to emit light of exactly the same color. Atoms and molecules usually absorb light, not emit it, so somehow we've got to pump them up in energy so that they're more inclined to emit than absorb. This is called a "population inversion." Once we've got a population inversion, there are some tricky things to do with mirrors to make a laser out of it—but we don't have to go into everything.

So how do we pump up the molecules to get that population inversion? One good way to do it is to use electrical energy, like we do in a fluorescent light. That's what you might use to light up a dark closet to look for your missing sneaker. And it works fine as long as you have a long enough extension cord. But think of looking for your jack in the back of your car when you have a flat out on a dark highway. That's where a flashlight comes in handy.

In a flashlight, the energy comes from a chemical reaction. That's what the batteries are all about. Could we use a chemical reaction to pump a laser? If so, it would be a "chemical laser," our laser flashlight. But that would require a chemical reaction that produces a population inversion. Trying to find out whether there are such reactions led chemists to the discovery of the first chemical laser. Of course, it's hardly news that chemical reactions can emit light. Candles do it all the time. And think about a firefly—he (or she) can do it without an extension cord. These emissions show that a reaction follows special pathways, and when energy is released, these preferred pathways might be a super way to get a population inversion.

The surprise, though, was that chemical lasers were not discovered by looking at bright flames or by copying the firefly. Chemical lasers were discovered to operate best in the infrared, where the human eye is blind. This spectral region is where molecular vibrations can cause molecules to absorb or emit light. From these lasers we learned that quite a few reactions prefer reaction pathways that put most of the available energy into vibrational motions of the final products. Why this happens still isn't clear, but we're working on it. In the meantime, we have a whole bunch of fine chemical lasers. They can be very efficient, which has made at least one chemical laser a candidate to be the match to light the nuclear fire of nuclear fusion. If that worked, chemical lasers would help us get "clean" nuclear energy for the rest of time. Chemical lasers can also be very intense, as shown with the fluorine-hydrogen flame laser. What good is that? Well, that gets us back to Buck Rogers, Star Trek, and Star Wars. If you want a laser out in space, you'll either need a chemical laser or a mighty long extension cord.

So what's next? We're still after that firefly.

V-A. Instrumentation for Study of Chemical Reactions

Section IV-A indicated how the chemist's use of the most modern instrumentation is making it possible to investigate even the fastest chemical processes in intimate detail. We are witnessing a quantum jump ahead in our understanding of the factors that control the rates of chemical reactions. Among the tools responsible for this rapid advance are lasers, computers, molecular beams, synchrotrons, and, on the horizon, free-electron lasers. We will consider each in turn.

LASERS

Chemical lasers have been discussed in "A Laser Flashlight" on page 168. For a laser to work, it needs a "population inversion" in which there are more molecules that have enough energy to emit light than there are molecules ready to absorb light. To maintain such an inversion, energy must be injected somehow. Some energy-releasing reactions do this (resulting in chemical lasers), but the energy can be injected in other ways. The simplest way is through irradiation with a conventional light source. However, electrical energy input is probably the most convenient way to establish a population inversion. The apparatus need not be much different from a fluorescent light fixture.

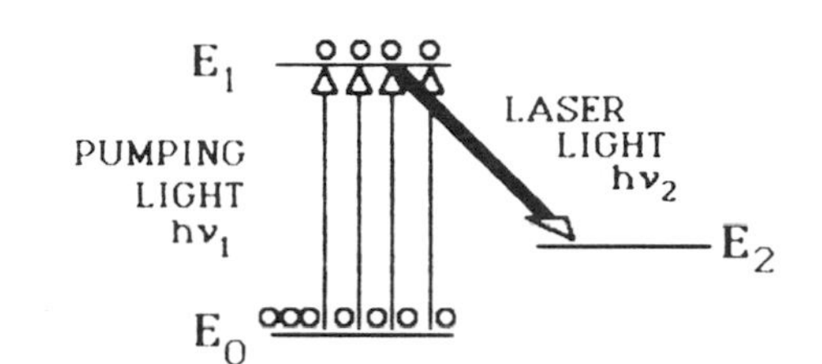

LIGHT ABSORPTION CAN ESTABLISH A POPULATION INVERSION

Whatever the manner of energy input (the "pumping" method), the special qualities of laser light arise from "stimulated emission," which can be regarded as the inverse of light absorption. A photon of light of the exact energy needed to excite a molecule from one energy level to another, higher level can stimulate emission of a second photon from a molecule already in the higher level. The second photon that is thus produced turns out to be perfectly in-phase ("coherent") with the electromagnetic wave of the first photon that started it all. This coherence gives lasers their distinctive character. It accounts, for example, for the pencil-sharpness that permitted us to reflect a laser "searchlight" beam off a mirror placed on the Moon by the Apollo astronauts.

The remaining feature of a laser is a set of accurately focussed mirrors that cause any stimulated emission to go back and forth many times through the population inversion. These mirrors are called an optical cavity; they permit and cause the buildup of the special qualities of laser light.

Lasers bring to mind a brilliant beam of light cutting through a sheet of steel or shining deep into space. But to a scientist, the beauty of the laser lies in its ability to deliver light of extremely *high intensity,* extremely *high power,* extremely *high spectral purity,* and/or extremely *short duration.* For a given experiment, laser design is dictated by the one of these features of greatest value to the experiment at hand, and usually at some sacrifice in the others. Some of this trade-off is

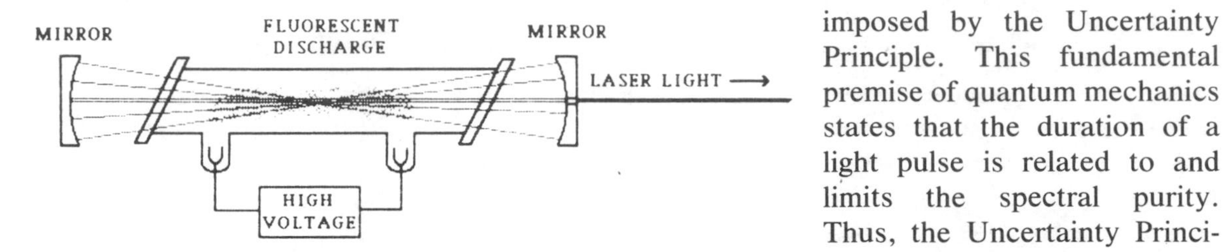

A LASER NEED NOT BE COMPLICATED

imposed by the Uncertainty Principle. This fundamental premise of quantum mechanics states that the duration of a light pulse is related to and limits the spectral purity. Thus, the Uncertainty Principle tells us that if a pulse is as short as one picosecond (10^{-12} seconds), there will be an uncertainty in the frequency (the color) at least as large as 5 cm^{-1}. With this much frequency spread, most information is lost about molecular rotations of gaseous molecules. On the other hand, if a line width of 0.005 cm^{-1} is needed to detect individual rotational states, then the molecule of interest must be examined by a light pulse at least as long as one nanosecond (10^{-9} seconds). This limitation deprives us of time information about species or events with shorter lifetimes than the nanosecond probe.

PULSE DURATION		⇄	SPECTRAL PURITY
ONE MICROSECOND	= .000 001 SECOND	⇄	.000005 cm^{-1}
ONE NANOSECOND	= .000 000 001 SECOND	⇄	.005 cm^{-1}
ONE PICOSECOND	= .000 000 000 001 SECOND	⇄	5 cm^{-1}
ONE FEMTOSECOND	= .000 000 000 000 001 SECOND	⇄	5000 cm^{-1}

PULSE DURATION LIMITS FREQUENCY ACCURACY AND VICE VERSA

Developments in the Last Decade

There were three crucially important developments in laser technology that took place during the 1970s and they are having a great impact on chemistry. First, several types of tunable lasers were developed and became commercially available. A "tunable laser" is one whose color (wavelength) can be selected according to need. The wider the range of the spectrum that lasers can work over, the more valuable they are as a research tool. The most important of these was the dye laser, which gave continuous color tuning throughout the visible region of the spectrum and a bit beyond into the near-infrared and near-ultraviolet. Dyes are chemical compounds whose intense color causes them to absorb light efficiently so that they can then emit coherent laser light. Second, the invention of efficient ultraviolet lasers gave scientists access to the photochemically important ultraviolet region at wavelengths shorter than 300 mm. These include "excimer lasers" that are based upon light emitted from molecules formed from electronically excited reactants. An example is the krypton fluoride laser. Krypton is an inert gas that does not form bonds in its ground state. After one of its valence electrons is excited, however, the resulting krypton atom has the chemistry of rubidium. Thus, the molecule formed between Kr and F has the bond strength and stability of RbF. This is a desirable factor in building up concentration to reach a population inversion, so that it can emit laser light. The third development was the discovery of methods of laser operation that gave short-duration light pulses—one picosecond or less.

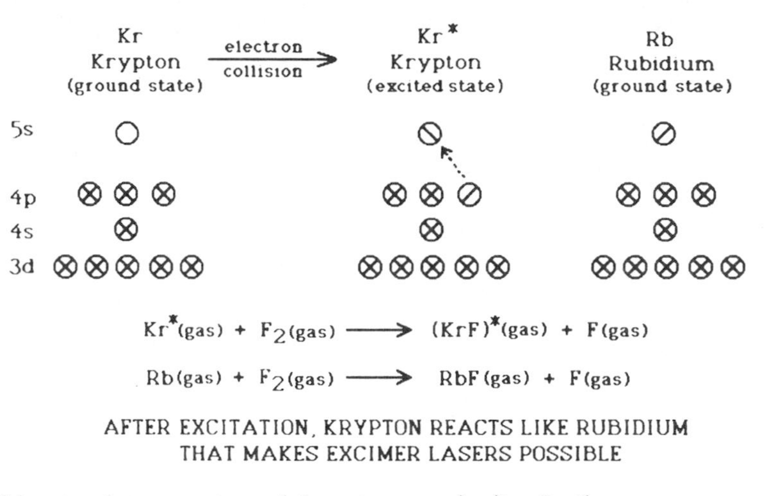

AFTER EXCITATION, KRYPTON REACTS LIKE RUBIDIUM THAT MAKES EXCIMER LASERS POSSIBLE

In 1970, the tunable dye laser did not exist except as a laboratory curiosity. In the early 1980s, almost every chemistry research laboratory had more than one tunable laser source. Tunable lasers can now be conveniently operated over the wavelength range from 4 microns in the infrared (40,000 Å) to 1,600 Å in the ultraviolet beyond the wavelength at which air becomes opaque (i.e., into the range called the "vacuum ultraviolet"). Already in the state-of-the-art stage are lasers that extend the wavelength range to beyond 20 microns (200,000 Å) in the infrared and to less than 1,000 Å in the vacuum ultraviolet.

Chemical Applications

Table V-A-1 lists many chemical applications of lasers. It is important to note that most of the more powerful lasers are not continuously tunable; they have only particular output wavelengths. They are most useful in the study of solid materials, which will usually absorb a wide range of wavelengths of light. For most chemical applications, tunable sources are critically important, and these lasers are often

TABLE V-A-1 Some Research Areas Utilizing the Laser

Area	Research Application	Laser Used
Photochemistry	Solar energy research, photosynthesis	Excimer, dye
Isotope separation	Uranium, plutonium isotope purification	Excimer, dye, TEA CO_2
Atomic absorption, fluorescence	Trace element analysis, environmental monitoring	Continuous ion, color center
Combustion diagnostics	Probing flames, explosions	Solid-state, dye
Atmospheric gas analysis	Monitoring industrial processes	Semiconductor diode
Biological cell sorting	Cell discrimination and separation	Ion laser
Cell bleaching	Photochemistry within biological cells	Dye laser
Microsecond kinetics $(1\text{-}100) \times 10^{-6}$ sec	Gas-phase deactivations, Chemical reactions	Flashlamp dye, TEA CO_2, chemical
Nanosecond kinetics 10^{-6} to 10^{-9} sec	Excited state lifetimes, very fast reactions	Solid-state, excimer
Picosecond kinetics 10^{-9} to 10^{-12} sec	Fast electronic state deactivation, coherence decay in liquids	Ion, solid-state
Subpicosecond kinetics $<10^{-12}$ sec	Vibrational deactivation in solids and liquids	Ion, solid-state

excited with another powerful, single-frequency laser. Having the best suited laser system is essential for work at many of today's most exciting chemical research frontiers.

COMPUTERS

The use of computers by chemists has paralleled the tremendous computer development of the last three decades. The size of this growth is reflected in the number of industrial installations of the largest IBM computers over this same time period. In the mid-1950s, there were 20 or 30 such machines (IBM 701s). By the mid-1960s, the much more powerful 7094 and 360 systems numbered about 350. Today, there are perhaps 1,700 industrial installations of IBM 3033s. This numerical growth has been accompanied by a phenomenal increase in computer power.

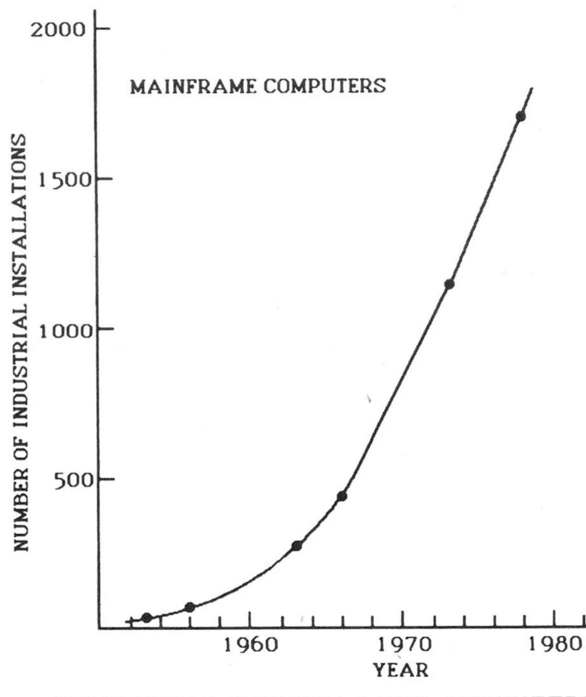

INDUSTRIAL USE OF LARGE COMPUTERS

The extent to which chemistry has benefited from this growth can be seen by comparing two landmark calculations. For polyatomic molecules, the first theoretical calculations based upon the Schroedinger Wave Equation without any simplifying assumptions (an *ab initio* calculation) appeared in the 1960s. Of special importance was the study of rotation around the carbon-carbon bond of ethane, C_2H_6. As the hydrogen atoms at one end rotate past the hydrogen atoms at the other end, the energy rises to a maximum. To learn the height of this internal rotation barrier, theoretical calculations ("self-consistent field" method) were based upon a basis set of 16 functions.

TABLE V-A-2 Relative Computing Speeds of Computer Levels

Computing	Example	Relative Speed
Superminicomputers	DEC VAX 11/780	(1)
Mainframes	IBM 3033	10-15
Supercomputers	CRAY 1S	80-120

This can be contrasted with a recent and similar study of decamethyl ferrocene, $[C_5(CH_3)_5]_2Fe$. This calculation used a basis set of 501 functions. Since such studies require computing effort proportional to the fourth power of the number of basis functions, the decamethyl ferrocene computation involves $(501/16)^4$ or one million times more computation than the ethane problem!

Superminicomputers

This level of computer has become a workhorse in chemistry. Instruments like the DEC VAX 11/780 are comparable to the world's largest mainframe computers available in the late 1960s. They have revolutionized computing in chemistry because of their substantial capacity, high speed, and lowered cost, which is now in the range $300,000 to $600,000.

The last 20 years have also seen three important development phases for the use of computers in chemical experiments. In the first, *computerization* phase, advances in both hardware and software greatly improved our ability to accumulate measurements (data acquisition). Then an *automation* phase increased the possibilities for experiment control through continuous monitoring of critical parameters. Finally, a *"knowledge engineering"* phase ushered in an era in which computers perform high-level tasks to interpret collected information.

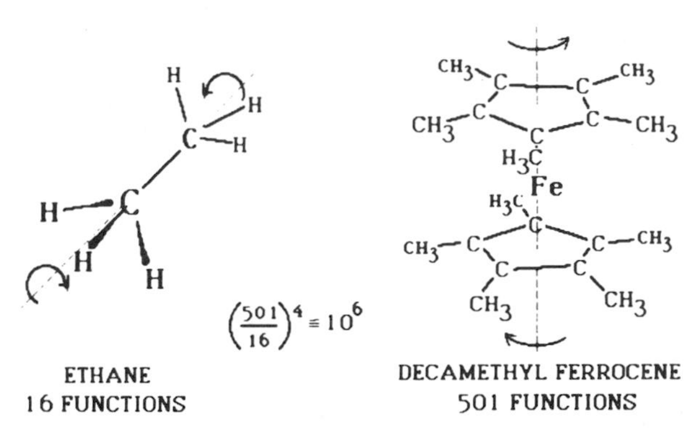

A MILLION-FOLD ADVANCE IN TWO DECADES

An excellent example is the Fourier Transform algorithm which permits us to record spectral data over a long time period, thereby to achieve high spectral resolution. Because this allows detection of quite weak signals, this algorithm is now routinely used to record ^{13}C NMR signals and to transform infrared interferograms. Because of the success of these instruments, the Fourier Transform algorithm is now being incorporated into all sorts of equipment: electrochemical, microwave, ion cyclotron resonance, dielectric, and solid-state NMR instrumentation.

Mainframe and Supercomputers

Some needs for computation in chemistry can be met only with the greater capacity and capability of the largest scientific computers (Cray/M and X-MP or CYBER 205), coupled with specialized resources such as software libraries and graphics systems. This is most notably true for electronic structure studies for many atom molecules beginning with the complete Schroedinger Equation and without approximations (*ab initio* calculations).

Another area that will benefit from supercomputers is computational biochemistry. Most dynamical simulation procedures applicable to biological molecules require calculation of the simultaneous motions of many atoms. A conventional 100-picosecond molecular dynamics simulation of a small protein in water would require about 100 hours on a DEC VAX 11/780 or 10 hours on an IBM 3033. Calculations of the rate constant for a simple activated process require a sequence of dynamical simulations to determine the free energy barrier, and additional simulations to determine the nonequilibrium contributions; the times can now reach 1,000 hours on a DEC VAX 11/780. More complicated processes or longer simulations become impossible without the much higher speeds of supercomputers.

MOLECULAR BEAMS

The advances of vacuum technology over the last three decades have made it possible to reduce the pressure in an experimental apparatus to a point at which molecular collisions become quite improbable (e.g., at pressures below 10^{-9} torr). Under these conditions, molecules that enter the vacuum chamber stream to the opposite chamber wall without deflection. Such a situation is called a "molecular beam." This provides a special opportunity to study chemical reactions. The most obvious application is to cause two such molecular beams to intersect. When a molecular collision does occur, it is almost always in this intersection zone. If the collision causes a chemical reaction, the product fragments leave the reaction zone with energies and directions that provide information about the reactive collision. By measuring the spatial distribution and fragment energies, we can learn intimate details about single-collision chemistry.

Capabilities

A typical, crossed molecular beam apparatus can contain as many as eight differentially pumped regions provided by various high-speed and ultrahigh-vacuum pumping equipment. It may be necessary to maintain a pressure differential from one atmosphere of pressure behind the nozzle of the molecular beam source to 10^{-11} torr at the innermost ionization chamber of the detector. What is glibly called the "detector" is likely to be an extremely sensitive mass spectrometer with which to measure the velocity and angular distributions of products. By replacing one of the beams by a high-power laser, molecular beam systems are now giving new kinds of information on the dynamics and mechanism of primary photochemical processes.

In the past 5 years, molecular beam experiments have played a crucial role in advancing our fundamental understandings of elementary chemical reactions at the microscopic level. These advances provide deeper insights with which to build our explanations of macroscopic chemical phenomena from the information gathered in microscopic experiments. The pervasive importance of these deeper insights was recognized in the award of the 1986 Nobel Prize in Chemistry to those responsible for bringing molecular beams into chemistry.

SYNCHROTRON LIGHT SOURCES

Characteristics of Synchrotron Sources

The most intense, currently available source of tunable radiation in the extreme ultraviolet and X-ray region is synchrotron radiation, which is produced when energetic electrons are deflected in a magnetic field. That happens, of course, all the time in a synchrotron, which is an instrument that accelerates electrons to very high energies for particle physics studies. To reach these high energies, the electrons must be "recycled" through the accelerating zone many, many times. "Recycling" requires bending their trajectories through four successive 90-degree turns. At each of these turns, the acceleration needed to change direction causes intensive radiation over the entire spectral range from the far-infrared to the X-ray range. This has, in the past, been looked on as an irritating energy loss.

Now, however, synchrotrons are running out of things to do in high-energy physics. Hence, attention has turned from synchrotrons as accelerators (with radiation seen as undesired energy loss) toward snychrotrons as sources of light. Devices are placed inside the accelerator that increase the number of sharp bends in the electron trajectories to increase these radiative properties. These devices are descriptively called "wigglers" or "undulators." They show potential for intensity increases by several powers of 10 over the already bright radiation emitted by an ordinary synchrotron. Principal current use of tunable synchrotron radiation falls in the X-ray energy range, 1-100 keV.

Applications of Synchrotron Sources in Chemistry

Extended X-ray Absorption Fine Structure (EXAFS) has been one of the more fruitful applications of synchrotron radiation to solid substances. When one of an atom's inner-shell electrons is excited by an X-ray photon, the atom emits light that is then diffracted by neighboring atoms. The result is a diffraction pattern that contains information about the interatomic spacings of these neighbors. Much attention has been directed toward crystal structures of inorganic solids, some of it seeking information on oxidation state when other methods are not definitive. Since heavy atoms are most readily detected, EXAFS has been usefully employed to learn the immediate chemical environment of transition metal atoms as they occur in biologically important molecules, including maganese in chlorophyll.

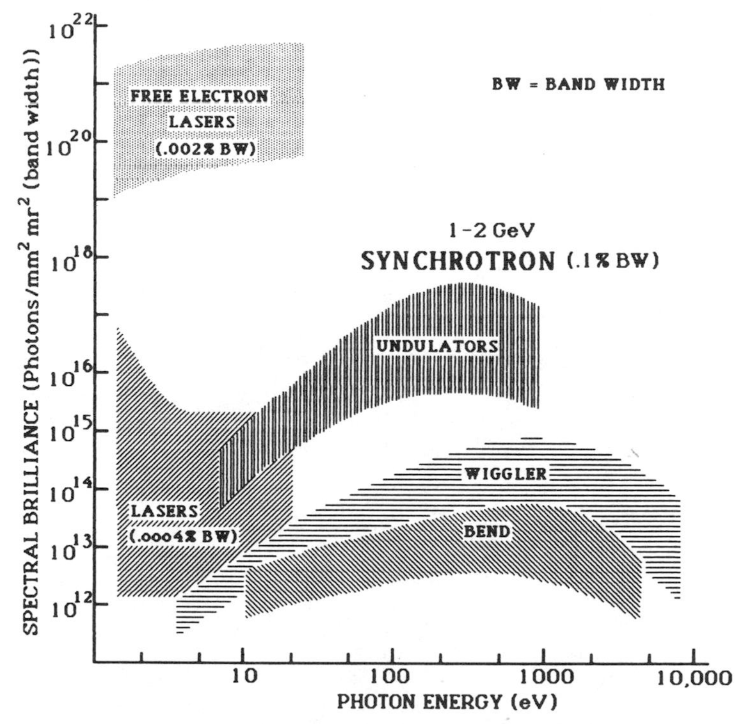

DESIGN GOALS ARE AMBITIOUS - AND PROMISING

FREE-ELECTRON LASERS

When a beam of electrons with velocities near the speed of light moves through a periodically alternating magnetic field (a wiggler), light is emitted in the direction of the electron beam. The wavelength of the light is determined by the period of the wiggler field and the energy of the electrons. This causes it to behave like a population inversion; i.e., if it is placed between the mirrors of a conventional laser, stimulated emission can occur to produce laser light. Such a device is called a Free-Electron Laser (FEL).

Potential Capabilities

Experience to date indicates that high-efficiency wavelength tunability and high average and peak power will all be forthcoming over a wavelength range extending from microwave frequencies through the infrared and visible to the vacuum ultraviolet spectral ranges. Average brightnesses several powers of 10 greater than those provided by conventional tunable lasers or synchrotron sources might be possible, particularly in the ultraviolet. An FEL has been operated at Los Alamos National Laboratory, based on a linear accelerator 2 or 3 meters long. Once a second, the device provides a train of pulses of tunable infrared radiation—currently, in the 9- to 11-micron wavelength range, with 30-picosecond pulses, peak power of 5 megawatts, and 50-nanosecond spacing between pulses. Such performance extended over the mid-infrared spectral region (4 to 50 microns) would open the way to many novel applications in chemistry. Examples are vibrational relaxation, multiphoton excitation, nonlinear processes in the infrared region, fast chemical kinetics, infrared study of adsorbed molecules, and light-catalyzed chemical reactions. As the wavelength is moved through the visible and toward the ultraviolet, a variety of novel chemical applications could be explored in photochemistry and fast chemical kinetics, as well as multiple photon and other nonlinear processes.

SUPPLEMENTARY READING

Chemical & Engineering News

"Laser Vaporization of Graphite Gives Stable 60-Carbon Molecules" by R.M. Baum (C.&E.N. staff), vol. 63, pp. 20-22, Dec. 23, 1985.

"Imaging Method Provides Mass Transport" (C.&E.N. staff), vol. 63, p. 29, Sept. 23, 1985.

"Computers Gaining Firm Hold in Chemical Labs" by P. Zurer (C.& E.N. staff), vol. 63, pp. 21-31, Aug. 19, 1985.

"Supercomputers Helping Scientists Crack Massive Problems Faster" by R. Dagani, vol. 63, pp. 7-14, Aug. 12, 1985.

"Spectroscopic Methods Useful in Inorganic Labs" (C.&E.N. staff), vol. 63, pp. 33-39, Jan. 14, 1985.

"Technique Allows High Resolution Spectroscopy of Molecular Ions" by R.M. Baum (C.&E.N. staff), vol. 62, pp. 34-35, Feb. 20, 1984.

"Extreme Vacuum Ultraviolet Light Source Developed" by R.M. Baum (C.&E.N. staff), vol. 61, pp. 28-29, Feb. 7, 1983.

"Synchrotron Radiation" by K.O. Hodgson and S. Doniach, vol. 56, pp. 26-27, Aug. 21, 1978.

The Ant That Doesn't Like Licorice

While touring through the Costa Rican jungle recently I stumbled on a terribly wide path completely devoid of plant life. The path must have been 6 feet wide, and as I strolled along it I tried to keep out of the way of the native ants who were bustling past me. Each of the ones going the other way was carrying a big piece of leaf overhead; the whole bunch looked like a fleet of Chinese junks sailing along.

Suddenly, I was overtaken by this very attractive native ant. "Hi there!" I introduced myself, "My name is Red Ant. What's your name?" Blushing, she answered,"My last name is Formicidae, but they call me Leafcutter." "Say, that's a pretty name. Why do they call you that?" Giggling, she said, "Everyone knows why—it's because that's my business." She gracefully pointed an antenna at a pitiful-looking tree up the path. "See that?" she asked, "My sisters and I did that. We cut every one of the leaves off that tree in only 5 days. Enough to feed the whole family for 2 months."

She turned to leave. "Don't go!" I exclaimed, "I'll get you a leaf from this tree right here." Reaching toward a lush tree that all the other ants were passing by, I pulled off a leaf and presented it to Leafcutter. "Pew!" she said, holding her nose, "Take it away—I hate licorice." Sure enough, the leaf I held smelled just like licorice. I wondered what was wrong with licorice. Leafcutter explained "I'm not sure why, but Mama doesn't like it when we bring leaves that smell like that into the anthill." I was still puzzled so I asked her if she'd show me her home.

Leafcutter lived in this gorgeous anthill along with her 5 million sisters, 500 brothers, her Mama, and, believe it or not, a fungus! Her lazy brothers never lifted a feeler to bring in even one leaf—all they seemed to do was amuse Mama. And guess what? The ants didn't even eat all those leaves they brought home at all—the fungus did!

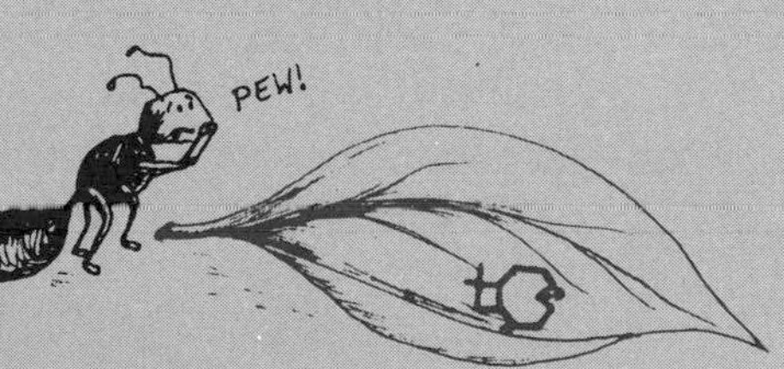

Apparently, the ants don't have the right enzymes to metabolize carbohydrates. But the fungus thrives off those leaves, and in gratitude to the ants for supplying them, it converts their carbohydrates into delicious sugars that the ant family lives on. "Mama says we're symbiotic," Leafcutter explained.

Scientists have also taken an interest in Leafcutter and her family. They have concentrated on the leaves that Leafcutter doesn't like, trying to find out what protects these leaves over the others. Using liquid chromatography, they've extracted 10 to 15 milligrams of about 50 different compounds from great piles of the rejected leaves. Then they've worked to purify and identify these compounds. NMR studies have shown that every one of those trees that Leafcutter dislikes contain compounds with molecular structures like that of carophyllene oxide, the compound that gives licorice its flavor.

They've also got this notion that it's the fungus that gets sick on those leaves. And when the fungus gets sick there's no sugar for the ant family. So it looks as though the licorice-flavored trees have learned to synthesize their own fungicide to protect themselves from the Leafcutters. The next step for those scientists will be to try to synthesize some similar compounds to combat harmful fungi elsewhere. The next step for me is into a cozy little anthill with Leafcutter—we're getting hitched in the spring.

V-B. Instrumentation Dealing with Molecular Complexity

The chemical identification and synthesis of complex molecules ultimately depends upon the chemist's ability to bring about a chemical change and then ascertain the composition and three-dimensional structures of the products. The fact that chemists are now active in the biological arena shows the capability that now exists. It permits us to expect to understand the chemistry of life processes at the molecular level. All this is within reach because of diagnostic tools invented by physicists and sharpened by chemists to meet the analytical and structural challenges presented by extremely complex molecules. Foremost among these tools are nuclear magnetic resonance, X-ray diffraction, and mass spectrometry.

NUCLEAR MAGNETIC RESONANCE (NMR)

The nucleus of an atom carries electrical charge, and its behavior in a magnetic field shows that it acts like a tiny magnet. To help explain the existence of this magnetic property, we attribute to the nucleus a rotational movement, or spin. If the electrical charge of the nucleus is distributed over the volume of the nucleus, then nuclear spin implies that some of this charge would move in a circle around the axis of rotation. Such a charge movement would generate a magnetic field, so the spin concept "explains" why the nucleus acts like a tiny magnet. When placed between the poles of a large magnet, the nuclear magnet will, like a compass, try to align itself parallel to the field. It will then require an input of energy to flip the magnet to an orientation opposing the field.

Through fine spectroscopic measurements, scientists have found that both the nuclear spin and the energy of interaction between the nuclear magnet (the "magnetic moment") and an external field are "quantized," as are all atomic properties. In contrast to the behavior of macroscopic magnets, only particular values of nuclear spin are found in nature, and these values determine sharp "energy levels." These discrete energy levels provide the basis for a nuclear spectroscopy called *nuclear magnetic resonance,* or NMR.

Spectroscopic studies guide us in assigning quantum numbers to the spin of a particular nucleus. Thus, electrons and protons are found only with spin quantum numbers of $+1/2$ or $-1/2$. A deuteron (a nucleus containing a proton and a neutron) has a spin of 1. The nuclei of ^{12}C and ^{16}O each have spins of zero (0) (i.e., they have no nuclear magnetic moment). In contrast, the isotopic nuclei ^{13}C, ^{14}N, ^{15}N, and ^{17}O have nuclear spins, respectively, of 3/2, 1, 1/2, and 1/2.

These nuclear spins determine the number of energy levels that will be seen if the nucleus is placed between the poles of a large magnet. A nuclear spin of zero (0) means that there will be no interaction with the field (i.e., ^{12}C and ^{16}O will be invisible). A spin of 1/2 implies two energy levels which correspond to the nuclear magnet oriented either parallel to the field ($+1/2$) or opposing the field ($-1/2$). A spin of 1 implies three energy levels that correspond to the nuclear magnet parallel to ($+1$), opposing (-1), or perpendicular (0) to the field. In general, if the spin is S, there are $(2S + 1)$ energy levels. As usual in spectroscopy, we can sense and measure these energy levels through the absorption of light.

The spacing of the energy levels depends, first, upon the magnitude of the applied magnetic field. It also depends upon the magnitude of the nuclear magnetic moment, which is not fixed by the spin quantum number but is dependent upon the nuclear structure. For a given nuclear moment, the energy level spacing can be increased by increasing the applied field. That makes the energy spacing easier to measure and improves the resolution. Hence, progression in the field of NMR spectroscopy has been connected with, and limited by, our ability to produce very high and very uniform magnetic fields. Present-day NMR performance takes advantage of the compactness of superconducting magnets to produce magnetic fields of tens of thousands of gauss (10^{-15} tesla).

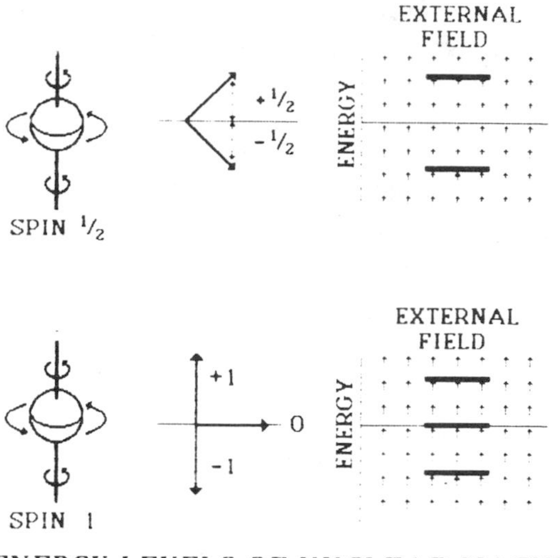

ENERGY LEVELS OF NUCLEAR MAGNETS IN AN EXTERNAL MAGNETIC FIELD

But in the 1950s, long before superconducting magnets, physicists began measuring the magnetic properties of the nucleus to learn about its structure. Their precision was sufficiently great that the physicists discovered, a bit to their dismay, that the measured nuclear resonance frequency depended not only on the magnetic properties of the nucleus but also upon the chemical environment the nucleus found nearby. Chemists were elated, however, since they saw the method as a new probe of molecular structure to supplement the rapidly developing infrared spectroscopic methods. Instrument developers quickly responded to the many opportunities seen for applications in chemistry. The outcome surpassed the most extravagant dreams. Today, NMR is surely one of the most important diagnostic tools used by chemists. It has had momentous impact in such diverse areas as synthetic chemistry, polymer chemistry, mechanistic chemistry, biochemistry, medicinal chemistry, and even clinical diagnosis. For example, we are now able to distinguish the chemical neighborhoods of the hydrogen atoms in molecules as complex as segments of DNA.

Solution NMR

Thus far, most of the chemical applications of NMR have involved liquid solution samples. This is because differences in chemical environments are sharply revealed because of averaging effects of the random motions in the liquid state. Performance has been limited by the uniformity of the high magnetic fields required, which also limits sample size and sensitivity. Through the 1960s and 1970s, technological developments (including superconducting magnets) permitted steady increases in magnetic field intensity and uniformity. Now a barrage of new developments in other factors, including Fourier Transform methods, high-resolution solid-state techniques, and a variety of pulsed measurements, is opening new dimensions for NMR.

Fourier Transform NMR (FT NMR)

Modern computers make it possible to record data continuously for some time period and then to transform the accumulated information into a frequency spectrum (See Section V-A, Computers). This Fourier Transform method was first applied to NMR in 1966; because of the better performance it brings, virtually all commercial research instruments now use FT. For example, it permits detection of the ^{13}C isotopically labeled molecules in an organic compound based on the ^{13}C present in nature (1 in 100 carbon atoms is ^{13}C). At the same time, improvements in superconducting magnet technology raised the magnetic field intensity almost threefold (from 5 tesla in 1966 to 12–14 tesla in 1979). Together, those two improvements provided 100-fold increases in sensitivity and 10-fold increases in resolution. Chemists can now ascertain the proton positions in the molecular structure of the anti-Parkinson's disease drug L-dopa with as little as 5-10 milligrams of sample. NMR spectra of such complex molecules as insulin and abnormal hemoglobin (e.g., sickle cell) can be studied. Such instruments are now essential for research on all new pharmaceuticals, novel anticancer drugs, hormones, and products of recombinant DNA technology.

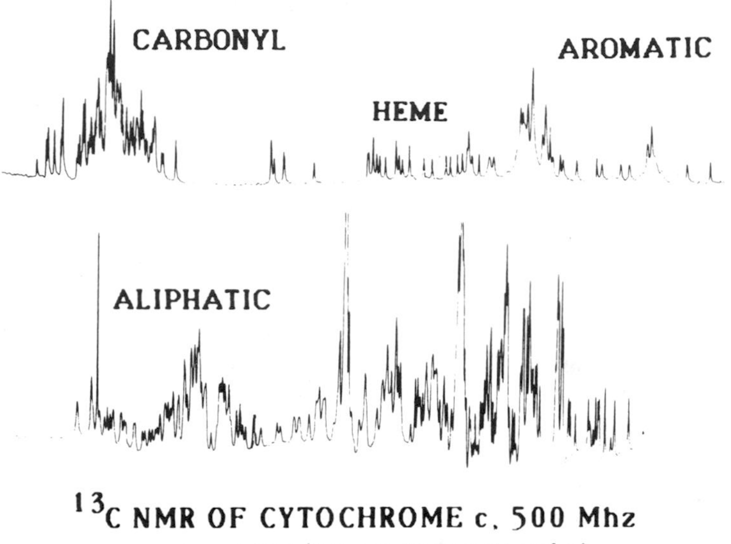

^{13}C NMR OF CYTOCHROME c. 500 Mhz
circa 1984 (NOT POSSIBLE IN 1969)

Solid-State NMR

In the late 1960s, a variety of pulsed NMR experiments was introduced which reawakened interest in obtaining high-resolution NMR spectra of solids, despite the fact that molecules in solids are fixed in position, so that the averaging effects of molecular motions in liquids are lost. Initially, abundant and sensitive nuclei (^{1}H, ^{19}F) were studied with resolution near one part per million. Then, in the period of 1972-1975, methods were developed in which the tube containing the sample is rapidly spun around an axis tilted relative to the magnetic field. Then, the spectrometer sees a "blur," an average, of the NMR spectra of all of the angles through which the spinning sample moves. The blurring effect is quantitatively calculated with an "averaging function," ($1-3 \cos^2 \theta$), where θ is the angle of tilt. If this tilt angle is fixed to be 54.7 degrees, the averaging function [$1-3 (\cos 54.7)^2$] is equal to zero. This angle is called "the magic angle." NMR spectra of solid samples spun at this magic angle provide band sharpening approaching those available for liquids. Today, both organic and inorganic solids can be studied at 0.01 parts per million resolution. Novel applications that have been made to inorganic samples include observations of quartz formed at meteor impact in which silicon atoms are found in unusual, six-coordinate crystal positions. Structures in rubbers, plastics, papers, coal, wood, semiconductors, and high-tech ceramics can be examined over wide temperature ranges, from 4K to 500K.

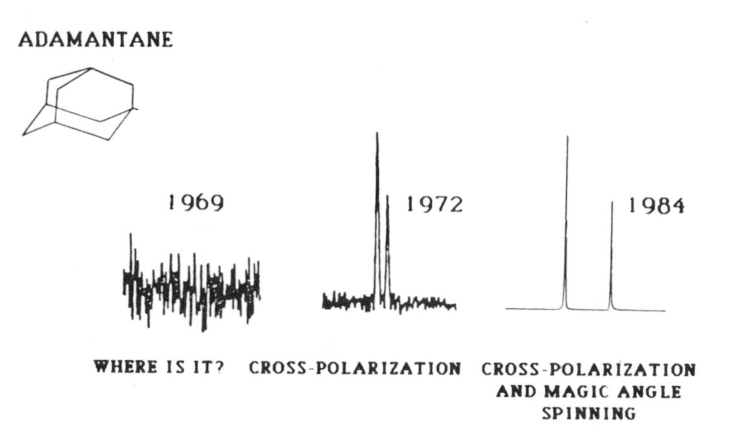

IN NMR, THINGS ARE GETTING BETTER!

Two-Dimensional NMR

By means of cleverly timed pulsed radiofrequency excitation techniques, it is now possible to observe multiple-quantum transitions and to record NMR spectra in "two dimensions." Such 2D spectra appear as contour maps in which different types of interaction spread out resonances along two axes. In addition to the characteristic frequency shifts caused by atoms in the immediate neighborhood (e.g., by which we can distinguish between CH_2 groups and CH_3 groups), the new dimension reveals more distant interactions. Thus, information about molecular shapes can be determined for complex molecules even when single crystals cannot be obtained (so X-ray techniques cannot be used). This is quite crucial for biological molecules because it gives access to conformational information under conditions close to the living conditions in which biological molecules actually function.

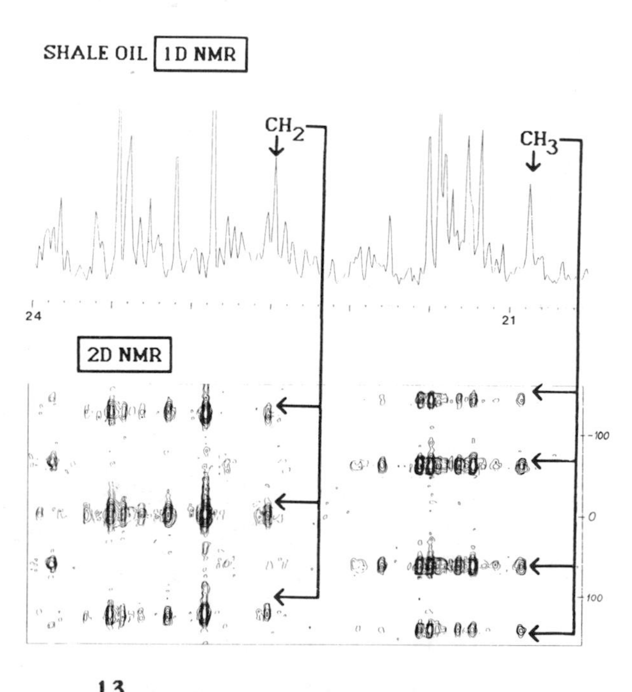

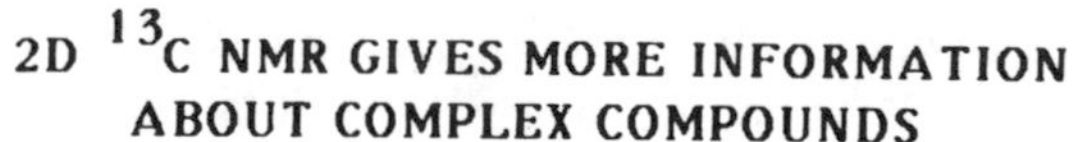

2D ^{13}C NMR GIVES MORE INFORMATION ABOUT COMPLEX COMPOUNDS

Imaging

In 1973 the first spatial resolution by NMR was reported by chemists. Today, there exist instruments capable of "mapping" in three dimensions the NMR chemical shifts and nuclear concentrations for objects as large as a human patient. Such NMR scanners, comparable in some respects to X-ray CT scanners, appear to have considerable potential for diagnosis of diseases, possibly including multiple sclerosis, muscular dystrophy, and malignant tumors. Most important, this diagnostic method does not require surgery or other invasive techniques. Further increases in field strength should permit real-time imaging of, for example, a beating heart. In a closely related, but invasive, medical application, NMR measurement coils have been surgically placed around intact and functioning animal organs. These have been used to study metabolism by measuring high- resolution phosphorus, carbon, and sodium NMR spectra in the organ while it is operating. These remarkable uses of NMR place before us the possibility of studying the chemistry of a living system truly *in vivo*.

NMR Performance, Availability, and Costs

Resolution and sensitivity of an NMR instrument depend upon the interplay among the magnetic field intensity, sample volume, and field uniformity over that sample volume. As chemists are working with more and more complex molecules, better resolution advances research capabilities as soon as it becomes technologically feasible. This can be seen in the steady rise in the magnetic fields available in commercial NMR instruments (as expressed in the proton NMR frequency, usually given in megahertz, MHz). Over the last 25 years, the highest field available has increased by a factor of about 1.5 every 5 years or so. Unfortunately, the resultant higher performance, coupled with other improvements, has exponentially increased the cost, hence, the availability, of the highest performance machines. Thus, the price of commercial NMR instruments has risen from about $35,000 in 1955 to $850,000 in 1985, a few percent per year faster than inflation.

The critical importance of state-of-the-art NMR instrumentation is reflected in the annual sales of NMR instruments, which, in 1984, totaled about $100 million. The most advanced NMR spectrometers were 500-MHz instruments, and about 70 such instruments had been produced worldwide. Many of these are in U.S. industrial laboratories; a number of them are in Europe, Japan, and the Soviet Union; and about 17 of them are placed in U.S. academic institutions. Magnet technology will presently permit commercial production of 600-MHz instruments at a cost of about $850,000, and on the horizon are 750-MHz instruments with an expected cost near $1.5 million.

Applications of NMR by chemists have revolutionized much of chemistry; and they are having profound influences on adjacent research fields in biochemistry, materials research, geochemistry, botany, physiology, and the medical sciences. Thus, while the costs of modern NMR instrumentation are high, the potential rewards are so great that we cannot afford to lose them.

MASS SPECTROMETRY (MS)

In a mass spectrometer, a molecule of interest is converted to a gaseous ion, and the ion is accelerated to a known kinetic energy with an electrical field. Then its mass can be measured, by tracking either its curved trajectory through a known magnetic field or its time of flight through a fixed distance to the detector. The first step, the production of molecular ions, causes some of the molecules to fragment and give a collection of ions whose masses are determined by the structural units in the original molecule. Thus, experience leads us to expect that the mass spectrum of CF_3-CH_3 will include a mass peak at 84 due to the "parent" ions $(CF_3\text{-}CH_3)^+$ but also a prominent peak at mass 69 due to $(CF_3)^+$ and another at 15 due to $(CH_3)^+$. Thus, the mass spectrum gives far more information than just the molecular weight of the parent molecule. Furthermore, the mass spectrometer can be coupled to other techniques, such as infrared spectroscopy or gas chromatographic fractionation, to add greatly to the significance of the mass spectrum. These coupling schemes are discussed in Section IV-C as a part of Analytical Chemistry.

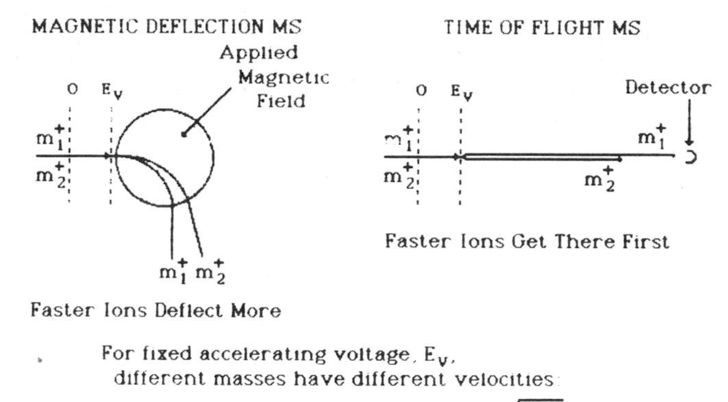

Applicability

Some scientists feel that gas chromatographic fractionation (see Section V-C) followed by mass spectrometric analysis provides the best general purpose analytical instrument for handling complex mixtures drawn from chemical, biological, geochemical, environmental, and crime lab applications. Until recently, however, such analytical use was limited to compounds that would vaporize at a temperature within their range of thermal stability. Now, over the last decade, applications of mass spectrometry are rapidly widening because of a recently developed series of related techniques using ion, neutral, and photon bombardment to desorb ions from solid samples (see Table V-B-1). These techniques dramatically increase the molecular weight range of mass spectrometry. Plasma desorption under bombardment by the fission fragments from the radioactive Californium isotope ^{252}Cf has given molecular ions of molecular weight 23,000 from the polypeptide trypsin, while Fast Atom Bombardment (FAB) has provided extensive structure information on a glycoprotein of molecular weight about 15,000. Laser and field desorption have produced molecular ion mass spectra displaying the oligomer distribution of sections of DNA. Now molecular weights of 20,000 can be measured, and mass resolution of one part in 150,000 is available in commercial instruments. Perhaps 5- to 10-fold higher resolution can be achieved with Fourier Transform techniques for relatively low-mass ions. Extremely high resolution can be quite useful to distinguish between the masses of one deuterium and two hydrogen atoms (seven parts per ten thousand) or between one ^{13}C atom and a ^{12}C plus a hydrogen atom (three parts per ten thousand). This becomes extremely important as we decipher the mass spectrum of a large molecule because both

TABLE V-B-1 Desorption Ionization Techniques for Analysis of High-Molecular-Weight Substances

Field Desorption (FD). Samples placed on a fine, carbon-coated wire are subjected to heat and high electric fields. Commercially available, somewhat erratic, but has been productively employed.

Plasma Desorption (PD). Samples placed on thin foil are bombarded with high energy fission fragments from radioactive californium (^{252}Cf) or ions from an accelerator. Not commercially available.

Secondary Ion Mass Spectrometry (SIMS). Solid samples are bombarded with kilovolt electrons. Low electron fluxes are used for molecular SIMS; high fluxes for inorganic analysis and depth profiling. Commercially available.

Electrohydrodynamic Ionization (EHMS). Samples are dissolved in a glycerol-electrolyte solvent. Desorption from solution occurs under high electric fields and without heating. Almost no molecular fragmentation! Not commercially available.

Laser Desorption (LD). Both reflection and transmission experiments and various sample preparations can be used. Tendency toward thermal degradation. Commercially available with time-of-flight mass analysis.

Thermal Desorption (TD). Sample is placed on probe tip which is heated to desorb ions (no ionization filament is used). Useful for inorganic analysis; recently applied to organic salts.

Fast Atom Bombardment (FAB). Samples in solution (usually glycerol) are bombarded with kilovolt-energy atoms. Fluxes higher than in SIMS. Wide applicability to biological samples, including pharmaceuticals. Commerically available.

deuterium and ^{13}C are present in nature. Consider, for example, that a molecular weight near 900 implies 60 or more carbon atoms. For such a molecule, the ^{13}C present in nature (1.1 percent) is sufficient that about half of the molecules will have at least one ^{13}C atom.

The breadth of use of mass spectrometry is shown by the fact that about $200 million worth of instruments are purchased each year. Several thousand people in the United States are engaged full-time in using them, more than double the number so employed 15 years ago. The chemical, nuclear, metallurgical, and pharmaceutical industries all make extensive use of mass spectrometry. Environmental regulations (particularly those covering organic compounds in water supplies) are written around mass spectrometry. Established and emerging methods of geologic dating and paleobiology are based on this technique. Research applications in chemistry are innumerable, ranging from routine analysis in synthetic chemistry to beam detection in a molecular beam apparatus.

Sensitivity and Selectivity

An unknown sample can be *identified* using MS with as little as 10^{-10} grams (100 picograms), while a specific compound with a known fragmentation pattern can be *detected* with as little as 10^{-13} grams (100 femtograms). As a striking example, a 0.1-milligram dose per kilogram of body weight of Δ9-tetrahydrocannabinol (an active drug from marijuana) can be tracked in blood plasma for over a week down to the 10^{-11} grams per milliliter level using combined gas chromatography and tandem mass spectrometry. As an example of specificity, in a simple MS examination of a coal sample containing a small amount of trichlorodibenzodioxin, interference by the great variety of similar compounds in the sample ("chemical noise") can completely hide the offending molecule. However, the parent mass of the desired compound (288) can be extracted from this background in a tandem MS/MS apparatus, in which two mass spectrometers are used in series. This extra step of separation produces a mass spectrum essentially identical to that of the pure compound.

Costs

Just as for NMR, costs of mass spectrometers have increased exponentially over the last few decades but, again, these increasing costs carry with them enormous increases in capability. For example, in 1950, for about $40,000, the best instrument available had a resolution of about one part in 300, and it could be used for molecular weights up to 150. Assuming an average inflation of 6 percent over the 30-year period, this same instrument would cost $230,000 in 1980 dollars. But in 1980, the best instrument available cost about $400,000, less than double this amount, but resolution had been raised 500-fold (to 150,000) and, at the same time, the mass limit had been raised over 10-fold (to 2,000). Along with these performance improvements, scanning speeds have been greatly increased, and data processing is done by built-in computers. Again, as for NMR, no first-rate research laboratory (academic or industrial) can operate without modern instrumentation of this type.

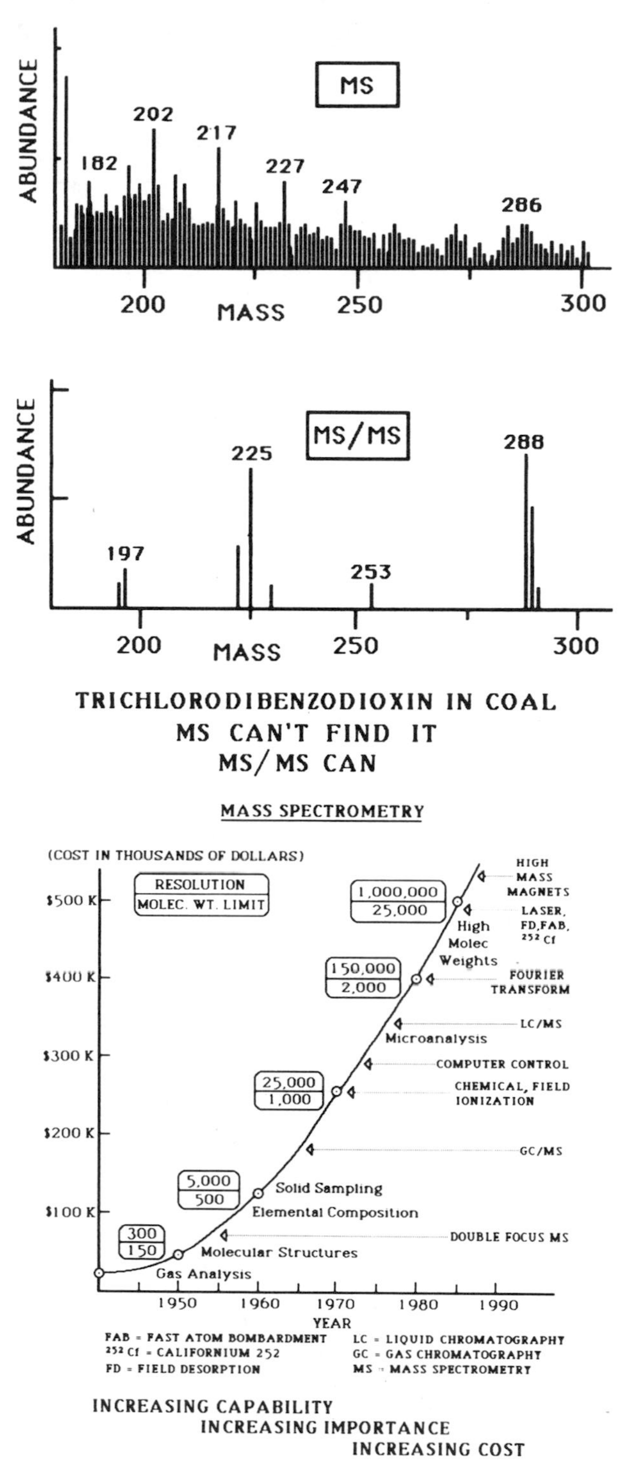

X-RAY DIFFRACTION

Molecular structure is concerned with the bond lengths, bond angles, and spatial placements of the atoms in a substance. Knowledge of such arrangements clarifies the physical and chemical properties of materials, it points to reaction mechanisms, and it identifies new compounds. At present, X-ray diffraction techniques offer the most powerful route to learning molecular structures for any substance that can be obtained in crystalline form.

When light shines on a mirror with regularly spaced straight lines scratched on it (a grating), the mirror still reflects somewhat. However, something special happens if the wavelength of the light, λ, is about the same as the spacing, *d*, between the lines on the mirror. Then the reflection pattern includes bright regions at special angles that are determined by the ratio of λ to *d*. This pattern is called a *diffraction pattern*. It arises from constructive and destructive interference between light waves, similar to that which occurs when two water waves merge. If the line spacing, *d*, is known, then the wavelength of the light, λ, can be determined by measuring the angles at which the bright regions of the pattern are found.

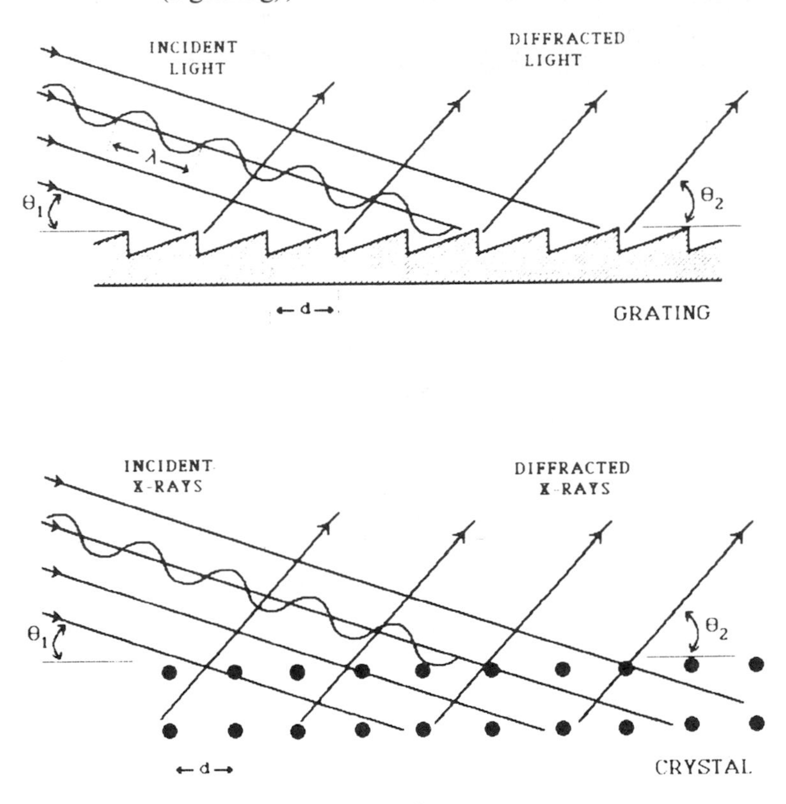

CRYSTALS, LIKE GRATINGS, DIFFRACT LIGHT

X-rays are light rays just like visible light, except that the human eye cannot see them and their wavelengths are only a few Å (green light has a wavelength of 5,500 Å—X-rays are around 2 Å). No machine shop can scratch a mirror with lines only a few Å apart to make a grating for X-rays. However, Nature provides us with excellent X-ray gratings in the form of natural crystals. The regular spacings of the atoms serve as regularly spaced scattering centers and so X-rays are diffracted by a crystal. In this case, we know the wavelength of the X-rays, so we use the angles at which bright regions appear (e.g., on a photographic plate) to determine the *atomic* spacings.

A crystal made up of single atoms (as in a pure metal) gives quite a simple diffraction pattern. The atomic spacings are regular, too, in a molecular crystal, like solid naphthalene, $C_{10}H_8$. But now, there are several types of spacings that will contribute to the diffraction pattern. First, there is the spacing between the centers of adjacent $C_{10}H_8$ molecules. In addition, there are the spacings determined by the fixed carbon-carbon and carbon-hydrogen bond lengths and the molecular bond angles. Now, the diffraction pattern becomes much more complex. Nevertheless, with precise instrumentation and modern computers, the complete molecular structure can be deduced from this pattern. Given a perfect crystal of the pure substance, this type of analysis can be used whether the crystal is an inorganic, organometallic, or organic substance, or a metal, a mineral, or a macromolecule of biological origin. The X-ray diffraction pattern reveals which atoms are bonded to which, the bond lengths and bond angles, and the molecular geometry; and it even indicates how the atoms are moving and how charges are distributed among them! It is as close as we can come to "seeing" the atoms in a molecule.

Applications

The X-ray diffraction technique has become an integral part of inorganic, metal-organic, and organic synthesis. Whenever an unknown substance can be crystallized, an X-ray structure determination is liable to reveal the identity, molecular structure, and conformation of the molecule. With present computer-automated data interpretation, molecular complexity is not a great obstacle. In fact, the requirement that the substance must be available in single-crystal form emerges as one of the major limitations to the range of applicability of this powerful technique. When single crystals can be obtained, even the most complex biological molecules can be examined.

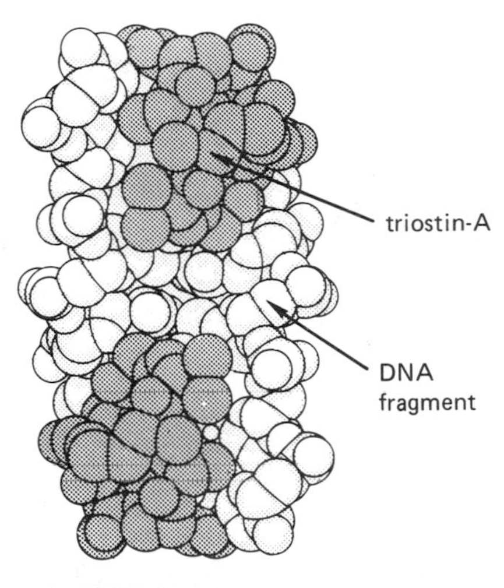

X-RAYS SHOW HOW A DRUG BINDS TO DNA

For example, X-ray structure analysis has become a vital tool for understanding the specific mechanisms for drug action. Such studies of molecular substrates, inhibitors, and antibiotics give information on the special geometry of the receptor site, a first step toward drug design. An example is the recent determination of the manner in which the beneficial drug triostin A attaches itself to a piece of DNA.

When a natural product has been shown to have useful biological properties, the molecular formula must be known before progress can be made toward chemical synthesis. If the active substance can be crystallized, X-rays can furnish that crucial information. Examples already mentioned in Section III-A extend from insect pheromones for pest control in agriculture and forestry to growth hormones to increase food, forage, and biomass production. In a similar way, the structures of toxins from poisonous tropical frogs, poisonous sea life, and poisonous mushrooms have advanced studies of nerve transmission, ion transport, and antitumor agents. Recently the seeds of *Sesbania drummondii,* a perennial shrub growing in wet fields along the Florida to Texas coastal plain, were found to yield a possible antitumor compound. The most active compound found in the seeds is present at only 1/2 a part per million, so 1,000 pounds of seed provided only milligram quantities. The structure of this molecule,

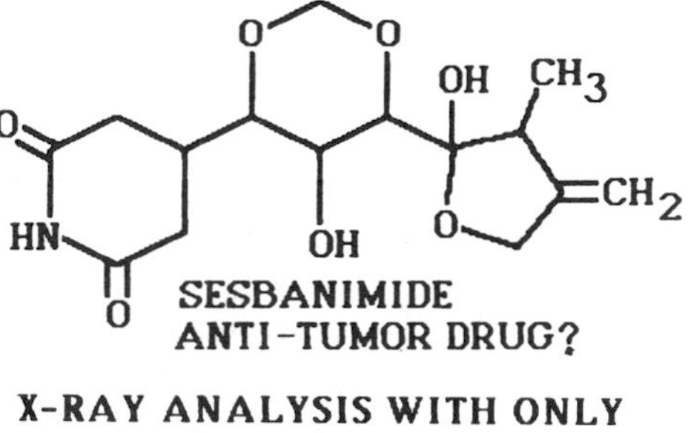

X-RAY ANALYSIS WITH ONLY TEN MICROGRAMS!

called sesbanimide, was determined by X-ray diffraction of a crystal weighing only 10 micrograms. This analysis displayed a novel tricyclic structure previously unknown either in nature or among synthetic organic compounds. With this knowledge, organic chemists have begun devising synthetic approaches to make sesbanimide and related compounds.

Molecular Graphics

For some time, computer-driven graphics programs have been used for modeling and fitting structures to X-ray-derived electron-density maps of molecules. In the last few years, however, new developments have appeared that greatly increase the ability to picture complex molecular arrangements. Computer-automated graphics units have recently become available that present the molecular structure in three dimensions, together with the capacity to rotate the molecule slowly and to highlight with color those molecular components of particular interest. Even an untrained eye can perceive three-dimensional spatial relationships that might go unnoticed without these instrumental features. As such capabilities become more widely available, they are sure to be regarded as an essential analytical tool for connecting molecular structure to molecular function, particularly for biological molecules.

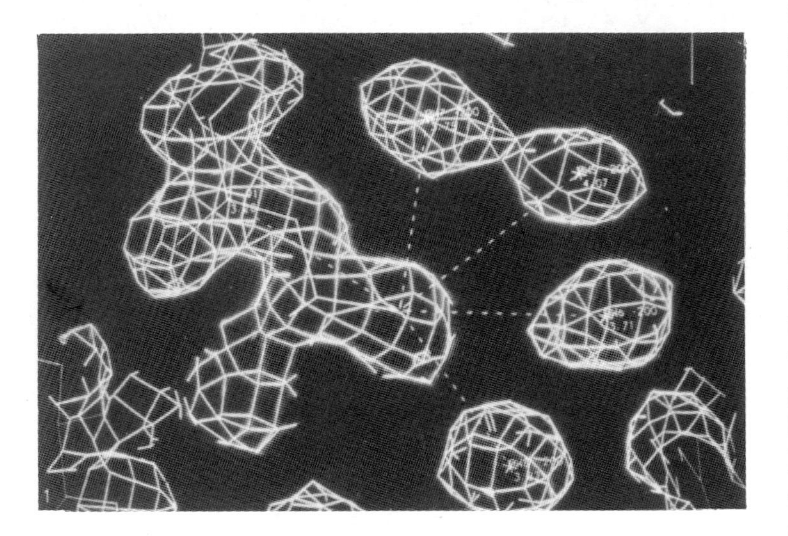

COMPUTER GRAPHICS SHOW MOLECULAR STRUCTURES IN 3D

NEUTRON DIFFRACTION

Complementary to X-ray diffraction and of considerable use to structural chemistry is neutron diffraction. Neutrons with room temperature velocities have wavelengths that are comparable to the atomic spacings in crystal lattices, so when they are scattered from crystalline materials, they give rise to diffraction patterns. To be practical, high-intensity neutron beams are needed, and these can be obtained only from nuclear reactors. If available, however, there are two unique advantages of neutrons over X-rays. First, their scattering from protons is of comparable intensity to that from heavier nuclei, so that neutron diffraction gives more precise information on positions and bonding of hydrogen atoms. Second, the neutron has a magnetic moment, so that neutron diffraction can be used to study magnetic structures.

Applications

Among the accomplishments of neutron scattering research in the past decade are the determination of structures of magnetic superconductors, determination of the spatial organization of macromolecular assemblies such as ribosomes, and the location of hydrogen atoms in the hydrogen bonds that determine protein structures.

ELECTRON SPIN RESONANCE

Most molecules contain an even number of electrons that occur in pairs with opposite spins. However, a reaction in which an electron is transferred can generate species with an unpaired electron (e.g., free radicals and radical ions). The unpaired electron gives the molecule magnetic properties that allow detection and characterization by the technique of electron spin resonance (ESR). The ESR instrument consists of a strong magnet, microwave equipment (originally based on radar technology), sensitive electronic apparatus, and, frequently, a dedicated computer.

Applications

Even though molecules with unpaired electrons tend to be reactive, they are important in many chemical and biological processes, usually as transient intermediates. For example, samples of photosynthetic materials give rise to ESR signals when they are irradiated. These signals arise from primary electron-transfer events initiated by the absorption of light by the photosynthetic pigments, and their study has been important in understanding the mechanism of photosynthesis. Organic radicals and radical ions produce a unique ESR spectrum that allows their identification. In addition, the pattern in the spectrum provides information about the electron-density distribution in the molecule.

SUPPLEMENTARY READING

Chemical & Engineering News

"Fourier-Transform Mass Spec Joins Analytical Repertoire" by S.C. Stinson (C.& E.N. staff), vol. 63, pp. 18-19, Mar. 18, 1985.

"Modern NMR Spectroscopy" by L.W. Jelinsky, vol. 62, pp. 26-40, Nov. 5, 1984.

"Field Flow Fractionation Used to Separate DNA" (C.& E.N. staff), vol. 62, pp. 23-25, Apr. 30, 1984.

"Potentiometric Electrode Aims to Measure Antibody Levels" by R.L. Rawls (C.& E.N. staff), vol. 62, pp. 32-33, Apr. 2, 1984.

"Zero-Field NMR Advances Molecular Structure Determinations" by R.M. Baum (C.& E.N. staff), vol. 61, pp. 23-24, Dec. 12, 1983.

"Multiple Quantum Technique Extends NMR" by R.M. Baum (C.& E.N. staff), vol. 61, pp. 30-31, Jan. 3, 1983.

"Mass Spectrometry/Mass Spectrometry" by R.G. Cooks and G.L. Glish, vol. 59, pp. 40-52, Nov. 30, 1981.

Science

"The Use of NMR Spectroscopy for the Understanding of Disease" by G. Radda, vol. 233, pp. 640-645, Aug. 8, 1986.

"Multiple Quantum NMR Spectroscopy" by M. Murowitz and A. Pines, vol. 233, pp. 525-531, Aug. 1, 1986.

"Two Dimensional NMR Spectroscopy" by A. Box and L. Lerner, vol. 232, pp. 960-967, May 23, 1986.

"The 1985 Nobel Prize in Chemistry" (for x-ray crystallography) by H.A. Hauptman and J. Karle, vol. 231, pp. 309-432, Jan. 24, 1986.

"High Resolution NMR of Inorganic Solids" by E. Oldfield and R.J. Kilpatrick, vol. 288, pp. 1537-1543, Mar. 29, 1985.

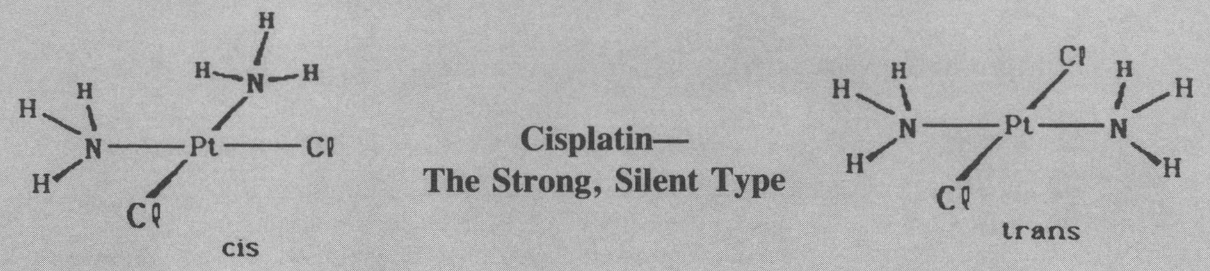

Cisplatin—The Strong, Silent Type

Cancer's a tricky dude to fight. Somehow it takes over a cell that's been functioning normally for years. Then, it fiddles with the control center of the cell—the nucleus—and causes it to reproduce more cancerous cells at an alarming and unhealthy rate. This enemy seems unstoppable. Luckily, scientists have found an ally in the war on cancer—a shy and unobtrusive molecule called cisplatin.

Cisplatin is merely one form of a platinum compound called diamminedichloroplatinum. In spite of its long name, it is a surprisingly simple molecule consisting of two ammonia groups (NH_3) and two chlorine atoms bound to a platinum atom. This compound comes in two shapes (*cis* and *trans*), with two very different personalities.

Cis-DDP, or cisplatin, is the form most effective in fighting cancer, even though *trans*-DDP seems to work in a similar manner. It was perplexing at first why one form of DDP would be so much more effective than the other when their mode of operation seems so similar. Cisplatin is very good at infiltrating enemy lines without detection; then, once inside the cell nucleus, cisplatin works undercover to block cell reproduction. *Trans* does basically the same thing but always manages to blow his cover before cell reproduction. He is recognized and removed from the nucleus before completing his mission.

How is this possible? Well, research has shed some light on the story. Apparently, both DDP molecules are easily taken into the cell and bind readily with DNA, forming what are called DNA "adducts."

The key seems to lie in the fact that *cis*-DDP binds consistently between two adjacent guanine rings in the DNA, while *trans*-DDP produces a variety of cross-links between bases with one or more nucleotides in between. All cells have a mechanism by which they notice and repair irregularities when possible. In the case of *trans*-DDP, it has made the mistake of being too obvious—the *trans* adducts are awkward and are recognized and removed within a few hours of having formed. But sneaky cisplatin is not noticed so easily and manages to stay in place, interfering with the DNA's attempts to replicate. We now believe this is precisely how cisplatin fights cancer growth.

Even though *cis*-DDP is more toxic to cancer cells than it is to normal cells, it, like many other forms of chemotherapy, carries with it serious side effects for the patient. But hopefully, inquiries into the workings of this simple drug will point the way toward similar antitumor agents that are just as effective but without the bummer side effects.

We'll get that tricky dude yet!

V-C. Instrumentation and the National Well-Being

As discussed in earlier chapters, sophisticated instrumentation has figured prominently in our discussions of environmental monitoring and economic applications of chemistry. The techniques of the surface sciences are of dominant importance to the advances being made in catalysis, upon which so many industries depend. Chromatography joins mass spectrometry and laser spectroscopy as an everyday tool in analytical chemistry. Infrared spectroscopy is typical of the several spectroscopic methods that are finding effective use in environmental monitoring as well as in research applications.

SURFACE SCIENCE INSTRUMENTATION

Surface science is a rapidly growing area. The development of powerful instruments that can reveal the atomic structure and chemical composition of surfaces has been largely responsible. The field has been stimulated, as well, by a wide range of important applications. For example, the electrical properties of surfaces and films are important in the miniaturization of semiconductor devices. Consequently, surfaces and thin films have attracted the interest of both physicists and chemists. They are investigating surface etching to permit removal of a layer a few atoms thick in an accurate pattern (a circuit). Another problem of current research interest is the growth of a semiconductor film (for example, a silicon film) when vapor is condensed on a cold surface. It is found that when silicon atoms are condensed on a crystalline surface, the electrical properties of the film can be fixed by the underlying crystal structure (epitaxial growth). And the prospect of understanding catalysis on a fundamental level is one of the most exciting and significant frontiers opened by these new instruments.

Instruments for the Study of Surfaces

The various techniques of surface science probe the surface with particles or with light (photons). Among the particles that have proven useful are electrons, ions, neutral atoms, neutrons, and electronically excited atoms. Photon probes extend from the X-ray region to the infrared. When particles are used, ultrahigh vacuum environments are essential (10^{-9} to 10^{-10} torr). In contrast, photon probes can be effective when the surface is in contact with a gas at high pressure or with a liquid, the conditions under which surface catalysis actually occurs.

A key question about chemistry as it takes place on a surface is the molecular structure of the molecules that have become attached to the surface. If each molecule is essentially intact (physisorbed) with its structure and bonding little changed, the surface is serving only as a site for reaction, immobilizing the reactant as it awaits its fate. But if the molecule reacts with the surface (chemisorbed), it acquires a new molecular identity with changed chemical behavior.

Table V-C-1 lists six types of surface science measurements that are the most informative about the structure of the adsorbed molecules. There are several other instruments that tell about the surface structure, composition, and bonding in the first few layers. Two or more complementary methods used together can greatly enhance the significance of any single measurement used alone.

TABLE V-C-1 Instrumentation Relevant to Chemistry on Surfaces

Method	Acronym	Bombard or Irradiate with:	Physical Basis	Information Obtained
Electron energy loss spectroscopy	EELS	Electrons, 1-10 eV	Vibrational excitation of surface molecules	Molecular structure, surface bonding of adsorbed molecules
Infrared spectroscopy	IRS	Infrared light	Vibrational excitation of surface molecules	Molecular structure of adsorbed molecules
Thermal desorption	TDS	Heat	Thermally induced desorption of adsorbates	Energy of surface binding
Auger spectroscopy	Auger	Electrons, 2-3 keV	Electron emission from surface atoms	Surface composition
Low-energy electron diffraction	LEED	Electrons, 10-300 eV	Back-scattering, diffraction	Atomic surface structure
Secondary ion mass spectroscopy	SIMS	Ions, 1-20 keV	Ejection of surface atoms as ions	Surface composition

Electrons are useful surface probes because their energies, and hence, their wavelengths, can be accurately controlled with their accelerating voltage. At low energy, near 25 electron volts, the wavelength of an electron is close to the atomic spacings in a metal, so a beam of such electrons reflected from the surface will show diffraction effects. Thus, low-energy electron diffraction (LEED) can play the same role in determining bond distances and bond angles in surface chemistry as X-ray diffraction plays in the structural chemistry of solids. LEED reveals the atomic structure of clean surfaces, as well as any regularity in the packing of atoms and molecules adsorbed on the surface.

In the Auger (pronounced ''Oh-jay'') effect, high-energy electrons (2,000-3,000 eV) striking an atom cause the atom to eject a secondary electron from an inner shell. The energy of the ejected electron is determined by the energy levels of the atom it came from, so measurement of the electron energy identifies the atom. Since the bombarding electrons do not penetrate deeply, these secondary electrons reveal, with high sensitivity, the composition of the first few surface layers. This information can be important because surface impurities and irregularities can dominate surface chemistry. Hence, the combination of Auger and LEED is used routinely to verify the cleanliness and perfection of the surface under study.

Electron energy loss spectroscopy (EELS) is of particular value because it detects the resonant vibrational frequencies of atoms and molecules bound to the surface. Chemists routinely use such vibrational frequencies for gaseous molecules to decide which atoms are hooked to which, how strong the bonds are, and their molecular geometry (see Infrared Spectroscopy later in this section). In EELS, an electron beam of known energy is bounced off the metallic surface into an energy analyzer. If the electrons hit an area where a molecule is adsorbed, the molecule can be left vibrating in one of its characteristic motions. The energy needed to do this, determined by the frequency of the motion, is taken away from the kinetic energy of the electron. The measurement of these electron energy losses of the reflected beam gives a vibrational spectrum of the adsorbed molecules.

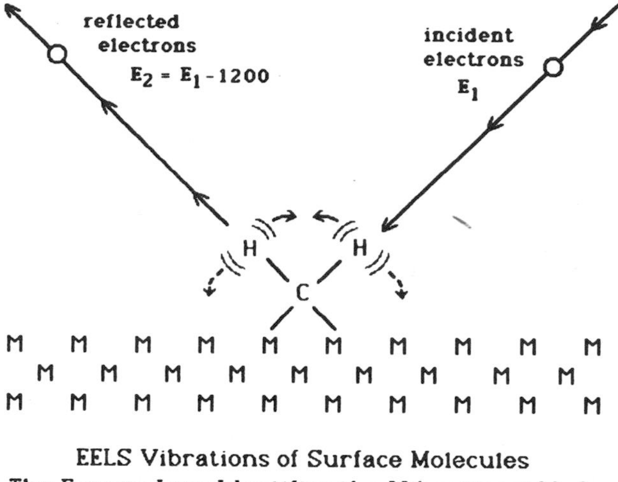

EELS Vibrations of Surface Molecules
The Energy Loss Identifies the Vibrational Mode

Ion scattering from surfaces has been used for surface composition analysis with great sensitivity, 10^9 atoms/cm^2. In secondary ion mass spectroscopy (SIMS), neutral and ionized atoms and molecular fragments are ejected by bombardment with high-energy (1-20 keV) inert gas ions. Ion scattering spectroscopy determines the surface composition by the energy change of inert gas ions upon surface scattering. Ion etching removes atoms from surfaces layer by layer. The combined use of ion etching and electron spectroscopy yields a depth profile analysis of the chemical composition in the near-surface region. This combination of instrumental methods is called "dynamic SIMS."

The availability of high-intensity laser sources is now awakening the development of a new set of surface-sensitive techniques. Surface infrared spectroscopy, laser Raman spectroscopy, and second harmonic generation surface spectroscopy all provide information about the surface chemical bonds of adsorbed atoms and molecules. All of these emerging surface science techniques will permit us to watch chemical reactions as they occur on well-characterized and clean surfaces. This is an important development in chemistry because surfaces provide the two-dimensional reaction domain that accounts for heterogeneous catalysis.

SURFACE ANALYSIS

Any sensitive measurement technique can be used as an analytical tool. This is the case in the surface sciences. Every one of the capabilities listed in Table V-C-1 can be put to analytical use in the pursuit of questions that may be only remotely connected to the surface sciences. As an example, a state-of-the-art laser microprobe device designed to desorb (remove) molecules from a solid surface can be used to detect the presence of a pesticide on the leaf of a plant. Such a capability was quite impossible only 10 years ago; today it permits us to contemplate tracking the amount, stability, weathering, and chemistry of a pesticide in field use. Of course, the analytical technique may just as well be concerned with monitoring or clarifying chemical changes that take place on a surface or with a surface. Many of these analytical studies relate to catalysis. In Section IV-C, examples were given of the use of EELS to determine the molecular structures that exist on a catalyst surface as it functions. Such applications have given rise to *surface analysis,* a new subdivision of analytical chemistry.

The effective sampling depth is a most important feature of any surface analytical technique. Sampling depth is important, because the measuring technique must be appropriate to the phenomenon under study. For example, bonding to the surface, wettability, and catalysis involve only a few atomic layers, whereas surface hardening treatments involve 10 to 1,000 atomic layers. Typical sampling depths for the primary surface analytical techniques are one or two atomic layers for low-energy ion scattering, 5 Å depth for SIMS, 20 Å for the Auger technique, and 100 Å for ion etching coupled with SIMS. Laser mass spectrometry, the Raman microprobe, and scanning electron microscopy (SEM) reach from 1,000 to 10,000 Å (i.e., to one micron). The shallower the sampling depth of the technique, the more finely it is able to define the surface composition of a sample.

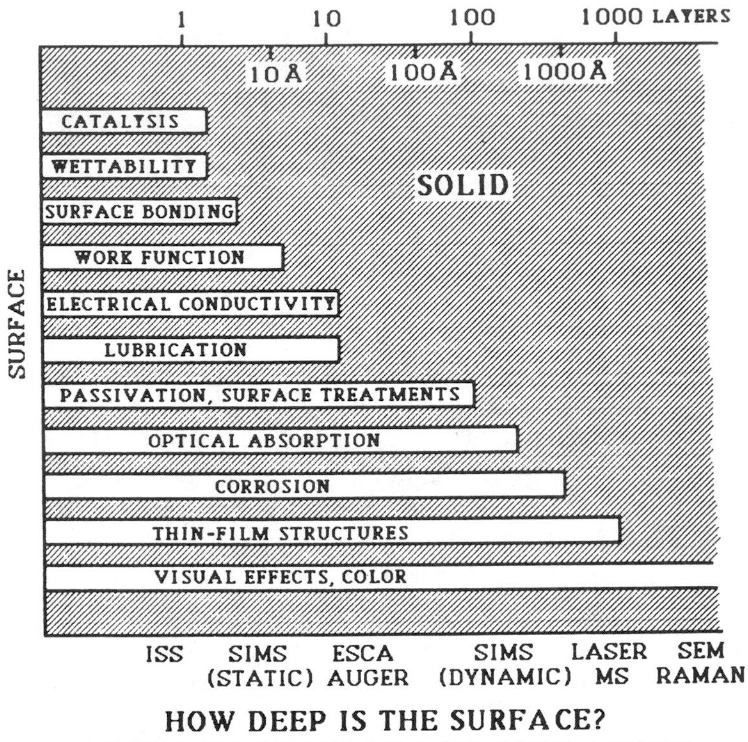

HOW DEEP IS THE SURFACE?
IT DEPENDS ON WHAT YOU CARE ABOUT

A major challenge in the development of surface analytical instrumentation is the reinforcement of its quantitative dimension. Most of the examples given have been concerned with *what* is there. We must also be able to determine *how much*. Another important problem is the development of microprobes which can provide both chemical and positional information about surface species. Currently, Auger and ion microprobes are useful in this respect for mapping elemental composition, as in revealing both the presence and location of the trace contaminants phosphorus and lead in silicon chips. However, they are not yet able to detect and map large organic molecules such as carcinogens or therapeutic drugs. Characterization of small particles is another important challenge for surface analysis; this is particularly important in environmental monitoring where the analysis of carcinogenic hydrocarbons on atmospheric dust and other particulates is a current problem.

CHROMATOGRAPHY

Chromatography separates molecules or ions by dividing species between a moving phase and a stationary phase. A liquid or a gas flowing continuously through a tube (called a "column") provides the moving phase. The stationary phase can be either small solid particles packed in the tube or, for a small-diameter tube (a capillary), the walls of the tube itself. If a pulse or a squirt of soluble substance enters the tube at one end, that substance will have some tendency to stick to the stationary surface, becoming adsorbed. However, the continuing flow of fresh solvent keeps acting to redissolve this adsorbed material, moving it forward in the tube. How fast the process of adsorbing and desorbing takes place depends sensitively on the composition and structure of the substance. Consequently, different substances that entered the tube together in the same pulse will move at different speeds through the tube, so they will exit at different times.

This separation technique takes advantage of small differences in properties such as solubility, absorbability, volatility, stereochemistry, and ion exchange, so that understanding the fundamental chemistry of these interactions is basic to progress in the field. Liquid chromatography has shown an impressive growth since 1970. The current $400 million annual sales are mainly by U.S. manufacturers. This growth has come through innovations such as high pressure and moving phases of changing composition (''gradient moving phases'') to give greater speed and resolution. ''Bonded-molecule'' stationary phases are chemically designed to increase selectivity and to extend the useful lifetime of a column. Detection also has improved with electrochemical, fluorometric, and mass spectrometric detectors, reaching sensitivities as low as 10^{-12} grams. Although gas chromatography is a more mature field by perhaps a decade, important advances continue to appear. High-speed separations can now be accomplished in a few tenths of a second; portable instruments the size of a matchbox are in use outside of the laboratory. A complex mixture can be separated into literally thousands of components, using fused-silica capillary columns that are a direct spin-off from optical fiber technology for communications. It is even possible to separate compounds that differ only in isotopic composition.

High-Performance Liquid Chromatography (HPLC)

During the 1970s, theoretical understandings of the complex flow and mass transfer phenomena involved in chromatographic separation helped perfect column design. During this same period, small-diameter (3-10-micron) silica particles with controlled porosity were introduced. Synthetic advances in silica chemistry led to the tailoring of particle diameter, pore diameter, and pore size distribution. Today, 15-cm columns with efficiencies exceeding 10,000 distillation steps (''theoretical plates'') are routine.

Still another major advance of the 1970s was the introduction of chemically bonded phases in which surfaces of porous silica are covalently coated with organic molecules containing silicon (organosilanes). Especially important is the use of hydrocarbon attachments (such as *n*-octyl and *n*-octadecyl) to make the surface look like an organic solvent. Then, the mobile liquid phase is typically an organic-aqueous mixture. This is called reversed phase chromatography (RPLC), and it currently provides well over 50 percent of all HPLC separations. It is especially well suited to substances that are at least partially soluble in water (drugs, biochemicals, aromatics, etc.).

Finally, the microprocessor/computer is playing an increasing role. ''Smart'' HPLC instruments are under development to program the performance. New detectors of greater sensitivity and selectivity are on the horizon. In particular, laser spectroscopy promises to yield highly sensitive devices for subpicogram detection (less than 10^{-12} grams).

Because of performance improvements, HPLC is having a major impact on diverse fields of biochemistry, biomedicine, pharmaceutical development, environmental monitoring, and forensic science. Today, peptide analysis and isolation requires HPLC because of its separating power and speed. Analysis of amino acids in protein/peptide sequencing is conventionally accomplished by RPLC. In clinical analysis, therapeutic drug monitoring can be accomplished by HPLC. The analysis of catecholamines (important as "neurotransmitters") is typically accomplished by RPLC with electrochemical detection. Isoenzyme analysis, which is important, for example, in assessment of heart damage after an attack, can be rapidly accomplished by HPLC. The analysis of polar and high-molecular-weight organic species in waste streams in sewage or factory treatment can be performed by HPLC, while the separation and analysis of phenols in water supplies by RPLC is recommended. Analysis of narcotics, inks, paints, and blood represent only a few of the forensic applications in police laboratories.

Capillary Chromatography

This version of chromatography uses an open capillary tube with a thin liquid layer on its inner wall. It began with capillary gas chromatography (GC), but the fragility of glass as an inert material for GC capillary columns discouraged many potential users. Now we have flexible, fused-silica capillaries with a polymer overcoat, a spin-off of fiber optics technology. These advances in capillary column technology led to intensive commercialization during the 1970s. Today's capillary columns exhibit efficiencies between 10^5 and 10^6 distillation steps ("theoretical plates") and are capable of separating literally hundreds of components within a narrow boiling point range. Direct introduction of samples at the nanogram (10^{-9} grams) levels has been developed, and much effort has been directed at perfection of gas-phase ionization detectors. Combined advances in the column and detector areas now make trace analytical determinations below 10^{-12}-gram levels practical by capillary gas chromatography.

Of particular note is the combination of capillary GC with powerful identification methods such as mass spectrometry and Fourier Transform infrared spectroscopy, as mentioned in Section IV-C. The combined techniques are now routinely capable of identifying numerous compounds of interest that are present in complex mixtures in only nanogram quantities. They have been used in identification of new biologically important molecules, as well as in drug metabolism studies, forensic applications, and identifications of trace environmental pollutants.

Every fluid has a characteristic temperature and pressure above which its gas and liquid phases become indistinguishable. Above these critical conditions, the "supercritical" fluid displays exceptionally low viscosity, and it can become a much better solvent. Consequently, the use of supercritical fluids in capillary chromatography has recently emerged as a promising approach to the analysis of complex nonvolatile mixtures. As the solute diffusion coefficients and viscosities of supercritical fluids are more favorable than those of normal liquids, chromatographic performance is substantially enhanced. Furthermore, the optical transparency of supercritical fluids makes them attractive for certain optical detection techniques.

Field-Flow Fractionation (FFF)

Chromatography becomes more difficult to apply as molecular size grows, and it becomes ineffective in separating macromolecules and colloidal particles in the size range 0.01 to 1 micron in diameter. A recent innovation, field-flow fractionation, may fill this need. In FFF, a liquid sample flows through a thin (0.1-0.3 mm), ribbon-like flow channel. A temperature difference or electric field is maintained across the ribbon. Each constituent in the sample distributes itself in a way that is determined by its diffusional properties and its response to the applied thermal or electrical field. Since flow through the channel is fastest near the middle of the ribbon, substances that are pulled close to the ribbon wall move more slowly than substances that reside near the middle of the flow channel. Separations are thus achieved. A useful aspect of this technique is that the strength of the applied field can be varied in a deliberate and programmed way by a computer during the course of the separation.

Such thermal gradients are effective in separating most synthetic polymers. The mass range of molecules and particles to which FFF has been applied extends from molecular weights of 1,000 up to 10^{18}, that is, up to particle sizes of about 100-micron diameters. FFF appears to be applicable to nearly any complex molecular or particulate material within that vast range. Applications have so far included macromolecules and particles of biological and biomedical relevance (proteins, viruses, subcellular particles, liposomes, artificial blood, and whole cells), of industrial importance (both nonpolar and water-soluble polymers, coal liquid residues, emulsions, and colloidal silica), and of environmental significance (waterborne colloids and the tiny particles called fly ash in smoke plumes).

INFRARED SPECTROSCOPY

A molecule can be pictorially, but accurately, viewed as a collection of wooden balls held together in a fixed geometry by springs. The masses of the balls are proportional to the atomic masses, and the strengths of the springs are proportional to the strengths of the chemical bonds. Such a "ball-and-spring" model will have resonant vibrational frequencies in which the wooden balls move back and forth in regular patterns. These frequencies are determined by the masses, the spring constants, and the geometry. A molecule is exactly the same. If measured, the resonant frequencies give direct information about the molecular architecture.

Consider, for example, the water molecule. This bent,

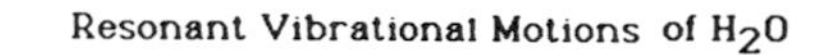

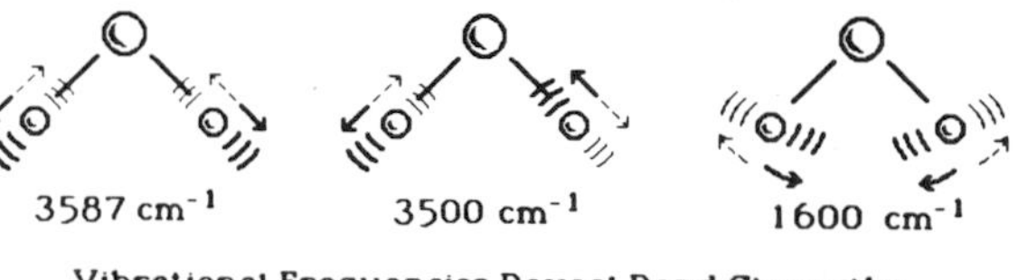

Vibrational Frequencies Reveal Bond Strengths and Bond Angles triatomic molecule has three resonant vibrations. In one of these, the two bonds stretch back and forth in phase, and in another, the two bonds both stretch, but this time out of phase. In the third characteristic vibration, the bond angle alternately opens and closes.

Such molecular vibrations do not break bonds, so they require little energy. Absorption of light is one way to excite these vibrations, but photons of appropriate energy are in the infrared spectral region, far beyond the visual sensitivity of the human eye. A typical molecular vibration, such as the bending motion of the water molecule, has a frequency of 4.8×10^{13} vibrations per second. This unwieldy number is usually brought into reasonable magnitude by dividing it by the speed of light, which changes the dimensions to 1/cm or cm^{-1} (‘‘reciprocal centimeters’’).

$$\frac{4.8 \times 10^{13} \text{ vibrations/sec}}{3 \times 10^{10} \text{ cm/sec}} = 1{,}600 \text{ cm}^{-1}.$$

Infrared vibrational frequencies are always expressed in reciprocal centimeters (cm^{-1}) (sometimes called ‘‘wave numbers’’). The measurement of these molecular frequencies is called vibrational spectroscopy or infrared spectroscopy.

These vibrational frequencies are so characteristic that they furnish a distinctive and easily measured ‘‘fingerprint’’ for each molecule. This spectral fingerprint, once measured for a particular molecule, can be used to determine whether that molecule is present in a sample and, if so, how much. The vibrational frequencies also reveal the molecular structure and bond strengths in the molecule, so they can be used to learn about the molecular architecture. When an unknown compound is under study, the infrared spectrum provides one of the easiest ways to decide what the compound is likely to be.

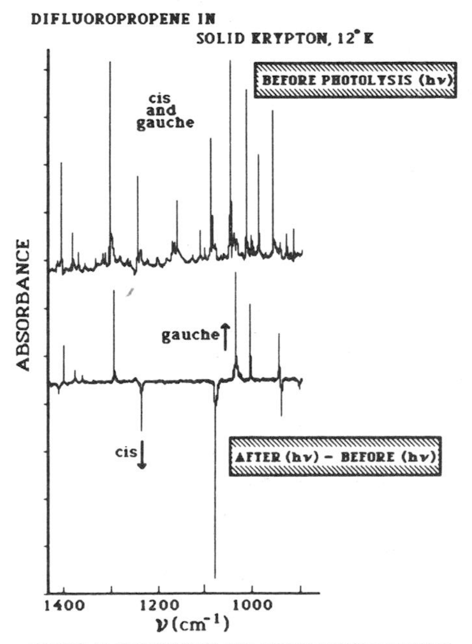

FTIR DIFFERENCE SPECTROSCOPY SHOWS ROTAMER INTERCONVERSION

Because infrared spectroscopy is so informative, it has become one of the routine diagnostic tools of chemistry. A large, research-oriented chemistry department might operate 5 to 10 infrared spectrometers with capabilities ranging from rugged, low-resolution instrument for instruction in an advanced first-year chemistry course, to high-resolution Fourier Transform Infrared Spectrometers (FTIR), suited to molecular structure determination and specialized research use.

Computer-Aided Spectrometers

Modern research infrared spectrometers incorporate computers to permit programmed operation, data collection, and data manipulation. The major impact of computers, however, has been their influence on the performance of Fourier Transform interferometers. The perfection of the

Fourier Transform algorithm (program), plus the reduction in accompanying computer costs, brought the interferometer from a trouble-plagued, research-only instrument to a routine, high-performance workhorse. A notable capability brought by the computer is the ease and accuracy with which one spectrum can be subtracted from another to emphasize small changes. This is called a difference spectrum. One important application relates to infrared spectra of biological samples in which evidence of a chemical change associated with a certain specific biological function can be completely covered up by the heavy infrared spectrum of the inactive substrate in which the sample is located. The digitized data permit precise spectral subtraction so that the background spectrum can be virtually eliminated to reveal the spectral changes of interest.

Another vivid display of the value of the difference capability is provided by photolysis of molecules suspended in a cryogenic solid ("matrix isolation"). If the digitized spectrum before photolysis is subtracted from the spectrum after photolysis, only the features that change are seen. Any molecule that is being consumed presents its spectral features downward, while spectral features of the growing product extend upward. This has been used, for example, to distinguish the two forms (*cis-* and *gauche-*) of the 2,3,-difluoropropene in the cluttered spectrum of a complex mixture. A laser tuned to an absorption frequency of one of the "rotamers," say, the *cis-* form, is used to irradiate the cold sample. Absorption of this light adds enough energy to the absorbing *cis-* molecule to permit it to convert to the *gauche-* form. Then in the difference spectrum, the *cis-* molecule spectrum appears as a negative spectrum and the *gauche-* molecule spectrum as a positive one. Absorptions due to other molecules do not change, so they simply do not appear at all.

Applications

The coupling of FTIR with gas chromatographic separations in a variety of analytical uses has been discussed. Also, as noted earlier, infrared spectroscopy is a specially effective method for monitoring and studying atmospheric chemistry. This is because gaseous molecules of low molecular weight are important, including formaldehyde, nitric acid, sulfur dioxide, acetaldehyde, ozone, oxides of chlorine and nitrogen, nitrous oxide, carbon dioxide, and the Freons. These substances are influential participants in photochemical smog production, acid rain, strato-

TABLE V-C-2 Additional Instrumental Techniques in Modern Chemistry

Instrument	Information Obtained
Ion cyclotron resonance	Reaction rates of gaseous molecular ions
Laser magnetic resonance	Precise molecular structures, gaseous free radicals
Laser Raman	Vibrational spectrum
Fluorimeter	Lifetimes, electronically excited molecules
Circular dichroism	Stereo conformations
Flow cytometer	Laser-activated cell sorter
Protein sequencer	Automated analysis of protein sequence
Oligonucleotide synthesizer	Automated synthesis of designed DNA segments
Electron diffraction	Molecular structure, gases
Scintillation counter	Tracking radiotracers

spheric disturbance of the ozone layer, and the "greenhouse effect." Infrared spectroscopy shows where they are and how much is there.

OTHER INSTRUMENTATION

In Sections V-A, V-B, and V-C, there has been detailed discussion of over a dozen different classes of instrumentation which are important in defining and advancing the current frontiers of chemistry. By no means, however, is the list all-inclusive. Table V-C-2 lists additional types of equipment and what kinds of chemical information each one provides. The length of this table is only one more signal of the crucial importance of instrumentation in modern chemistry.

SUPPLEMENTARY READING

Chemical & Engineering News

"Instrumentation '86—Optical Spectroscopy" (C.& E.N. staff), vol. 64, pp. 34-42, Mar. 24, 1986.

"Instrumentation '86—Chromatography" (C. & E.N. staff), vol. 64, pp. 52-68, Mar. 24, 1986.

"Instrumentation '86—Mass Spectrometry" (C.& E.N. staff), vol. 64, pp. 70-72, Mar. 24, 1986.

"Low Cost FTIR Microscopy Units Gain Wider Use in Microanalysis" (C.& E.N. staff), vol. 63, pp. 15-16, Dec. 9, 1985.

"Affinity Chromatography" by Parikh and P. Cuatrecasas, vol. 63, pp. 17-31, Aug. 26, 1985.

"GC Detector Uses Gold Catalyst for Oxidation Reactions" by W. Worthy (C.& E.N. staff), vol. 63, pp. 42-44, June 24, 1985.

"X-Ray Technique May Provide New Way to Study Surfaces, Films" by W. Worthy (C.& E.N. staff), vol. 63, pp. 28-30, April 8, 1985.

"Centrifugal Force Speeds Up Countercurrent Chromatography" by S.C. Stinson (C.& E.N. staff), vol. 62, pp. 35-37, Nov. 26, 1984.

Science

"A New Dimension in Gas Chromatography" by T.H. Maugh II (Science staff), vol. 227, pp. 1570-1571, Mar. 29, 1985.

"Ion Beams for Compositional Analysis" (SIMS), by A.L. Robinson (Science staff), vol. 227, pp. 1571-1572, Mar. 29, 1985.

Investigating Smog Soup

Air pollution is a visible reminder of the price we sometimes pay for progress. Emissions from thousands of sources pour into the atmosphere a myriad of molecules that react and re-react to form a "smog soup." We are already aware of some of the potential dangers of leaving these processes unstudied and unchecked: respiratory ailments, acid rain, and the greenhouse effect. Surprisingly, you and I are the principal culprits in generating much of this unpleasant brew—everytime we start our cars or switch on our air conditioning or central heating! Transportation, heating, cooling, and lighting account for about two-thirds of U.S. energy use, almost all derived from combustion of petroleum and coal.

Pinpointing cause and effect relationships begins, inevitably, with the identification and measurement of what is up there, tiny molecules at parts-per-billion concentrations in the mixing bowl of the sky. Finding out what substances are there, how they are reacting, where they came from, and what can be done about them are all matters of chemistry. The first two questions require accurate analysis of trace pollutants. Physical and analytic chemists have successfully applied to such detective work their most sensitive techniques. An example is the *Fourier Transform Infrared Spectrometer*. This sophisticated device can look through a mile or so of city air and identify all the chemical substances present and tell us their concentrations down to the parts-per-billion level. Recognizing a substance at such a low concentration is comparable to asking a machine to recognize you in a crowd at a rock concert attended by the entire U.S. population.

How does this superb device work? "Infrared" means light just beyond the red end of the rainbow visible to the human eye. Hence infrared light is invisible, though we can tell it is there by the warmth felt under an infrared lamp. But molecules can "see" infrared light. Every polyatomic molecule absorbs infrared "colors" that are uniquely characteristic of its molecular structure. Thus each molecular substance has an infrared absorption "fingerprint"—different from any other substance. By examining these fingerprints, chemists can identify the molecules that are present.

An example of what can be done is the measurement of formaldehyde and nitric acid as trace constituents in Los Angeles smog. Unequivocal detection, using almost a mile-long path through the polluted air, revealed the growth during the day of these two bad actors and tied their production to photochemical processes initiated by sunlight. Continuing experiments led to detailed characterization of the simultaneous and interacting concentrations of ozone, peroxyacetyl nitrate (PAN), formic acid, formaldehyde, and nitric acid in the atmosphere. These detections removed an obstacle to the complete understanding of how unburned gasoline and oxides of nitrogen leaving our exhaust pipe end up as eye and lung irritants in the atmosphere. This advance doesn't eliminate smog soup, but it is a big step toward that desirable end.

CHAPTER VI

The Risk/Benefit Equation in Chemistry

Gasoline is a mixture of hydrocarbons, mainly alkanes and alkenes, but with aromatics or tetraethyl lead added to improve combustibility. It is toxic to drink, the aromatics are carcinogenic, the tetraethyl lead can cause lead poisoning, the mixture is extremely flammable, and, when burned, it produces the noxious and toxic substances nitric oxide and carbon monoxide. Gasoline is probably the most dangerous compound that the average person will ever encounter on a daily basis. Yet this same average person stores 5 to 10 gallons at home (in the car's gas tank), and he or she might purchase some 500 gallons of this perilous liquid every year, and then combust it to release into the atmosphere 250,000 cu.ft. of oxides of nitrogen and carbon. Every city has dozens of repositories for this fluid, gasoline stations, each storing perhaps 10,000 gallons, always in a crowded neighborhood. These supplies must be regularly replenished by trucks that carry 20,000 gallons each through the city amongst the normal traffic.

This is risky business! But evidently the public (that's you and me) believes that the benefits are important enough to justify the risks. Every time you start your car, you are implicitly stating that the risk/benefit equation in our reckless use of gasoline comes down on the side of benefit.

Dichlorodiphenyl trichloroethane is an insecticide that has saved over 50 million human lives and prevented untold suffering. It did so by almost eliminating malaria. In 1959, at the height of its use, over 156 million pounds of this chemical, commonly called DDT, were produced in the United States alone. Many thousands of people were dusted with DDT, literally from head to toe, without apparent harm. Such programs reduced the number of malaria victims in India alone from 75 *million* in 1952 to less than 100 *thousand* by 1964. In Sri Lanka, the number of malaria cases dropped from about 3 million, with 12,000 deaths per year, to less than 100, with no deaths at all. But production of this lifesaving chemical has been sharply cut back, and in the United States it has almost gone out of use. The reason is that DDT has been judged to be potentially dangerous because of its persistence in the environment and because of its accumulation in the tissues of living organisms.

This is a dilemma. There is no doubt that human lives could be saved, but that ecological disturbances will accompany continued use of DDT. The ban against its use indicates that the public has decided that the risks exceed the benefits.

Gasoline and DDT are only 2 of the 70,000 or so chemicals that have come into widespread use. These chemicals range from aspirin to Vitamin C, flea powder to household detergents, Dacron shirts to Teflon-coated frying pans. Our quality of life is determined, sustained, and constantly improved by our ability to control chemical reactions and to make chemical compounds that are useful in everyday life. But gasoline and DDT are excellent examples because they point out vividly the fact that handling all of these chemicals must involve some risks. We have only recently realized that minimizing risks must be considered to be as important a dimension of progress as maximizing benefits from technological change. *We must learn to deal wisely with the risk/benefit equation.*

FEARS OF THE MODERN AGE

In a recent, highly publicized incident, the compound trichloroethylene, TCE, was found at several parts per million in the drinking water from 35 private wells near Palo Alto, California. A lawsuit caused these wells to be closed. Yet the available evidence implies that these wells are probably quite safe—if that is so, then why are they closed?

Part of the answer is that there is a growing alarm in this country over chemicals and their possible effects on individuals. The populace is concerned about chemical pollutants, additives, waste, by-products, residues—in short, any chemical that is the direct result of technological change. Some of this current fear stems from a general lack of information—a fear of the unknown; some is triggered by sensational or overzealous reporting in newspapers and on television. Many people feel a sense of helplessness, a feeling they have no say in the control of these new substances entering the environment. There is also a vague feeling of mistrust of the priorities and interests of those with a vested financial stake in producing, distributing, and utilizing these substances.

Yet all of these chemicals came into being to establish and maintain our high quality of life. For decades, we have been blithely enjoying the fruits of our technological success without thought of the possible incumbent hazards and undesired effects. Now, suddenly, our society has become "chemically aware"; we have become hypersensitive about all chemicals, no matter what the source, amount, degree of hazard, or intended purpose.

Unfortunately, such fear can lead us to overreact, so that situations presenting minimal danger can divert attention and resources from real dangers that must be corrected and eliminated. We must alleviate this chemical insecurity so that we can find the optimum balance in the risk/benefit equation. Then we can continue to enjoy the growing benefits of the chemical age while guaranteeing that we protect the health and well-being of our planet and its occupants.

WHAT IS TOXIC?

"Everything is poisonous. The dose alone determines the poison." Paracelsus, the sixteenth century chemist, physician, and healer, made this assertion, one that comes to mind when we read that a healthy diet should not contain too much salt, sugar, or butterfat. Nitrogen is 80 percent of the air we breathe, but too much nitrogen acts as an anesthetic, and it can give deep sea divers a dangerous feeling of euphoria called "rapture of the depths." Selenium is essential to human and animal health, but in excess can cause a variety of ailments. And if too much salt, sugar, butterfat, and nitrogen is unhealthy, perhaps we should believe Paracelsus—everything is poisonous if taken in too large amounts. This news can be a bit unnerving since everything around us and in us is chemical, including everything we eat and drink. But it is reassuring, too, because we, and all other life forms, evolved and thrived in the presence of these chemicals. Perhaps the persistence of life is evidence that the corollary to Paracelsus's remark is true as well—things that are poisonous in too large doses are not necessarily poisonous in small enough doses.

With that premise, we are faced with two fundamental questions. First, we must find out what hazard levels we face—*risk assessment*—and then we must decide what to do about them—*risk management*. The Environmental Protection Agency wisely recognizes these as separate dimensions of their responsibility. Risk assessment is mainly connected with the known scientific facts about a given possible hazard. Risk management requires choices among options, as well as consideration of costs and social consequences. We will discuss them in turn.

RISK ASSESSMENT

To begin with, there are two kinds of toxicity to worry about. A toxic chemical can cause illness soon after exposure, which is called *acute* toxicity. Another chemical can have no immediate effect, but it may be injurious much later, after continuous, long-term exposure; this is called *chronic* toxicity. For example, phosgene, Cl_2CO, is an acutely toxic gas that is accidentally produced by use of a carbon tetrachloride fire extinguisher to put out an electrical fire. A concentration of 5 parts per million (ppm) will cause eye irritation in a few minutes, whereas greater than 50 ppm would probably be fatal. On the other hand, benzene, C_6H_6, is chronically toxic—its vapor inhaled at the same level, 50 ppm, would cause no immediate effect, but if inhaled every day for many months or years, benzene can cause decreases in red blood cell count, hemoglobin level, and the number of leukocytes in the blood.

Unfortunately, it isn't easy to get such detailed information. The most definitive way, the hard way, is to have enough people exposed to a given chemical to show that it is safe or to show the exposure at which toxicity begins to appear. Plainly, it is hardest to learn about chronic toxicity. Very large populations must be exposed for long periods to give a statistical chance of establishing something useful. That is what *epidemiology* is all about.

What Is Epidemiology?

Historically, epidemiology is the study of epidemics, contagious diseases that spread rapidly. But today, epidemiology is also used in a statistical way to try to detect acute or chronic toxicity even when the effects on health are quite small. For example, vinyl chloride, CH_2CHCl, is known to be carcinogenic (a cause of cancer). The reason is that a very rare form of liver cancer, angiosarcoma, is found statistically to be concentrated in a small number of workers who have been continuously exposed over long periods to high levels of vinyl chloride, in the hundreds of ppm. Here we can reach the confident epidemiological conclusions that this chemical has a toxic effect and that the degree of hazard to the general public is extremely small.

What Causes What?

Unfortunately, epidemiological data can be misinterpreted, even when the statistics are firm. It is one thing to show (1) that colon cancer is much more prevalent in the United States than it is in India, and (2) that Americans eat more dairy products than people in India do. Before jumping to the conclusion that dairy products cause colon cancer, we must remember that colon cancer appears in older people and that U.S. citizens live a lot longer (on the average) than citizens of India. Thus, the opposite conclusion might be reached—that eating dairy products allows one to grow old enough for colon cancer to show up (from other causes). *Epidemiology can show "association" but not necessarily "causality."* The epidemiologists' joke is that the twentieth century growth of population in Western Europe has decreased at about the same rate as the decrease in the number of storks. Few of us would conclude that the human birth rate is going down because there aren't enough storks for the deliveries.

Animal Tests

These difficulties have driven us to the use of laboratory test animals as substitutes for humans. Without debating the ethics of such practice, we observe that among the animals so used are mice, rats, guinea pigs, monkeys, hamsters, dogs, cats, pigs, and even fish. Mice and rats are used most, probably because they are inexpensive, they breed rapidly, and their use is generally accepted.

In a typical study, groups of a few thousand mice might be exposed to two or three different doses of a particular chemical every day for 2 years (including a zero dose for a control group). Then these mice are killed ("sacrificed" is the term used) and every mouse is autopsied to search for tumors. The statistical differences between the control group (zero dose) and the exposed groups are taken as a measure of hazard. Such an experiment might show that one milligram of chemical X eaten every day by a half-pound mouse causes a 14 percent increase in stomach cancer. To decide what this means for us, we usually assume that a human weighing about 150 pounds, 300 times more than the average mouse, will need to eat about 300 milligrams per day of chemical X to have about the same 14 percent cancer probability. Thus, toxic doses are expressed in terms of milligrams per kilogram of body weight (mg/kg).

For acute toxicity, a substance legally earns the title "poison" if 50 percent of a test group of animals dies from a dose of 50 mg/kg or less. This dose—the one that causes a 50 percent death rate—is called the LD_{50} dose (lethal dose, 50 percent death rate). Thus, strychnine is a poison—it takes only 1.2 mg/kg body weight to kill 50 percent of a rat population. On the other hand, trichloroethylene (TCE) (recently found in water wells at ppm concentrations) is not called a poison since a rat has to eat 7,200 mg/kg of body weight to reach the LD_{50} level. Transferred to humans, this means that a 150-pound adult would have to eat about 3 pounds of TCE per day to receive this same dose, based on body weight. A 50-pound child would have to drink 4,000 gallons per day of well water containing 25 ppm of TCE to receive this dose.

Since we cannot deliberately use human populations to test poisons and possible cancer-causing chemicals (carcinogens), this use of animal subjects is at least a rational approach. Even then, it raises the difficult question of whether animal responses provide a reliable estimate of human responses. After all, we aren't really trying to protect mice with this testing—we have in mind the health of human beings, who come from an entirely different rung on the evolutionary ladder. The uncertainty comes when we have to rely only on this method to make decisions.

As an example, the compound 2,3,7,8-tetrachlorobenzo-*p*-dioxin (popularly called "dioxin") is extremely poisonous to guinea pigs. For these little fellows, the LD_{50} is only 0.6 *milli*grams per kg body weight. In astonishing contrast, it takes 10,000 times larger doses to reach the LD_{50} level for hamsters! From species to species, we find enormous variation in toxic responses to dioxin. For this substance, we have had many documented human exposures. Among 400 severe-exposure cases that occurred 20 to 35 years ago, painful skin lesions were the only definite injury that appeared, and no deaths attributable to the exposures have yet occurred. In this case, we can't even predict usefully the lethal dose of dioxin to a 50-gram hamster from measurements on a 200-gram guinea pig, let alone a 150-pound human. This type of uncertainty is part of every regulatory decision in which tolerance limits must be established for humans with only animal tests as a guide.

TABLE VI-1 Dioxin's Lethal Dose Varies from Species to Species

Animal	LD_{50} (mg/kg)
Guinea pig (male)	0.6
Rat (male)	22
Rat (female)	45
Mouse	114
Rabbit	115
Dog	>300
Bullfrog	>500
Hamster	5,000

Is There a Dose-Time Relationship?

There remains one further question to complicate this perplexing but crucial issue of risk assessment. It is natural to wonder, if a large dose of something is poisonous in a short time, whether a small dose of the stuff is also poisonous, but over a longer time? For example, suppose we want to eliminate disease-spreading rats with the fumigant ethylene dibromide, $C_2H_4Br_2$. Lab experiments show that rats are killed by breathing 3,000 ppm of this gas after 6 minutes of exposure. If that is so, how long would it take to have a lethal dose at 300 ppm? The simplest guess we can make is that at one-tenth the concentration, it takes just 10 times as long, 60 minutes, one hour. Actually, this *linear assumption* with its one-hour estimate is pretty close to what is found in lab tests. Does this, then, let us predict the lethal concentration for a 6-month exposure? Six months is 4,320 hours, so our linear model predicts lethality at the very low concentration of 0.07 ppm (300/4,320). This time, however, experiment shows that the 6-month LD_{50} exposure for rats is 50 ppm. For prolonged exposure, the rats can tolerate much more ethylene dibromide—700 times more—than we estimated. In this case, the linear assumption has failed.

This is not an isolated case. We have already mentioned selenium, which, at low concentrations, is essential for both human and animal health. At higher concentrations, selenium produces serious health effects. Evidently, this contradicts the linear model, which gives no clue to the beneficial effects of selenium but, instead, would lead us to expect selenium to be toxic at any level if exposure is long enough. Carbon monoxide, a treacherous poison, provides another clear example. In the blood, CO bonds to hemoglobin and renders it useless as an oxygen carrier. If about one third of the hemoglobin is so tied up, the victim dies. This would happen to the average person after one hour of exposure to 4,000 ppm of CO (a partial pressure of 3 torr). From this evidence, the linear model would predict that 1 ppm would be lethal in about 4,000 hours, i.e., in about 6 months. However, the natural atmosphere we breathe all of our lives always contains about 1 ppm of CO and is clearly not lethal.

There are examples in the opposite direction as well. Liquid mercury has a rather low vapor pressure, about one millitorr, and breathing it continuously has no immediate effect on health. However, the body cannot efficiently eliminate mercury, once ingested, so it accumulates over time. After many years of continuous exposure, various undesirable symptoms begin to appear, including unsteadiness, inflammation of the gums, general fatigue, and headaches. In this case, the absence of short-term effects gives no warning about the chronic, long-term danger of continuous exposure to the vapor of liquid mercury.

All of these examples suggest that we must be very careful when trying to extrapolate data concerning potentially harmful substances. It is not possible to predict with confidence the long-term, low-exposure toxicity of a given chemical just from evidence about toxicity at high exposures for short times.

Summary

We see that risk assessment is a difficult business. Nevertheless, it is an essential part of maintaining a healthy environment. We want to enjoy the benefits of technological advances, so we must learn to evaluate possible undesired side effects. We cannot afford to ignore the possibility that something might be hazardous, but at the same time, we cannot be paralyzed by indecision or fear.

RISK MANAGEMENT

Our everyday lives are full of risks, but this is not a new situation. We evolved in a threatening environment. Nature provides many risks free of charge: tornados, hurricanes, avalanches, earthquakes, floods, fires, volcanic eruptions. But many of the risks were "invented" by the human race—from catching one of the plagues which were spawned in the Middle Ages by the growth of cities, to falling off your horse while riding off to war. Our modern list is getting longer and longer—it includes automobile accidents, airplane crashes, ferryboats sinking, muggings in Central Park, smoggy air, and catching colds in the subway. Some risks we choose to take, like risking skin cancer while getting a suntan; others we may choose to avoid, like smoking cigarettes. Some risks we prefer not to take, but cannot find a way to avoid, like living under the ominous threat of nuclear war.

Thus, the sheer volume of risks we face seems to be steadily increasing with time. Yet a look at one particular measure, life expectancy, reassures us that we are not only holding our own in this battle, but perhaps actually winning. In the United States, life expectancy has steadily risen throughout this century, and it continues to rise at a rate of 3 years of additional age per decade. This impressive statistic can ease some of the anxiety about modern, technological risks, and steady our resolve to deal with them in an attentive, prudent, and rational way. Dealing with them begins with thinking about what is an "acceptable risk."

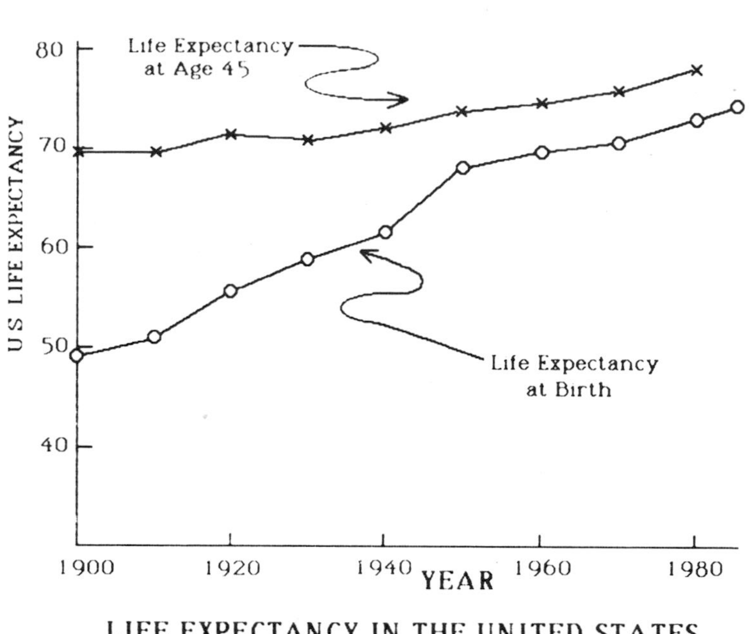

LIFE EXPECTANCY IN THE UNITED STATES
THREE YEARS INCREASE EVERY DECADE

Acceptable Risk

Notice that we propose to "think about," not to "decide," what is an acceptable risk. That is because risk acceptance is a highly personal and subjective thing. It can be arbitrary and even contradictory. An individual can decide to avoid the dangers of hang gliding but regularly drive an automobile at 60 mph with seat belt unfastened. Another individual may smoke cigarettes but vigorously shun marijuana—or vice versa. Some risks we take unthinkingly because they are familiar. Californians are used to earthquakes, Nebraskans to tornados, and Floridians to hurricanes. Everyone who can afford it goes from here to there in airplanes equipped with life rafts but not parachutes. What factors are at work as each of us decides what to worry about?

While the answer to this question varies enormously from person to person, there seems to be one factor that is generally important. *Most people are highly sensitive about taking a risk involuntarily;* they like to have a choice in the matter. The same individual who may choose to try skydiving, smoke cigarettes, consume caffeine, take birth control pills, or just cross a busy street in the middle of the block is likely to object strongly to the news that a pesticide might be found on his or her fresh fruit. People just feel better making their own choices. The current nervousness about chemicals probably stems largely from the feeling that chemical risks are being taken involuntarily and that they are increasing.

On the other hand, it is difficult for most people to deal with the magnitude of a risk. How does one evaluate the news that eating peanut butter might involve a risk of one in 500,000 that its 2 parts per billion (ppb) aflatoxin content might result in cancer? The numbers tend to blur in the face of the news that a dreaded disease may be involved. But there is one way to evaluate risks that most people can appreciate and use as a guide in decision making. It is to compare the risk for an unfamiliar hazard with another risk that is usual and similar in kind. That type of comparison calibrates for each of us what is an acceptable risk.

Comparable Risks

Let us begin by examining quantitatively the magnitude of a risk almost everyone takes voluntarily and daily: riding in an automobile. Ample statistics indicate that the chance of being killed while driving 3 miles is about one in a million. That means that a city dweller who travels that much every day for a year takes an annual risk of about 4 in 10,000. Over such an individual's 60-year lifetime, the chances of a fatal accident in an automobile are about one percent. This single example says that we regularly expose ourselves to an ''acute risk'' of one in a million and probably without realizing it, a ''chronic risk'' of one in a hundred.

Now let's look at some comparisons between similar risks. For example, 10 years ago, careful estimates were made of the possible effect of spray can propellants and refrigerator fluids on the stratospheric ozone. The outcome at that time was that the ozone would be lowered in the next few decades by at most 20 percent if chlorofluorocarbon use were continued. Since stratospheric ozone shields us from ultraviolet radiation, this depletion, if it were to occur, would cause a predictable increase in the number of nonfatal skin cancers in the U.S. population. To what could this new risk be compared? Is it large or small? Here we have a firm basis for an answer. The increased probability that an individual might end up with such a skin cancer is about the same as the increased probability of skin cancer associated with moving from San Francisco to the more sunny Los Angeles or from Baltimore to Miami.

That comparison does not *decide* the question of whether the chlorofluorocarbons should or should not be used. That depends upon the benefits to be lost as well. But it gives a person with a nonscientific background a tangible measure of what is at stake.

The Palo Alto water wells contaminated with trichloroethylene provide a second example. These wells, used for drinking water, were found to contain up to 3 ppm of trichloroethylene (TCE). At this concentration level, TCE is known to be a weak carcinogen through laboratory animal tests. Is there a comparison risk we can use to guide us? Tests show that the TCE contaminant is 1,000 times *less* hazardous than drinking an equal volume of cola, beer, or wine, each of which also contain weak carcinogens. Cola and beer, for example, both contain the carcinogen formaldehyde, cola at 8 ppm and beer at 0.7 ppm. These can be compared, in turn, to human blood, which also contains formaldehyde at about 3 ppm from normal metabolism. These contrasts allow a lay person to compare, in everyday terms, just how big (or small) the TCE risk is in those water wells.

What about pesticides? Man-made pesticide residues are found in our food at about 0.1 ppm, and most of them are classified as noncarcinogenic. Here, a useful comparison can be made to the presence of *natural* pesticides that are also present in our foods, but at 10,000 times higher concentrations than the man-made pesticides. These natural pesticides, present in every plant, are toxic chemicals that Nature evolved to protect the plant from fungus, insect, and animal predators. Some of these have been tested on rats or mice and found to be carcinogenic at sufficiently high dosages: estragole in basil, safrole in various herbs, psoralins in parsley and celery, hydrazines in mushrooms, and allyl isocyanate in mustard. That is useful and relevant evidence to be considered when we are considering the current hazard posed by agricultural pesticides.

We see that comparable risks provide us with an easily understood measuring stick that can help us decide which potential hazards are serious enough to warrant corrective or restraining action. This kind of evidence can help us sort out those hazards small enough to be ignored so that we can concentrate our efforts (and resources) on the hazards that deserve and require attention.

Who Is at Risk?

Risk management is connected with getting general agreement on the balance between risks and benefits. But sometimes the people who see themselves at risk are different people than the ones who benefit. Thus, the society at large wishes to dispose of radioactive waste by storing it safely in some remote area. Unfortunately, it is necessary to transport it there on trucks that pass through many small towns on the way. The people who live in these towns are the ones at risk that a truck might overturn right downtown at the intersection of 1st and Main. They are likely to agree with Los Angeles that the wastes should be stored in some desert, but not be ready to endorse carrying that dangerous material through the town. As another example, lots of people in Logan, Utah, depend for their livelihood on their jobs in the smelters, and people all over benefit from the useful materials that come from these smelters. However, the smelters make their contribution to the air pollutants that generate acid rain a thousand miles away in the Northeast and in Canada. The people in the regions suffering from acid rain see the issue entirely differently than those in the region whose economy depends upon the vitality of those industries.

There is no easy answer to such dilemmas except to say that the interests of both groups must be considered in shaping public policy. Recognizing the nature of the problem helps, though, in seeing why diametrically opposed positions might be persuasively argued by reasonable and sincere people on each side.

TABLE VI-2 Media Treatment of Three Chemical Spills

Incident	Newspaper Headline
10,000 gallons of toluene leaked into bay	Chemical Scare Blocks Estuary Toluene—the "T" in TNT
White solid substance spilled on roadway	Giant Traffic Jam Scare Closes Bay Bridge
"Mysterious white substance" found in roadway	Chemical Scare on Gate Bridge

MEDIA TREATMENTS OF CHEMICAL SPILLS

Table VI-2 lists three incidents that occurred during 1983 in the San Francisco Bay Area. In the first of these, both the television and the local newspapers opened their reports with the misleading remark: "Toluene—that's the T in TNT." What the press did not report was what all this toluene was doing in the area. Toluene is widely used as a solvent for many useful products such as paints and lacquers. Because of its wide use, large volumes of toluene are regularly transported over large distances. It has come into use as a substitute for benzene because toluene is safer: it is less flammable than benzene, it has a lower vapor pressure, and benzene is considered to be carcinogenic. These facts and the outcome—that no one was injured—add perspective to this undesirable accident, but they were not effectively communicated to the public.

The second incident caused the closure of the Bay Bridge during the peak commute hours, trapping 20,000 autos and disrupting the plans of their 40,000 occupants heading to work, to the airport, to the hospital, or to visit the San Francisco Art Museum. The spilled white substance proved to be lime, used in making concrete, handled daily by construction workers. The third incident in Table VI-2 was reason for closing the Golden Gate Bridge for 3 hours. This "chemical scare" was associated with a big bag of cornstarch that must have dropped off the back of a truck. These are two of the five closures of these bridges due to "chemical spills" since 1980. As in the three examples in Table VI-2, the other three closures were reported in the press and on television with emphasis on the worst-case scenario. These three chemicals were iron oxide (used as a pigment; it has the composition of rust), calcium phosphate (a fertilizer and component of detergents), and talcum powder. The talcum powder case closed the bridge for 10 hours! In none of the five bridge closures was there a newspaper follow-up article reassuring the public that not one individual had been harmed and, indeed, that no one was ever in danger.

What do we learn from these examples? The first lesson to be learned is that reporters rushed to the scene of an accident cannot be expected to be chemists. They will be reporting information received from other individuals, such as police, who are also unlikely to be chemists but whose responsibility is to act in the public interest in the face of the limited information they have. These latter officials can do nothing other than assume the worst possible situation. *Consequently, media reports of chemical spills will usually overestimate the danger.*

We might hope, though, that the reporters would feel enough responsibility to avoid unjustifiable, fear-laden expressions (the " 'T' in TNT"). We might expect them to tell us what the chemicals are used for (once identified) and to let the public know, in retrospect, that this incident, at least, did not pose any real public hazard.

As for the officials, they did what they had to do. In our social climate, they must act as though every white powder is as lethal as, say, sodium cyanide. What we can do to help them is to look ahead to avoidance of repeats. How might these spills have been handled after the first example? First, a plastic sheet held down over the spill would reduce wind-blown distribution. Then, bridge personnel wearing fine-pore dust masks could sweep the solid material into a pile for later removal by a cleanup truck equipped with a vacuum cleaner.

Turning from bridges to the larger context, it will be, more often than not, fire personnel and police who will deal with the immediate consequences of a chemical spill. Such personnel must (and many already do) have specific training in how to deal with chemical spills. There should be particular emphasis on the industrial feedstocks and chemical products in local use. A college chemistry course with a laboratory would be of immeasurable value and should be a normal criterion for advancement in a fire department.

LARGE-SCALE USES OF CHEMICALS

Any large-scale human activity carries with it a special consideration. While an unexpected and undesired outcome may have a quite low probability, the fact that very large numbers of people may be affected must influence our thinking. This special consideration is obviously applicable to nuclear war, reactor meltdowns, and genetic engineering; it also is awakened by large-scale uses of particular chemical substances. We have already discussed in Chapter II the possible global impact of the widespread use of chlorofluoromethanes as spray-can propellants and air-conditioner refrigerants. The worldwide use of DDT provides a second, informative case history and will be discussed here. Large-scale industrial accidents fall in this category as well.

Most chemical spills are well-handled and contained, but there have been, and will continue to be, rare but serious industrial accidents in which there is the possibility of catastrophe. While the potential dangers do not approach the centuries-long and worldwide impact of a Chernobyl reactor meltdown, we are reminded by Bhopal and the recent chemical spills into the Rhine River that large-scale industrial operations pose real public risks. Two large-scale accidents occurred within the last 10 years in which large populations were put at risk. One of these took place at Seveso, Italy, in 1976, and the other at Bhopal, India, in 1984. These two catastrophic events deserve review.

Seveso and Dioxin

A Swiss-Italian chemical firm, Industrie Chemiche Meda Societa, Anonima (ICME-SA), manufactured the effective herbicide 2,3,5-trichlorophenoxyacetic acid (2,4,5-T) in large quantities at its plant in northern Italy, near the town of Seveso. The herbicide, used worldwide to increase food supply, is made from the simpler compound 2,4,5-trichlorophenol (TCP). In manufacturing TCP, an undesired impurity is formed in small quantity. The impurity, 2,3,7,8-tetrachlorodibenzo-*p*-dioxin, popularly called "dioxin," has been discussed earlier in this section because of its extreme toxicity for certain small animals and the strong species dependence of its toxicity (see Table VI-1).

This large-scale accident at the ICME-SA plant began in July 1976, when cooling water was turned off to a chemical reactor making TCP. The temperature and pressure rose until a safety valve opened, releasing the reactor contents into the atmosphere over a densely populated area. The reactor was estimated to contain several pounds of the dioxin impurity.

Among the toxic chemicals that have received notoriety during the last 20 years, dioxin may well be the one for which there exists the largest amount of and most systematic epidemiological data. Since 1949, there have been eight large industrial accidents, two of them involving U.S. companies. Table VI-3 shows that 823 workers were exposed, two thirds of whom suffered chloracne, a very unpleasant skin lesion, as a result. Much smaller numbers suffered liver dysfunction, elevated

TABLE VI-3 Industrial Accidents Causing Dioxin Exposure[a]

Date	Number of Workers Exposed	Location	Number of Chloracne Cases	Deaths: Number Expected *without* Exposure	Deaths: Number Observed
1949	250	Nitro, West Virginia	132	46.4	32
1953	75	Ludwigshaften, West Germany	55	~18	17[b,c]
1963	106	Amsterdam, The Netherlands	44	13	8[c]
1964	61	Midland, Michigan	49	7.8	4
1965-1969	78	Prague, Czechoslovakia	78	?	5[a,b,d]
1966	7	Grenoble, France	21	?	?
1968	90	Derbyshire, United Kingdom	79	?	1
1976	156	Seveso, Italy	134	?	Normal[a,b,c,d]
Total	823		592		

[a]All at plants manufacturing TCP; dioxin was an undesired impurity.
[b]Impaired mobility; fatigue, neurological symptoms.
[c]Liver damage.
[d]Elevated blood cholesterol levels.

lipid and cholesterol levels in the blood, and neurological damage. All of these conditions gradually recover. Astonishingly, for the 492 workers exposed before 1964 (22 years ago), the number of individuals who have died (61) is 30 percent *lower* than expected from normal causes. No one concludes, of course, that dioxin increases longevity, but it is difficult to conclude that it has a lethal effect on humans.

To summarize the outcome of the widely reported Seveso accident, hundreds of townspeople and ICME-SA workers were evacuated, many of these having received severe exposure to the chemicals released. An estimated 37,000 people received minor exposure. The health of 500 people who received the largest exposures is being carefully monitored. To date, there have been no known deaths, involuntary abortions, or birth defects attributable to the exposure. Many small animals and rodents were killed in the town of Seveso. The disposal of 41 drums of toxic waste from the cleanup was in the European news for the next few weeks. A French waste disposal company contracted to move the drums from Italy to an authorized waste disposal site in West Germany. Then began a strange odyssey for the 41 drums occasioned by the fact that everyone wanted to get rid of the wastes but no one wanted to have them pass through their town, let alone be stored nearby.

Dioxin has received much attention, both in the courts and in the press. Certainly, a factor has been the extreme and well-documented toxicity for guinea pigs and mice, coupled with the fact that there are some well-established, though most often temporary, human ailments caused by severe exposure. It does help to show that the "chronic risk" to any individual, even those living near TCP chemical plants, is negligible compared with the chronic risks involved in driving a car, smoking cigarettes, eating peanut butter sandwiches, or drinking beer or wine. However, these are familiar and *voluntary* risks. As we have said before, the public is extraordinarily sensitive to any risks to which it is exposed involuntarily.

Bhopal and Methyl Isocyanate

Near midnight on Sunday, December 2, 1984, the impoverished residents of the squatter community called J.P. Nagar slept, unaware of the tragedy about to strike them. They occupied huts and hovels in a crowded shantytown built in the safety zone surrounding the Union Carbide plant just outside of Bhopal. Bhopal is a city of 800,000 and the capital of the agricultural State of Madhya Pradesh, the largest State in India. The plant was sited near Bhopal to manufacture pesticides, an essential element of the "green revolution" in a country trying to come to grips with its most critical national problems, starvation and malnourishment.

This plant had three large, underground storage tanks containing the volatile and toxic liquid methyl isocyanate, which is an immediate precursor to several effective herbicides. Late in the evening of December 2, the weekend plant personnel found that the pressure in one of these tanks, number 610, was abnormally low. Then, the temperature and pressure in 610 began to rise, this dangerous development being accentuated by the fact that the protective refrigeration unit may have been switched off. The plant personnel panicked as the temperature began to rise precipitously. The vapor pressure of the volatile liquid soared upward until it ruptured first a safety disk and then a relief valve designed to alleviate such an emergency. However, the vent line to the flare tower, where such releases are burned to harmless products, was closed off for repairs. The torrent of gas passed into, and overwhelmed, chemical scrubbers intended to neutralize any methyl isocyanate release not handled by the (inoperative) flare tower. Pressurized sprinklers designed to form a "water curtain" over such a release did not function because the water pressure was too low.

In an accelerating calamity, tank 610 discharged 41 tons of lung-searing methyl isocyanate gas on the people of J.P. Nagar. The wind carried the lethal cloud south toward the Railway Station, which had its own shanty community. Before that terrible night ended, about 14,000 of the 800,000 inhabitants of Bhopal had been seriously exposed. Perhaps 1,500 men, women, and children died within the first few hours. Without question, the world had seen the worst mass exposure to toxic chemicals since the deliberate chemical warfare of World War I. The repercussions of this tragic event are still with us, as both societies and chemical industries around the world work to ensure that it will not happen again.

The Chemistry Behind Bhopal

Methyl isocyanate (MIC) is a volatile, reactive, toxic, and flammable liquid. It boils at 39°C, and its vapor pressure is almost half an atmosphere at 20°C. It is shipped only in stainless steel or glass-lined containers under a slight overpressure of dry nitrogen to prevent the entry of atmospheric moisture. In bulk storage, it should be cooled, preferably to 0°C.

It is toxic to rats with an LD_{50} of 21 ppm for 2 hours of exposure and 5 ppm for 4 hours of exposure. In 1965 (when it was still permitted), four human volunteers in West Germany were exposed to low levels of MIC. At 0.4 ppm, none of the subjects detected it, but at 2 ppm, there was nasal irritation and their eyes watered.

At 21 ppm, the irritation became extreme and the test was terminated. There were no lasting aftereffects.

When water comes into contact with MIC, it reacts rapidly to form methyl amine and carbon dioxide. The reaction releases heat, so if there is no cooling, the temperature rises and the reaction speeds up. As the temperature rises (or in the presence of catalysts such as Fe, Cu, Sn, or Zn), MIC reacts with itself to form a trimer, again liberating heat so that the temperature rises still more and reactions accelerate. These undesired reactions have the potential for a thermal "runaway," so safe handling of MIC requires careful temperature control, avoidance of moisture, and meticulously clean containers (to avoid catalysts).

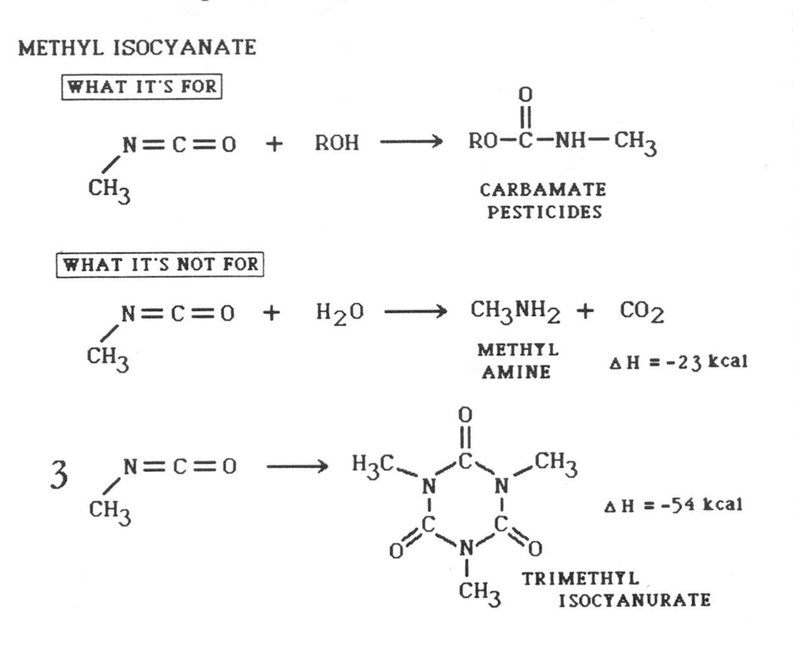

With all of these hazardous properties, why does anyone use MIC, let alone store 40 tons in a single storage tank? The value of MIC is that it readily reacts with alcohols to form carbamates, which are extremely effective pesticides. It is used by Union Carbide to make the pesticide Sevin® (1-naphthyl-*N*-methyl carbamate), by Shell to make Nudrin® (methonyl), by DuPont to make Lennate®, and by FMC to make Furudon® (carbofuran).

Union Carbide has its biggest MIC production plant at Institute, West Virginia (10 times bigger than the Bhopal plant). Because of the importance of pesticides in raising the food supply for the 700 million population of India, Union Carbide established near Bhopal the most advanced Research and Development Center for pesticides in all of Asia. In addition, Union Carbide built the plant at Bhopal so that pesticides for use in India could be manufactured in India by Indian personnel.

The Victims

One year after the Bhopal accident, the official government death toll was placed at about 1,800, although it is possible that this estimate is too low by as much as 500 to 1,000. Among the most exposed victims, lung damage was persisting, eye damage was generally recovering, and liver dysfunctions were prevalent (due, in part, to complications caused by the drugs administered). In the exposed population, with a normal death rate of 250 per month, deaths were being registered at 265 per month. No birth defects had been associated with the exposure.

Lessons To Be Learned

Governments in developing countries are generally enthusiastic, sometimes insistent, that manufacturing of essential products be conducted within the country's borders. Some countries require majority local ownership, local engineering and construction, and local operating and maintenance staff. These requirements can affect safety adversely because of cultural differences in work attitudes, understanding of industrial concepts, and response to training. Some of these factors may have contributed to the magnitude of the Bhopal disaster. Whether that is so or not, these are real problems that must be recognized and handled effectively while giving proper attention to national sensitivities.

More generally, this catastrophe draws acute attention to the importance of safety in chemical operations. *There should be an enforced safety zone around a chemical plant, and care should be exercised in the placement of chemical plants.* There is now an increased awareness in communities near chemical plants, and in this country, many chemical industries have responded to this awareness with active communication programs that directly involve the local citizenry. These efforts are leading to improved readiness plans for various emergency situations that might develop.

Perhaps the most important single lesson of Bhopal is, however, that *particularly dangerous intermediate chemicals should not be stored in unnecessarily large quantities*. Processes must be designed to manufacture these intermediates at the time of use and only in the quantity needed. This is a time-honored principle in any chemistry research laboratory, and it is even more important when the lives and health of many people are involved.

The DDT Story

It all began in 1939 when a Swiss chemist, Paul Müller, synthesized dichlorodiphenyl trichloroethane (DDT) during a systematic exploration for new insecticides. At the outset DDT appeared to be a miracle compound; it was extremely potent against a wide range of insect pests, and it did not have the acute human toxicity problems associated with the lead and arsenic compounds widely used at the time.

The Benefits

The United States first used DDT extensively in 1944 during World War II to counter a growing typhus epidemic among troops and the civilian population in Italy. Typhus is carried by body lice, and thousands of people were liberally dusted from head to toe with DDT to eliminate these pests. The epidemic was stopped, preventing a potentially devastating loss of human life.

In the light of this massive success, DDT was put into service against the *Anopheles* mosquito, which spreads malaria across many parts of the globe. Before the use of DDT, malaria was responsible worldwide for 2 million-3 million deaths per year and recurrent, periodic suffering for a much larger number. After a decade of use, malaria has been removed as the primary scourge of human existence in several countries. In India, the number of malaria cases was lowered from 75 million in 1952 to 100,000 in 1964. In the Soviet Union, the number of cases dropped from 35 million in 1956 to 13,000 in 1966. In Sri Lanka over this same time period, the malaria toll went from 12,000 deaths per year to zero! The World Health Organization of the United Nations has credited this wonder chemical with saving possibly 50 million lives from malaria alone. For his accomplishments, Dr. Paul Müller was awarded the 1948 Nobel Prize in Medicine.

The Risks

Unfortunately, the story does not end here. In 1972, the Environmental Protection Agency placed a ban on the use of DDT in the United States. How this came to pass is, in itself, a success story since it displays the importance of monitoring the environment as we watch for unexpected side effects of widely used chemical substances.

Already in 1946, scientists realized that DDT is stored in the fatty tissues of the body and remains there for inordinately long periods. Animals (including humans) and fishes are predominantly water systems; they transport and eliminate substances via aqueous fluids. But chlorinated hydrocarbons like DDT have quite low solubility in water (about 2 ppb), so they preferentially dissolve and concentrate in fatty tissues. For example, DDT shows up readily in the butterfat of a mother's breast milk. The EPA properly responded to this disturbing information by seeking a limit for acceptably safe levels of DDT in cows' milk and other foods. Exercising caution, the EPA first set the "safe" limit at zero. However, setting any "safe" limit at zero raises a problem of another kind. It would mean that for a milk sample to be judged safe, no DDT should be detected down to the detectability limit of the most sensitive measuring technique. Hence, as detection methods improved, the meaning of the "safe" level changed. *A zero limit always ties safety to detection techniques rather than to the best estimate of the hazard.* For this and other reasons, the zero limit was unattainable, and the EPA changed the "acceptably safe" limit to 0.05 ppm.

Then, as time passed, it began to be clear that DDT is not easily decomposed in the environment. As chemical detection methods became more refined, it was possible to estimate that, a decade after use, perhaps only 50 percent of the pesticide would be gone—either degraded or moved somewhere else.

Finally, evidence accumulated about how DDT tended to concentrate moving up the food chain. When elm trees were sprayed, it resulted in 100 ppm of DDT in the surrounding soil, 140 ppm in earthworms in that soil, and more than 400 ppm in the robins feeding on those worms. In birds, especially the larger, predatory species, the results of this concentration was quite detrimental. Apparently, DDT interferes with reproductive capability by causing the birds to produce eggs with dangerously thin shells. Some species, such as the bald eagle and the peregrine falcon, rapidly neared extinction as this new danger was added to other human encroachments on their habitats.

At its peak, the U.S. production alone of DDT reached 156 million pounds (in 1959). Since its use began, it has been used so extensively around the globe that no part of the earth remains untouched. It has been detected in the fat of native Alaskans in remote areas, as well as in penguins and seals in the Antarctic. In addition, some insects and pests became resistant to DDT after its prolonged use, and some beneficial insects were locally exterminated unintentionally.

The Risk/Benefit Equation

Here we see the evolution of a classic risk/benefit case. At the outset, it is clear that short-term benefits are great (in this case, the saving of human lives), and there are no known costs to be weighed against these benefits. But, despite the realization of the anticipated benefits, vigilant monitoring revealed environmental disturbances that were too pervasive to be ignored. Even though no human ailment has ever been connected with exposure to DDT, it is plain that some of its properties are incompatible with our desire to protect the world around us—DDT's incredible stability, its mobility, and its affinity for living systems. However, at the same time that these special problems worked against its continued use, they defined the properties needed for a substitute. These now exist—insecticides that are much more species specific, that are also nontoxic for human exposure, and that degrade in the environment after a few days or a few weeks. While DDT was saving millions of human lives, it was also guiding us to better solutions to the risk/benefit equation.

CONCLUSION

The most resounding message that emerges is that *risk assessment is a difficult business*. Paracelsus told us: "Everything is poisonous. The dose alone determines the poison." Yet it is extremely difficult to determine the dose. Tests with humans are not permitted, and animal tests have questionable applicability to humans. Epidemiology shows association but not necessarily causality.

There are strong subjective elements, too. One person's negligible risk is another's unacceptable hazard. Worse yet, often the group at risk is different from the group that benefits. Finally, everyone is sensitive about any risk to which the exposure is involuntary.

Despite these difficult and sometimes perplexing aspects, risk/benefit trade-offs have become a common element of countless decisions that affect us all. Some of these are decided for us by our elected officials in our state capitols and in Washington, D.C. Some of them we decide for ourselves in voting booths. Wherever these decisions are made, they should reflect both the common good and the common will. To make this possible, we need to improve scientific literacy throughout our population. Plainly, this must begin early in our schools; science education must receive more attention.

In conclusion, we should be reminded that our quality of life and steadily increasing longevity are directly attributable to our technological advances in chemistry. An approach to chemical hazards based on unreasonable fear can deprive us of health-restoring drugs, essential sources of energy, increased food supplies, useful commodities, and industrial productivity. To avoid paralysis and the loss of these benefits, we need calm, wise, and rational decisions in deciding when and how much regulation is needed. *We can achieve this by dealing wisely with the risk/benefit equation.*

SUPPLEMENTARY READING

Chemical & Engineering News

"Bhopal" by W. Lepkowski (C.&E.N. staff), vol. 63, pp. 18-32, Dec. 2, 1985.

"Stringfellow Cleanup Mishaps Show Need to Alter Superfund Law" by L.R. Ember (C.& E.N. staff), vol. 63, pp. 11-21, May 27, 1985.

"Bhopal, A C. and E.N. Special Issue" (C.& E.N. staff), vol. 63, pp. 14-63, Feb. 11, 1985.

"Dioxin, A C. and E.N. Special Issue" (C.& E.N. staff), vol. 61, pp. 20-64, June 6, 1983.

"Acid Pollutants: Hitchhikers Ride the Wind" by L.R. Ember (C.& E.N. staff), vol. 59, pp. 20-31, Sept. 14, 1981.

"William Lowrance: Probing Societal Risks," Interview, W. Lowrance, vol. 59, pp. 13-20, July 6, 1981.

Science

"Risk Assessment and Comparisons: An Introduction" by R. Wilson and E.A.C. Crouch, vol. 236, pp. 267-270, April 17, 1987.

"Ranking Possible Carcinogenic Hazards" by B.N. Ames, R. Magow, and L.S. Gold, vol. 236, pp. 271-289, April 17, 1987.

"Perception of Risk" by P. Slovic, vol. 236, pp. 280-285, April 17, 1987.

"Risk Assessment in Environmental Policy-Making" by M. Russell and M. Gruber, vol. 236, pp. 286-290, April 17, 1987.

"Health and Safety Risk Analysis: Information for Better Decisions" by L.B. Lave, vol. 236, pp. 291-295, April 17, 1987.

Libraries into Space

Unbelievable though it sounds, we may have to place whole libraries in a space-like environment over the next decade! This strange proposal is not made because our orbiting astronauts need more reading material, but because if we don't do that or something similar, most of our books won't be around very long for the rest of us to read. An alarming and little know problem faces mankind today—the vast majority of books, those printed since the 1850s, are relentlessly yellowing and crumbling to dust. The library at the University of California at Berkeley alone stands to lose 60,000 books and periodicals per year to decomposition. This is not because of air pollution; the source of the destruction lies in the very paper on which the books are printed. Now, some clever chemists have discovered that, surprisingly, a trip into an environment similar to space provides at least one solution to this vexing problem.

Papermaking processes used since the 1850s universally employ an alumrosin sizing to keep ink from "feathering" or spreading on the paper. Slowly, this papermaker's alum—aluminum sulfate—combines with moisture in the pages and in the air to form sulfuric acid. This aggressive substance, in turn, facilitates attack on the cellulose fibers in the paper, breaking them into smaller and smaller fragments and, ultimately, to dust. Between 75 and 95 percent of the deterioration in "modern" paper is caused by such acid attack.

In recent years, chemists have developed a number of acid-neutralizing processes for books. One of these, developed in the Library of Congress research laboratory, suggests that the chemical diethyl zinc may be ideal for the job. Diethyl zinc is a gas, so its molecules can easily permeate even a closed book. Once inside, the substance deacidifies each book and then looks ahead to the future by leaving an alkaline residue of zinc oxy-carbonate. This residue, uniformly distributed throughout the paper fibers, protects the book from any future acid attack.

Ironically, this life-saving agent, diethyl zinc, bursts into flame on contact with air and explodes when it touches water. How does a chemist work with a compound that cannot be exposed to air or water? In a deep space environment, of course. A suitable location was found at NASA's Goddard Flight Center, where 5,000 books from the Library of Congress took a simulated flight, not on a rocket into space, but in a laboratory space-simulating vacuum chamber.

First, the books were thoroughly dried by warming under vacuum for about 3 days. Then, with all oxygen removed from the chamber, gaseous diethyl zinc was introduced and allowed to diffuse into the books. As the neutralizing reaction proceeds, harmless ethane gas is produced and pumped away. Then the protective zinc oxy-carbonate is formed. The results have been extremely promising, and, as the technology is perfected, libraries across the nation will be looking to install huge deacidification facilities. These countermeasures, coupled with the new "alkaline reserve" papers now used in modern printing, promise that the precious heritage of the world's libraries, including the enormous Library of Congress, will be preserved for future generations to enjoy, and profit from, just as we do today.

CHAPTER VII

Career Opportunities and Education in Chemistry

Chemistry, as a central science, helps us understand the universe around us, see our place in that universe, and respond to the needs of human society. Furthermore, chemistry figures importantly in the economic fabric of our country. Hence, the pursuit of chemistry provides a fulfilling and rewarding career for young people interested in science and in service to humankind. We shall discuss here these career opportunities and the educational pattern associated with chemistry as a profession.

CHEMISTRY: AN ACTIVITY OF CREATIVE INDIVIDUALISTS

Today's public image of science is still heavily influenced by the reverberating impact of the World War II Manhattan Project that brought us the atomic bomb and the Apollo Project of the 1960s that let us set foot on the Moon. But embedded in this glamorous, highly organized, and well-publicized setting, there are several scientific disciplines that have somehow maintained the highly personal characteristics of classical human creativity. (How many poets were needed to write *Hamlet*? How many artists to paint the *Mona Lisa*? How many scientists to propose relativity?) Chemistry is one of these disciplines. Somehow it has remained an individualistic and highly competitive activity that depends upon prolonged individual initiative and personal creativity. Scientific publications in the field generally involve only two or three authors.

Chemistry has remained, worldwide, an innovative "cottage industry" that has been remarkably productive. Its continuing success is shown by the increasing rate of discovery of new compounds (see Chapter I, p. 2), despite the fact that at any given moment the molecules easiest to synthesize have already been made; the harder ones remain. This evidence shows that chemistry in the small project mode is an extremely effective enterprise, both here and abroad. Thus, the term cottage industry describes a highly individualistic and personally creative activity rather than a group one. These characteristics impart a healthy competitiveness and a liberating freedom from accepted dogma. They make chemistry an ideal field in which to nurture a young scientist's originality and initiative. He or she can be intimately involved and in control of every aspect of an investigation, selecting the question, deciding on the approach, assembling and personally operating the

TABLE VII-1 Employed Scientists and Engineers in Selected Field (1980)

Employer	Chemists	Chemical Engineers	Mathematicians	Biological Scientists	Physicists and Astronomers
Business/Industry	86,640	63,710	42,190	39,350	22,400
Academia (Ph.D. granting)	26,940 (7,800)	3,980 (1,665)	52,230 (9,140)	95,240 (28,135)	24,110 (7,995)
Federal government	9,075	2,025	12,580	16,160	6,585
State and local government	7,940	1,015	4,985	13,685	1,175
Other nonprofit organizations	7,660	580	4,510	22,620	3,115
Military	1,560	510	1,190	1,520	590
Other	1,985	580	1,185	1,525	835
Total	141,800	72,400	118,870	190,100	58,810

SOURCES:
U.S. Scientists and Engineers 1980, NSF Report No. 82-314, Table B-12.
Academic Science: Scientists and Engineers, January 1981. Washington, D.C.: National Science Foundation.
Detailed Statistical Tables, NSF Report No. 82-305, Table B-5. 1981. Washington, D.C.: National Science Foundation.
Science, Engineering, and Humanities Doctorates in the United States: 1981 Profile. 1982. Table 1.5A. Washington, D.C.: National Academy of Sciences.

equipment, collecting and analyzing the data, and deciding on the significance of the results.

The contention that chemistry responds to the needs and desires of our society is strikingly verified by the statistics on the number of professional chemists employed by industry. Table VII-1 compares the number of scientists and engineers employed in various fields. The first line shows that in 1980, business and industry employed almost one and a half times more chemists and chemical engineers than the sum of the mathematicians, biological scientists, physicists, and astronomers. Of course, this pattern is the sum of many individual hiring decisions by industries that exist and survive only if they market products needed by the people of the world. These figures imply that a young person thinking of entering a professional career in the chemical sciences can be assured that there is "somewhere to go."

This contrast is equally significant if we look at employment of professional scientists at the Ph.D. or doctoral level. In 1981, business and industry employed 24,320 Ph.D. chemists, more than the sum of Ph.D. mathematicians, biological scientists, physicists, and astronomers combined. This figure indicates that industry employed 56 percent of the 43,200 working Ph.D. chemists. (The corresponding percentage for the four disciplines above is 21 percent.) Academic institutions were the next largest employer; in 1981, 14,775 doctoral chemists were so employed, 34 percent of the total.

Because of the clear potential for positive economic return from chemical research, the chemical industry invests heavily in its own in-house research. In 1982, the Chemical and Allied Products industries invested about $4.2 billion in corporate research and development, of which about $380 million might be classified as basic research. The rest is applied research and development of new products. These statistics again indicate that research in chemistry pays off in future processes and products used by society. They also show that industrial laboratories furnish an important arena for chemical research.

THE BACHELORS DEGREE IN CHEMISTRY (AB OR BS)

College preparation for a professional career in chemistry begins with a 4-year degree leading either to a Bachelor of Arts (AB degree) or a Bachelor of Science (BS degree), with a major in chemistry in either case. The former degree tends to place more emphasis on the humanities and to carry somewhat more flexibility. Both of these characteristics are of significant value, as discussed below.

Because of the basic character of chemistry and its centrality among the sciences, introductory chemistry classes are not dominated by majors in chemistry but, rather, by students thinking of careers in fields adjacent to chemistry. A knowledge of the atomic makeup of the world around us is a necessity in most advanced courses to be taken by the student entering the health and biological sciences, physics, engineering, geology, oceanography, and even astronomy. This implies that the course content encountered in the first 2 years of chemistry tends to be general and suited to a wide range of student interest. This is undoubtedly an advantage to every individual taking the introductory courses. One of the problems of modern higher education is the tendency to force specialization too early. The college curriculum should permit easy movement toward more suitable career goals as the student's breadth of experience and maturity provide a firmer basis for these important life choices. Introductory chemistry courses tend to permit such mobility.

Of course, the last 2 years of a major in chemistry provide the focus needed to give personal experience with the major areas in chemistry. Laboratory courses occupy a special place in this inductive science, and access to modern instrumentation (including computers) is a crucial element. These laboratory activities also furnish a fascinating exposure to the challenging puzzles that are day-to-day fare in chemistry, as well as the colorful changes that take place in flasks and in nature. Next, it is important that the budding scientist be well grounded in the principles that guide a chemists's thinking: molecular structure and bonding, based in quantum mechanics, and the driving force for chemical change, based in chemical thermodynamics. Finally, there should be opportunity for participation in undergraduate research.

However, it is important to recognize that we are in a period of increasingly rapid change in which boundaries within science are disappearing. Each student should ensure that his or her curriculum leaves ample flexibility to engage in studies of adjacent disciplines such as biology, molecular biology, solid-state physics, geochemistry, and the environmental sciences. Equally important is the need for time reserved for courses in the humanities. No single remark is heard more often from experienced scientists (and employers) than the observation that ability to communicate—to write and to speak clearly—is as important as any other component of a scientific education.

THE DOCTORAL DEGREE IN CHEMISTRY (Ph.D.)

There is no room for doubt that the higher levels of professional activity in chemistry depend directly on the educational experiences embodied in the Ph.D.

program. The dependence is rooted in the rapid pace of scientific progress over the span of a professional chemist's career. This pace requires ability to cope with and develop new ideas—the heart of Ph.D. thesis work in chemistry.

Graduate education in chemistry provides a valuable, career-molding interaction with a mature scientist who is working productively at an active research frontier. There is a significant one-on-one aspect to the research director–graduate student interaction. In a highly personalized way, the faculty member will encourage individuality and creativity while directing the student toward problems likely to be solvable, interpretable, and significant to the advancement of existing frontiers. As the student matures, he or she assumes more and more responsibility for selecting the next question to be addressed and the experimental approach to be followed, for eliminating obstacles as they appear, and for interpreting results as they are obtained.

At the same time, the typical chemistry graduate student will be a member of a group working with the same research director on related problems based on similar experimental and theoretical techniques. This group might include several other graduate students and postdoctoral students. The transfer of ideas and techniques within this peer group is another vital and rewarding part of graduate study in chemistry.

Currently, a large proportion of Ph.D. degree recipients continue their educational preparation by conducting one or two years of postdoctoral study at another University, a National Laboratory, or in industry. This, too, has become an important part of the chemist's career development. It lets the student broaden horizons by venturing into a field different from the thesis work, by interacting with other productive researchers at a different locale, and by assuming more complete responsibility for the course of the research program. The combination of close collegial collaboration with a research-active professor, followed by more independent postdoctoral research work, identifies chemistry as an excellent prescription for the encouragement and nurturing of individual creativity in talented young scientists.

Chemistry Doctorates in U.S. Education

Table VII-2 shows the number of U.S. degrees awarded in chemistry for the period 1960 to 1980. It is not to be assumed that most of the Ph.D.s have progressed through the Master's degree; quite the opposite, the M.S. is for many their final graduate degree, usually received 2 to 3 years after the Baccalaureate. The larger fraction of the Ph.D. candidates enter graduate school with a 4-year Bachelor's degree, and they complete the Ph.D. between 4 and 5 years later.

TABLE VII-2 Number of Degrees Awarded in Chemistry, 1960-1980

Year	Bachelors	Masters	Ph.D.s
1960	7,603	1,228	1,048
1964	9,724	1,586	1,301
1968	10,847	2,014	1,757
1972	10,721	2,259	1,971
1976	11,107	1,796	1,623
1980	11,446	1,733	1,551

Table VII-2 shows that in recent years about 1/7th of those receiving Bachelor's degrees continue on to receive the Ph.D. For Chemical Engineering this fraction would be 1/12th, for Biological Sciences, 1/13th, and for Mathematics, 1/27th. The larger fraction for chemistry reflects the direct value of and need for graduate education in the chemistry profession.

The trend in the annual number of Ph.D. degrees awarded has changed dramatically over the last two decades. During the 1960s, the number of Ph.D.s in chemistry doubled, peaking at 2,200 Ph.D.s in 1970. Then there was a decline that seemed to level off by the end of the 1970s at about 1,500 Ph.D.s per year. Now, it is rising again. These long-range trends are difficult to interpret because they span a period of complicated demographic, social, and economic changes. They do, however, indicate that the decline in Ph.D.s during the 1970s has ended, and Ph.D. entry into chemistry is again rising, presumably in response to positive career expectations.

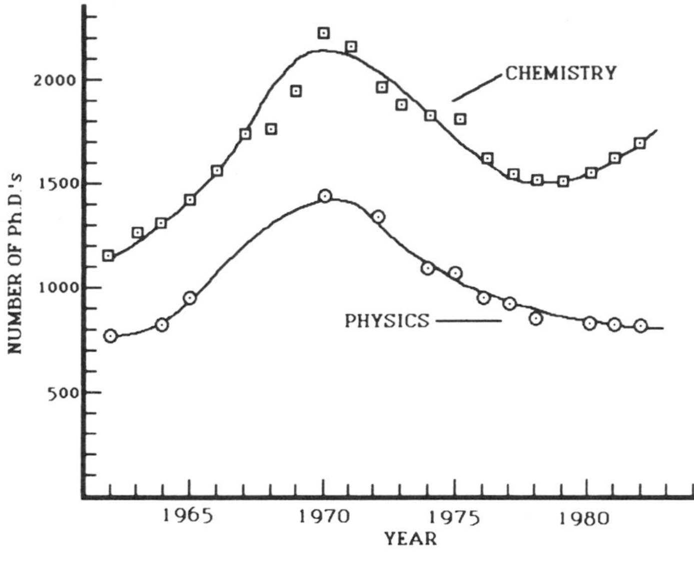

Ph.D. DEGREES IN CHEMISTRY AND PHYSICS

Post-Baccalaureate Educational Patterns for Chemists

While considerable variation exists, a typical chemistry Ph.D. graduate experience involves three essential elements: teaching, course work, and thesis research. In many graduate schools, teaching is required for one year, sometimes including fellowship holders. The rationale for this element has several components: teaching is a valuable educational experience for the graduate; it helps him or her evaluate an academic career as a career goal, it provides financial support, and it aids chemistry departments in meeting their large role in undergraduate education for related fields. From the point of view of financial support, teaching can thus provide approximately 20 percent of the support usually received by a chemistry graduate student.

There are several qualifying steps that may be required for successful completion of doctoral study in chemistry: entrance examinations, course grades, cumulative examinations, preliminary examinations, thesis submission, and final defense of thesis. Few schools would use all of these, and of those used, there is considerable variation in relative importance. Generally, the most significant are cumulative examinations taken during the first 2 years (if used) and the preliminary examination taken during the second or third year. Of course, the ultimate completion of Ph.D. study depends upon submission of a suitable research-based thesis. A thesis is a written account detailing substantial research accomplished by the graduate student. Almost always, portions of the thesis are published in the research literature.

In addition to payment for teaching duties (as Teaching Assistants or TAs), most chemistry graduate students have either won fellowship financial aid (National Science Foundation, National Institutes of Health, etc.) or they receive Research

Assistantship (RA) financial aid (''stipends''). A number of these stipends are supported by industrial grants, but the majority are drawn from federal grants to an individual faculty member to support the graduate students under his or her direction. At the major research universities, *essentially all of the chemistry graduate students receive continuous stipend and tuition support throughout their graduate study*. For the national average over the period 1974 to 1980, the best available data indicate that between two-thirds and three-fourths of U.S. doctoral students in chemistry currently receive either TA or RA stipends.

CAREER DIRECTIONS

A chemistry degree provides entry to a variety of fulfilling and rewarding careers. Many undergraduates choose the chemistry major to obtain a good foundation for employment and/or advanced studies in a variety of adjacent fields. Chemists are needed in such fields as environmental protection, the health sciences (including toxicology), the biological sciences (including genetic engineering), transportation industries (including aviation), and the semiconductor industry. Of course, the chemical industry offers a wide variety of jobs to help it produce and market its products and to help it discover new products needed by the public.

A second career goal of great social importance is in teaching. The need for science teachers at the high school and middle school levels is probably greater than in any other teaching area. An individual with a baccalaureate degree in chemistry who goes on to obtain a teaching credential (usually one more year of advanced study) is assured of a choice among teaching jobs.

Research is the major career avenue pursued by those who go on to an advanced degree (MA or Ph.D.). Research in chemistry is carried out in various arenas: industrial laboratories, private (not-for-profit) laboratories, national or other federal laboratories, and in our Universities and Colleges. Progressively through this sequence, research tends to be increasingly directed toward the fundamental understanding of nature and less toward practical or goal-oriented problems. In the United States, more than anywhere else in the world, the most fundamental research is conducted in the Universities, thus coupling the basic research function to the education of the next generation of scientists. Thus, it continuously renews our pool of scientific personnel with young scientists whose thesis research work has probed the edges of our knowledge.

SUPPLEMENTARY READING

ACS Information Pamphlets

''Futures Through Chemistry: Charting a Course,'' 12 pages, March 1985.

''Careers in Chemistry: Questions and Answers,'' 4 pages, May 1984.

''Chemical Careers in the Life Sciences,'' 18 pages, 1984.

''Careers in Chemical Education,'' 13 pages, Spring 1982.

''Graduate Programs in Chemistry,'' 39 pages, 1983.

Pamphlets available from:
American Chemical Society
Educational Division
1155 16th Street, NW
Washington, DC 20036

Index

major discussions; **definitions**

D

major discussions; **definitions**

major discussions; **definitions**

major discussions; **definitions**

major discussions; **definitions**

major discussions; **definitions**

Q

R

major discussions; **definitions**

T

major discussions; **definitions**